Project Management

Systems Innovation Series

Series Editor:

Adedeji B. Badiru

Air Force Institute of Technology (AFIT) – Dayton, Ohio

Systems Innovation refers to all aspects of developing and deploying new technology, methodology, techniques, and best practices in advancing industrial production and economic development. This entails such topics as product design and development, entrepreneurship, global trade, environmental consciousness, operations and logistics, introduction and management of technology, collaborative system design, and product commercialization. Industrial innovation suggests breaking away from the traditional approaches to industrial production. It encourages the marriage of systems science, management principles, and technology implementation. Particular focus will be the impact of modern technology on industrial development and industrialization approaches, particularly for developing economics. The series will also cover how emerging technologies and entrepreneurship are essential for economic development and society advancement.

Technomics

The Theory of Industrial Evolution

H. Lee Martin

Computational Economic Analysis for Engineering and Industry

Adedeji B. Badiru and Olufemi A. Omitaomu

Industrial Project Management

Concepts, Tools, and Techniques

Adedeji Badiru, Abidemi Badiru, and Adetokunboh Badiru

Systems Thinking

Coping with 21st Century Problems

John Boardman and Brian Sauser

Triple C Model of Project Management

Communication, Cooperation, and Coordination

Adedeji B. Badiru

Knowledge Discovery from Sensor Data

Edited by Auroop R. Ganguly, Joao Gama, Olufemi A. Omitaomu, Mohamed Gaber, and Ranga Raju Vatsavai

Handbook of Military Industrial Engineering

Edited by Adedeji B. Badiru and Marlin U. Thomas

STEP Project Management

Guide for Science, Technology, and Engineering Projects

Adedeji B. Badiru

Conveyors
Application, Selection, and Integration
Patrick M McGuire

Inventory Management
Non-Classical Views
Edited by Mohamad Y. Jaber

Social Responsibility
Failure Mode Effects and Analysis
Holly Alison Duckworth and Rosemond Ann Moore

Moving from Project Management to Project Leadership
A Practical Guide to Leading Groups
R. Camper Bull

Innovations of Kansei Engineering
Mitsuo Nagamachi amd Anitawati Mohd Lokman

Kansei/Affective Engineering
Edited by Mitsuo Nagamachi

Industrial Control Systems
Mathematical and Statistical Models and Techniques
Adedeji B. Badiru, Oye Ibidapo-Obe, and Babatunde J. Ayeni

Project Management
Systems, Principles, and Applications, Second Edition
Adedeji B. Badiru

Statistical Techniques for Project Control
Adedeji B. Badiru and Tina Agustiady

Carbon Footprint Analysis
Concepts, Methods, Implementation, and Case Studies
Matthew John Franchetti and Defne Apul

Sustainability
Utilizing Lean Six Sigma Techniques
Tina Agustiady and Adedeji B. Badiru

Quality Tools for Managing Construction Projects
Abdul Razzak Rumane

Handbook of Industrial and Systems Engineering
Edited by Adedeji B. Badiru

Communication for Continuous Improvement Projects
Tina Agustiady

Culture and Trust in Technology-Driven Organizations
Frances Alston

Managing Projects as Investments
Earned Value to Business Value
Stephen A. Devaux

Cellular Manufacturing
Mitigating Risk and Uncertainty
John X. Wang

Kansei Innovation
Practical Design Applications for Product and Service Development
Mitsuo Nagamachi and Anitawati Mohd Lokman

Total Project Control
A Practitioner's Guide to Managing Projects as Investments, Second Edition
Stephen A. Devaux

Total Productive Maintenance
Strategies and Implementation Guide
Tina Kanti Agustiady and Elizabeth A. Cudney

Global Manufacturing Technology Transfer
Africa-USA Strategies, Adaptations, and Management
Adedeji B. Badiru

Guide to Environment Safety and Health Management
Developing, Implementing, and Maintaining a Continuous Improvement Program
Frances Alston and Emily J. Millikin

Project Management Simplified
A Step-by-Step Process
Barbara Karten

A Six Sigma Approach to Sustainability
Continual Improvement for Social Responsibility
Holly A. Duckworth and Andrea Hoffmeier

Project Management for Research
A Guide for Graduate Students
Adedeji B. Badiru, Christina F. Rusnock, and Vhance V. Valencia

Handbook of Construction Management
Scope, Schedule, and Cost Control
Edited by Abdul Razzak Rumane

Project Feasibility
Tools for Uncovering Points of Vulnerability
Olivier Mesly

Essentials of Engineering Leadership and Innovation
Pamela McCauley

Learning Curves
Theory, Models, and Applications
Edited by Mohamad Y. Jaber

Handbook of Emergency Response
A Human Factors and Systems Engineering Approach
Edited by Adedeji B. Badiru and LeeAnn Racz

Project Management for the Oil and Gas Industry
A World System Approach
Adedeji B. Badiru and Samuel O. Osisanya

Profit Improvement through Supplier Enhancement
Ralph R. Pawlak

Work Design
A Systematic Approach
Adedeji B. Badiru and Sharon C. Bommer

Additive Manufacturing Handbook
Product Development for the Defense Industry
Adedeji B. Badiru, Vhance V. Valencia, and David Liu

Global Engineering
Design, Decision Making, and Communication
Carlos Acosta, V. Jorge Leon, Charles R. Conrad, and Cesar O. Malave

Handbook of Measurements
Benchmarks for Systems Accuracy and Precision
Edited by Adedeji B. Badiru, LeeAnn Racz

Design for Profitability
Guidelines to Cost Effectively Manage the Development Process of Complex Products
Salah Ahmed Mohamed Elmoselhy

Introduction to Industrial Engineering
Avraham Shtub and Yuval Cohen

Company Success in Manufacturing Organizations
A Holistic Systems Approach
Ana M. Ferreras and Lesia L. Crumpton-Young

Quality Management in Construction Projects
Abdul Razzak Rumane

Productivity Theory for Industrial Engineering
Ryspek Usubamatov

Defense Innovation Handbook
Guidelines, Strategies, and Techniques
Edited by Adedeji B. Badiru and Cassie Barlow

For more information about this series, please visit: https://www.crcpress.com/Systems-Innovation-Book-Series/book-series/CRCSYSINNOV

Project Management
Systems, Principles, and Applications

Second Edition

Adedeji B. Badiru

CRC Press
Taylor & Francis Group
Boca Raton London New York

CRC Press is an imprint of the
Taylor & Francis Group, an **informa** business

CRC Press
Taylor & Francis Group
6000 Broken Sound Parkway NW, Suite 300
Boca Raton, FL 33487-2742

First issued in paperback 2020

© 2019 by Taylor & Francis Group, LLC
CRC Press is an imprint of Taylor & Francis Group, an Informa business

No claim to original U.S. Government works

ISBN-13: 978-1-138-74086-0 (hbk)
ISBN-13: 978-0-367-77973-3 (pbk)

Library of Congress Cataloging-in-Publication Data

Names: Badiru, Adedeji Bodunde, 1952- author.
Title: Project management : systems, principles, and applications / by Adedeji B. Badiru.
Description: Boca Raton : Taylor & Francis, a CRC title, part of the Taylor & Francis imprint, a member of the Taylor & Francis Group, the academic division of T&F Informa, plc, 2019. | Series: Systems innovation series | Includes bibliographical references and index.
Identifiers: LCCN 2019004121 | ISBN 9781138740860 (hardback : alk. paper) | ISBN 9780429282829 (e-book : alk. paper)
Subjects: LCSH: Project management.
Classification: LCC HD69.P75 B328 2019 | DDC 658.4/04—dc23
LC record available at https://lccn.loc.gov/2019004121

Visit the Taylor & Francis Web site at
http://www.taylorandfrancis.com

and the CRC Press Web site at
http://www.crcpress.com

To the memory of Omolade Bisola in her perpetual state of being a bud.

Contents

Preface xxi
Acknowledgments xxiii
Author xxv
Quick Reference Formulas for Project Management xxvii

1. **Systems Approach to Project Management** 1
   Systems Control Framework 1
   Diversity of Organizational Performance 2
   Definition of Systems Engineering 4
   Project Systems Logistics 5
   Systems Constraints 5
   Systems Influence Philosophy 7
   Systems Value Modeling 8
   Example of Value Vector Modeling 10
   Management by Project 11
   Integrated Systems Implementation 12
   Critical Factors for Systems Success 13
   Early Systems Engineering 14
   DODAF Systems Architecture for Project Management 14
   Grand Challenges for Engineering 15
   Systems View of the Grand Challenges 16
   Body of Knowledge Methodology 17
   Components of the Knowledge Areas 18
      Step-by-Step and Component-by-Component Implementation 20
   Project Systems Structure 22
      Problem Identification 23
      Project Definition 23
      Project Planning 23
      Project Organizing 23
      Resource Allocation 23
      Project Scheduling 24
      Project Tracking and Reporting 24
      Project Control 24
      Project Termination 24
   Project Systems Implementation Outline 25
      Planning 25
      Organizing 26
      Scheduling (Resource Allocation) 26
      Control (Tracking, Reporting, and Correction) 27
      Termination (Close, Phaseout) 27
      Documentation 28
      Value of Lean Times 28
   Systems Decision Analysis 28
      Step 1. Problem Statement 28

Step 2. Data and Information Requirements 29
Step 3. Performance Measure 29
Step 4. Decision Model 29
Step 5. Making the Decision 30
Step 6. Implementing the Decision..31
Group Systems Decision-Making Models..31
Brainstorming...32
Delphi Method 32
Nominal Group Technique..34
Interviews, Surveys, and Questionnaires...35
Multivote 36
Systems Hierarchy 36
D-E-J-I Model for Project Execution..41
Design Stage of D-E-J-I Systems Model 41
State Transformation in Project Systems 43
Evaluation Stage of D-E-J-I 44
Half-Life Computation for Learning Curves...45
Justification Stage of D-E-J-I 47
Integration Stage of D-E-J-I..49
Computational Example 52
Exercises..53
References 54

2. Systems-Wide Project Planning 55
Project Planning Objectives 55
Time–Cost–Performance Criteria for Planning...60
Systems Level of Planning..60
Components of a Plan 61
Motivating the Project Team 62
Axiom of Theory X 63
Axiom of Theory Y 64
Maslow's Hierarchy of Needs..64
Hygiene Factors and Motivators..65
Management by Objective...66
Management by Exception 66
Project Feasibility Study 67
Scope of Feasibility Analysis..68
Contents of Project Proposals 69
Proposal Preparation...70
Proposal Incentives..72
Budget Planning 72
Top-Down Budgeting...73
Bottom-Up Budgeting 73
Zero-Base Budgeting 73
Project Work Breakdown Structure 74
Legal Systems Considerations 75
Systems Information Flow...76
Cost and Value of Information..77
Triple C Model..79

Communication..79
   Complexity of Multiperson Communication...85
Cooperation...87
Coordination...89
Resolving Project Conflicts with Triple C 90
Classical Abilene Paradox..91
Exercises..93
Reference...94

3. Project Systems Organization..95
Environmental Factors in Project Organization...95
Issues in Social and Cognitive Domains 95
   Leadership Development...95
   Preparing the Project Personnel 96
   Project Office..96
Organizational Breakdown Structure 97
   Selecting an Organization Structure..97
Formal and Informal Structures..98
   Span of Control...99
Functional Organization...100
Product Organization..101
Matrix Organization Structure 103
Mixed Organization Structure 105
Alternate Organization Structures..106
   Bubble Organization Structure..106
   Market Organization Structure 106
   Chronological Organization Structure..107
   Sequential Organization Structure 107
   Military Organization Structure..108
   Political Organization Structure...109
   Autocratic Organization Structure..109
   Project Transfer Organization..110
Organizing Multinational Projects 112
Exercises..115

4. Project Systems Scheduling 117
Fundamentals of Network Analysis 117
Critical Path Method 121
   Rule 1 122
   Rule 2 122
   Rule 3 122
   Rule 4 122
   Rule 5 122
   Rule 6 122
   Rule 7 123
   Rule 8 123
CPM Calculation Example...123
   Forward Pass 123
   Backward Pass 124

Determination of Critical Activities...125
    Using Forward Pass to Determine the Critical Path.............................128
    Subcritical Paths...128
Gantt Charts...129
    Gantt Chart Variations...131
    Activity Crashing and Schedule Compression.....................................134
PERT Network Analysis...140
    PERT Estimates and Formulas..140
    Beta Distribution...143
    Triangular Distribution..144
    Uniform Distribution...145
    Distribution of Project Duration...146
    Central Limit Theorem..146
    Probability Calculation..147
    PERT Network Example...148
Precedence Diagramming Method..149
    Anomalies in PDM Networks...156
    Complexity of Project Networks..157
    Example of Complexity Computation...161
    Evaluation of Solution Time...162
    Performance of Scheduling Heuristics..163
Formulation of Project Graph...164
    Depth-First Search Method...165
    Procedure to Detect Cycles in a Network $G = (N, A)$ Using the dfs Method.........165
    Activity-on-Arc Representation..165
Exercises..168

5. Resource Allocation Systems...175
Resource Allocation and Management...175
Resource-Constrained Scheduling..176
Resource Allocation Examples...177
    Longest-Duration-First...178
Resource Allocation Heuristics..179
    Activity Time..179
    Activity Resource..180
    Time Resources..180
    GENRES..180
    Resource over Time...180
    Composite Allocation Factor...181
    Resource Scheduling Method..182
    Greatest Resource Demand...182
    Greatest Resource Utilization...183
    Most Possible Jobs..183
    Other Scheduling Rules...183
    Example of ACTIM...184
    Comparison of ACTIM, ACTRES, and TIMRES....................................185
Quantitative Modeling of Worker Assignment...188
Takt Time for Activity Planning..191
    Tips for Using Takt Time...192

Resource Work Rate..193
Work Rate Examples...195
Resource-Constrained PDM Network..196
Critical Resource Diagram 198
    Resource Management Constraints 198
    CRD Network Development...198
    CRD Computations...199
    CRD Node Classifications...200
RS Chart 200
    CRD and Work Rate Analysis 201
Resource Loading and Leveling...202
    Resource Leveling...203
    Resource Idleness Graph...204
Probabilistic Resource Utilization..204
Learning Curve Analysis..207
    The Log-Linear Model..207
        Average Cost Model...207
        Unit Cost Model...211
    Graphical Analysis...212
    Multivariate Learning Curves..215
        Bivariate Example...217
Interruption of Learning...219
Learning Curves in Health-Care Projects...219
Exercises..221
References 226

6. Project Control System 227
Elements of Project Control..227
    Control Steps..230
    Formal and Informal Control...230
    Schedule Control..232
    Project Tracking and Reporting...233
    Performance Control...236
    Continuous Performance Improvement 237
    Cost Control..239
Information for Project Control 240
    Measurement Scales 241
    Data Determination and Collection 242
    Data Analysis and Presentation..245
        Raw Data 245
        Total Revenue..247
        Average Revenue 247
        Median Revenue...250
        Quartiles and Percentiles 251
        Mode..251
        Range of Revenue...251
        Average Deviation...252
        Sample Variance 253
        Standard Deviation...254

Control Charts..255
Statistical Analysis for Project Control..261
    Sampling Techniques..262
        Sample...262
    Diagnostic Tools...266
Probabilistic Decision Analysis...269
    Normal Distribution...270
Decision Trees...273
Project Control through Rescheduling...280
Experimental Analysis for Project Control...281
    Personnel Interactions for Experimentations.............................281
    Need for Project Experimentation..282
    Experimental Procedure...283
    Types of Experimentation..285
    Hypothesis Testing...285
    One-Tailed versus Two-Tailed Hypothesis Testing....................286
    Producer's Risk versus the Consumer's Risk..............................288
Control through Termination...288
    Project Control Verification and Validation.................................289
        What Is Verification?..290
        What Is Validation?..290
        What to Validate...290
        How Much to Validate...290
        When to Validate..290
    Verification and Validation Stages..290
    Factors Involved in Validation..291
    How to Evaluate the System..292
    Sensitivity Analysis for Project Control.......................................292
Exercises...293

7. **Modeling for Project Optimization**...297
Project Modeling..297
General Project Scheduling Formulation...297
Linear Programming Formulation..298
Activity Planning Example...300
Resource Combination Example..302
Resource Requirements Analysis..304
LP Formulation...305
Integer Programming Approach for Resource Scheduling.............306
Time–Cost Trade-Off Model..309
    Maximum Flow Procedure...310
    Time–Cost Trade-Off Procedure..311
    Sensitivity Analysis for Time–Cost Trade-Off............................318
    Knapsack Problem...319
    Knapsack Formulation for Scheduling...320
    Example of Knapsack Activity Scheduling..................................322
    Transportation Problem for Project Scheduling..........................324
    Balanced versus Unbalanced Transportation Problems.............326
    Initial Solution to the Transportation Problem...........................327

Northwest-Corner Technique...327
Transportation Algorithm 329
Example of Transportation Problem 330
Transshipment Formulation..335
Assignment Problem in Project Optimization 335
Example of Assignment Problem..337
Traveling Resource Formulation...339
The 2-Opt Technique 341
Shortest-Path Problem 341
Goal Programming..344
Resource Allocation Using Simulated Annealing 347
Exercises 350
References 354

8. Project Cost Systems 355
Economic Analysis Process 355
Simple and Compound Interest Rates 355
Investment Life for Multiple Returns 357
Nominal and Effective Interest Rates 359
Cash-Flow Patterns and Equivalence 361
Compound Amount Factor 362
Present Worth Factor 363
Uniform Series PW Factor 363
Uniform Series Capital Recovery Factor 364
Uniform Series Compound Amount Factor...364
Uniform Series Sinking Fund Factor 365
Capitalized Cost Formula...366
Permanent Investments Formula 367
Arithmetic Gradient Series..368
Increasing Geometric Series Cash Flow 369
Decreasing Geometric Series Cash Flow...371
Internal Rate of Return 372
Benefit–Cost Ratio...373
Simple Payback Period 374
Discounted Payback Period..375
Fixed and Variable Interest Rates 377
Amortization of Capitals 378
Equity Break-Even Point 380
Analysis of Tent Cash Flows 382
Special Application of AGS 383
Design and Analysis of Tent Cash-Flow Profiles 383
Derivation of GTE 388
Multiattribute Project Selection 390
Utility Models...390
Additive Utility Model...393
Multiplicative Utility Model..395
Fitting a Utility Function 395
Polar Plots 400
Analytic Hierarchy Process..406

Cost Benchmarking 411
Exercises 412
References 414

9. Advanced Forecasting and Inventory Modeling 415
Forecasting Techniques 415
    Forecasting Based on Averages 416
        Simple Average Forecast 416
        Period Moving Average Forecast 416
        Weighted Average Forecast 416
        Weighted *T*-Period Moving Average Forecast 417
        Exponential Smoothing Forecast 417
    Regression Analysis 418
        Description of Regression Relationship 418
        Prediction 419
        Control 419
    Procedure for Regression Analysis 419
    Coefficient of Determination 420
    Residual Analysis 421
    Time Series Analysis 422
    Stationarity and Data Transformation 424
        Moving Average Processes 427
        Autoregressive Processes 428
Classical Inventory Management 430
    Economic Order Quantity Model 430
    Quantity Discount 432
Calculation of TRC 433
    Evaluation of the Discount Option 433
    Sensitivity Analysis 435
    Wagner–Whitin Algorithm 436
    Notations and Variables 437
    Propositions for the W–W Algorithm 438
        Computational Example of W–W Algorithm 438
        Silver-Meal Heuristic 440
Seasonal Pattern Modeling 442
    Modeling Approach 443
        Standard Cyclic PDF 443
        Expected Value 445
        Variance 445
        Cumulative Distribution Function 445
        General Cyclic PDF 445
        Application Examples 446
    Exercises 451

10. Multiresource Scheduling 453
Scarcity of Project Resources 453
Notations Used in the Methodology 453
Analysis of Project Resources 454
Resource Modeling Background 455

Multiresource Methodology 456
    Representation of Resource Interdependencies and Multifunctionality 456
Modeling of Resource Characteristics 458
Resource Mapper 460
Activity Scheduler 462
Model Implementation and Graphical Illustrations 468
    Sequence-Based Scheduling and Function-Based Scheduling 471
References 472

11. Case Examples and Applications 473
Case Example: Project Systems View of World Economy 473
Global Systems View of the World 475
Economic Interdependence Model 475
Hybridization of Cultures 476
Market Integration as a Factor of Deterrent 477
Pursuit of Global Awareness for the Sake of Deterrent 477
Questions for Overseas Economic Engagement 477
Questions for Assessing Local Conditions 478
Questions for Economic–Cultural Nuances 478
Questions for Geographic Awareness 478
Questions for General Assessment 479
Labor as a Vehicle of Mutual Development 479
Awareness of Overseas Workforce Constraints 480
Hierarchy of National Needs 481
Prevention Is Better than Correction 482
Empathy, Sympathy, and Compassion Instead of Apathy 482
    Educate 483
    Engage 483
    Empower 483
Sustainability of Global Alliances 483
Summary of the Case Example 483
References 490

12. Emerging Roles of Artificial Intelligence in Project Management 493
Artificial Intelligence in Project Management 493
Background of AI 493
The First AI Conference 496
Branches of AI 499
Neural Networks 500
Expert Systems 501
Embedded Expert Systems 502
Conclusion and Projection 503
References 503

Index 505

Preface

A systems view of the world is the premise of this book. The book's emphasis is the belief that there is a better way (systems approach) to accomplishing goals and objectives in managing projects. This second edition has been updated to incorporate more recent developments in systems models, principles, and applications of project management. Of particular interest is the expanded inclusion of case studies. In general, project management is the process of managing, allocating, and timing resources to achieve a given goal in an efficient and expeditious manner. The objectives that constitute the specified goal may be in terms of time, cost, or performance expectations. Performance can be in terms of output quality or compliance with project requirements. A project can be simple or complex. In each case, proven project management processes must be followed with a world systems view of the project environment. While on-the-job training is possible for many of the project management requirements, formal education must also be utilized. This second edition covers contemporary tools and techniques of project management, from an established pedagogical perspective. Specific programs that will be interested in the book include Industrial Engineering, Systems Engineering, Construction Engineering, Operations Research, Engineering Management, Business Management, General Management, Business Administration, Mechanical Engineering, Civil Engineering, Production Management, Industrial Management, and Operations Management. The book contains ample graphical representations throughout to clarify the concepts and techniques presented. The end-of-chapter exercises help to reinforce the topics covered in each chapter. The World Systems View used in this book is essential for working across countries and across cultures, which is a desirable accomplishment in the globalized and intertwined economies of our present day.

Adedeji B. Badiru
Beavercreek, Ohio
January 2019

Acknowledgments

Several individuals contributed to the completion of this book, both in its first edition as well as this second edition. The preparation of the manuscript spanned several years of arduous work, assisted by students, colleagues, and professional associates. Many people pitched in with intellectual contributions here and there. The passage of time between the first edition and second edition saw the emergence of new knowledge, principles, and techniques, which are incorporated into this updated manuscript. As in the past, notable editorial comments, guidance, typing, graphics development, suggestions, and administrative support were provided by Iswat Badiru, TJ Badiru, Abi Badiru, Ade Badiru, Mack Everly, Anna Maloney, Annabelle Sharp, Songmi Berarducci, LeeAnn Racz, Jinan Andrews, Luke Farrell, Jesse Peterson, Mark Skouson, Christina Akers, and Sharon Bommer. I express my profound gratitude to them for their contributions over the past several years. I thank venerable Cindy Renee Carelli, my executive editor at CRC Press, and her entire team for the unflinching commitment, not only to the quality of the output but also to the expeditious processing of the book manuscript. The direct involvement of so many people is a good demonstration of a systems approach to project management, which is the premise of this book. An author does not have all the answers, but supported by several people from a variety of angles, he or she can deliver a marvelous manuscript.

Author

Adedeji B. Badiru is a professor of systems engineering and dean of the Graduate School of Engineering and Management at the Air Force Institute of Technology (AFIT). He was previously professor and head of systems engineering and management at AFIT, professor and department head of industrial engineering at the University of Tennessee, Knoxville, and professor of industrial engineering and dean of University College at the University of Oklahoma, Norman. He is a registered professional engineer (PE), a certified project management professional (PMP), a fellow of the Institute of Industrial & Systems Engineers, and a Fellow of the Nigerian Academy of Engineering. He is also a program evaluator for ABET. He holds a leadership certificate from the University Tennessee Leadership Institute. He has BS in industrial engineering, MS in mathematics, and MS in industrial engineering from Tennessee Technological University, and PhD in industrial engineering from the University of Central Florida. Badiru is a member of the Project Management Executive Forum, based in Cincinnati, Ohio.

Quick Reference Formulas for Project Management

(More comprehensive formulas are provided in Appendix A)

PERT time: $t_e = \dfrac{a + 4m + b}{6}$

PERT time variance: $s^2 = \dfrac{(b-a)^2}{36}$

Earliest start (ES) time for Activity i: $ES(i) = \underset{j \in P\{i\}}{\text{Max}}\left\{EC(j)\right\}$

Earliest completion (EC) time of Activity i: $EC(i) = ES(i) + t_i$

Earliest completion time of a project: $EC(\text{Project}) = EC(n)$, n is the last node.

Latest completion (LC) time of a project: $LC(\text{Project}) = EC(\text{Project})$, with no deadline

Latest completion (LC) time of a project: $LC(\text{Project}) = T_p$ (deadline)

Latest completion (LC) time for activity i: $LC(i) = \underset{j \in S\{i\}}{\text{Min}}\{LS(j)\}$

Latest start time for activity i: $LS(i) = LC(i) - t_i$

Total Slack (TS): $TS(i) = LC(i) - EC(i)$; $TS(i) = LS(i) - ES(i)$

Free Slack (FS): $FS(i) = \underset{j \in S(i)}{\text{Min}}\left\{ES(j)\right\} - EC(i)$

Interfering Slack (IS): $IS_i = TS_i - FS_i$

Takt Time: $T = \dfrac{\text{Available work time} - \text{Breaks}}{\text{Customer demand}}$

Crew Size: $\text{Crew size} = \dfrac{\text{Sum of manual cycle time}}{\text{TAKt time}}$

Efficiency: $e = \dfrac{\text{Output}}{\text{Input}} = \dfrac{\text{Result}}{\text{Effort}}$

Project Network Density: $D = \displaystyle\sum_{i=1}^{N} \text{Max}\{0, (p_i - s_i)\}$

$\text{Expected \% completion} = \dfrac{\text{Workdays for activity}}{\text{Workdays for planned}}$

$\text{Planned project \% completion} = \dfrac{\text{Workdays completed on project}}{\text{Total workdays planned}}$

Permutations: $P(n, m) = \dfrac{n!}{(n-m)!}$, $(n \geq m)$

Combinations: $C(n,m) = \dfrac{n!}{m!(n-m)!}, \quad (n \geq m)$

Failure rate: $q = 1 - p = \dfrac{n-s}{n}$

Expected Value: $\mu = \sum \left(x f(x) \right)$

Variance: $\sigma^2 = \sum (x-\mu)^2 f(x) \quad \text{or} \quad \sigma^2 = \displaystyle\int_{-\infty}^{\infty} (x-\mu)^2 f(x) dx$

1

Systems Approach to Project Management

A systems view of a project makes the project execution more agile, efficient, and effective. A system is a collection of interrelated elements working together synergistically to achieve a set of objectives. Any project is, in actuality, a collection of interrelated activities, people, tools, resources, processes, and other assets brought together in pursuit of a common goal. The goal may be in terms of generating a physical product, providing a service, or achieving a specific result. This makes it possible to view any project as a system that is amenable to all the classical and modern concepts of systems management.

Project management is the foundation of everything we do. Having knowledge is not enough; we must apply the knowledge strategically and systematically for it to be of any use. The knowledge must be applied to do something in pursuit of objectives. Project management facilitates the application of knowledge and willingness to actually accomplish tasks. Where there is knowledge, willingness often follows. But it is the project execution that actually gets jobs accomplished. From the very basic tasks to complex endeavors, project management must be applied to get things done. It is thus essential that project management be a part of the core of every academic curriculum in any discipline, whether it is in the liberal arts, medical science, business, retail, education, science, advanced technology, or engineering. The tools and techniques presented in this book are generally applicable to any project-oriented pursuit in business, industry, education, military, and government. This practically means everything that everyone does, because every pursuit can, indeed, be defined as a project. Even a national political process is amenable to a rigorous application of project management tools and techniques. In this regard, a systems approach is of utmost importance in any human pursuit.

Systems Control Framework

Classical control system focuses on control of the dynamics of mechanical objects, such as a pump, electrical motor, turbine, rotating wheel, and so on. The mathematical basis for such control systems can be adapted (albeit in iconical formats) for organizational management systems, including project management. This is because both technical and managerial systems are characterized by inputs, variables, processing, control, feedback, and output. This is represented graphically by input–process–output relationship block diagrams. Mathematically, it can be represented as

$$z = f(x) + \varepsilon$$

where

    z is the output

    $f()$ is the functional relationship

    ε is the error component (noise, disturbance, etc.)

For multivariable cases, the mathematical expression is represented as vector–matrix functions, as shown in the following:

$$Z = f(X) + E$$

where

    each term is a matrix

    Z is the output vector

    f(·) is the input vector

    E is the error vector

Regardless of the level or form of mathematics used, all systems exhibit the same input–process–output characteristics, either quantitatively or qualitatively. The premise of this book is that there should be a cohesive coupling of quantitative and qualitative approaches in managing a project system. In fact, it is this unique blending of approaches that makes systems application for project management more robust than what one will find in mechanical control systems, where the focus is primarily on quantitative representations.

Diversity of Organizational Performance

Organizational performance is predicated on a multitude of factors, some are quantitative while some are qualitative. Systems engineering efficiency and effectiveness are of interest across the spectrum of the diversity of organizational performance under the platform of project management. Project analysts should be interested in having systems engineering serve as the umbrella for improvement efforts throughout the organization. This will get everyone properly connected with the prevailing organizational goals as well as create collaborative avenues among the personnel. Systems application applies across the spectrum of any organization and encompasses the following elements:

1. Technological systems (e.g., engineering control systems and mechanical systems)
2. Organizational systems (e.g., work process design and operating structures)
3. Human systems (e.g., interpersonal relationships and human–machine interfaces)

A systems view of the world makes everything work better and projects more likely to succeed. A systems view provides a disciplined process for the design, development, and execution of complex projects, both in engineering and nonengineering organizations. One of the major advantages of a systems approach is the win–win benefit for everyone.

A systems view also allows full involvement of all stakeholders of a project, beyond mere rhetoric, as can be deduced from the sayings of Confucius:

> Tell me and I forget;
> Show me and I remember;
> Involve me and I understand.

CONFUCIUS, CHINESE PHILOSOPHER

For example, the pursuit of organizational or enterprise transformation is best achieved through the involvement of everyone, from a systems perspective. Every project environment is complex because of the diversity of factors involved. There are differing human personalities, technical requirements, expectations, and environmental factors. Each specific context and prevailing circumstances determine the specific flavor of what can and cannot be done in the project. The best approach for effective project management is to adapt to what each project needs. This requires taking a systems view of the project. The project systems approach presented in this book is needed for working across organizations and countries, cultures, and unique nuances of each project. This is an essential requirement in today's globalized and intertwined project goals. A systems view requires a disciplined embrace of multidisciplinary execution of projects in a way that each component complements other components in the project system. Project management represents an excellent platform for the implementation of a systems approach. Project management integrates various technical and management requirements. It requires control techniques, such as operations research, operations management, forecasting, quality control, and simulation to deliver goals. Traditional approaches to project management use these techniques in a disjointed fashion, thus ignoring the potential interplay among techniques. The need for integrated systems-based project management worldwide has been recognized for decades. In 1993, the World Bank reported that a lack of systems accountability led to several worldwide project failures. The bank, which has loaned more than $300 billion to developing countries over the last half century, acknowledged that there has been a dramatic rise in the number of failed projects around the world. A lack of an integrated systems approach to managing the projects was cited as one of the major causes of failure. Unfortunately, the 1993 World Bank assessment is still applicable today. More recent reports by other organizations point to the same flaws in managing global projects and the need to apply better project management to major projects. Press headlines in April 2008 highlight that "Defense needs better management of projects." This was in the wake of a government audit that revealed gross inefficiencies in managing large defense projects. In a national news release on April 1, 2008, it was reported that auditors at the Government Accountability Office (GAO) issued a scathing review of dozens of the Pentagon's biggest weapons systems, citing that "ships, aircraft, and satellites are billions of dollars over budget and years behind schedule." According to the review, "95 major systems have exceeded their original budgets by a total of $295 billion; and are delivered almost two years late on average." Further, "none of the systems that the GAO looked at had met all of the standards for best management practices during their development stages." Among programs noted for increased development costs were the "joint strike fighter and future combat systems." The costs of those programs have risen "36% and 40%, respectively," while C-130 avionics modernization costs have risen to 323%. And, while "Defense Department officials have tried to improve the procurement process," the GAO added that "significant policy changes have not yet translated into best practices on individual programs." A summary of the report of the accounting office reads

Every dollar spent inefficiently in developing and procuring weapon systems is less money available for many other internal and external budget priorities, such as the global war on terror and growing entitlement programs. These inefficiencies also often result in the delivery of less capability than initially planned, either in the form of fewer quantities or delayed delivery to the warfighter.

In as much as the military represents the geopolitical–economic landscape of a nation, the aforementioned assessment is representative of what every organization faces, whether public or private. In systems-based project management, it is essential that related techniques be employed in an integrated fashion so as to maximize the total project output. One definition of systems project management offered here is stated as follows:

Systems project management is the process of using systems approach to manage, allocate, and time resources to achieve systems-wide goals in an efficient and expeditious manner.

The definition calls for a systematic integration of technology, human resources, and work process design to achieve goals and objectives. There should be a balance in the synergistic integration of humans and technology. There should neither be an overreliance on technology nor an overdependence on human processes. Similarly, there should not be too much emphasis on analytical models to the detriment of commonsense human-based decisions.

Definition of Systems Engineering

Systems engineering is growing in appeal as an avenue to achieve organizational goals and improve operational effectiveness and efficiency. Researchers and practitioners in business, industry, and government are all clamoring collaboratively for systems engineering implementations. So, what is systems engineering? Several definitions exist. The following is one quite comprehensive definition:

Systems engineering is the application of engineering to solutions of a multifaceted problem through a systematic collection and integration of parts of the problem with respect to the life cycle of the problem. It is the branch of engineering concerned with the development, implementation, and use of large or complex systems. It focuses on specific goals of a system considering the specifications, prevailing constraints, expected services, possible behaviors, and structure of the system. It also involves a consideration of the activities required to assure that the system's performance matches the stated goals. Systems engineering addresses the integration of tools, people, and processes required to achieve a cost-effective and timely operation of the system.

Project Systems Logistics

Logistics can be defined as the planning and implementation of a complex task, the planning and control of the flow of goods and materials through an organization or manufacturing process, or the planning and organization of the movement of personnel, equipment, and supplies. Complex projects represent a hierarchical system of operations. Thus, we can view a project system as a collection of interrelated projects serving a common end goal. Consequently, we present the following universal definition:

> Project systems logistics is the planning, implementation, movement, scheduling, and control of people, equipment, goods, materials, and supplies across the interfacing boundaries of several related projects.

Conventional project management must be modified and expanded to address the unique logistics of project systems.

Systems Constraints

Systems management is the pursuit of organizational goals within the constraints of time, cost, and quality expectations. The iron triangle model depicted in Figure 1.1 shows that project accomplishments are constrained by the boundaries of quality, time, and cost. In this case, quality represents a composite collection of project requirements. In a situation where precise optimization is not possible, there will have to be trade-offs between these three factors of success. The concept of iron triangle is that a rigid triangle of constraints encases the project. Everything must be accomplished within the boundaries of time, cost, and quality. If better quality is expected, a compromise along the axes of time and cost must be executed, thereby altering the shape of the triangle.

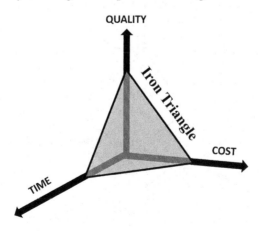

FIGURE 1.1
Systems constraints of cost, time, and quality within an iron triangle.

The trade-off relationships are not linear and must be visualized in a multidimensional context. This is better articulated by a three-dimensional view of systems constraints, as shown in Figure 1.2. Scope requirements determine the project boundary, and trade-offs must be done within that boundary. If we label the eight corners of the box as (a), (b), (c), ..., (h), we can iteratively assess the best operating point for the project. For example, we can address the following two operational questions:

1. From the point of view of the project sponsor, which corner is the most desired operating point in terms of combination of requirements, time, and cost?
2. From the point of view of the project executor, which corner is the most desired operating point in terms of combination of requirements, time, and cost?

Note that all the corners represent extreme operating points. We notice that point (e) is the do-nothing state, where there are no requirements, no time allocation, and no cost incurrence. This cannot be the desired operating state of any organization that seeks to remain productive. Point (a) represents an extreme case of meeting all requirements with no investment of time or cost allocation. This is an unrealistic extreme in any practical environment. It represents a case of getting something for nothing. Yet, it is the most desired operating point for the project sponsor. By comparison, point (c) provides the maximum possible for requirements, cost, and time. In other words, the highest levels of requirements can be met if the maximum possible time is allowed and the highest possible budget is allocated. This is an unrealistic expectation in any resource-conscious organization. You cannot get everything you ask for to execute a project. Yet, it is the most desired operating point for the project executor. Considering the two extreme points of (a) and (c), it is obvious that the project must be executed within some compromise region, within the scope boundary. Figure 1.3 shows a possible view of a compromise surface, with peaks and valleys representing give-and-take trade-off points within the constrained box. The challenge is to come up with some analytical modeling technique to guide decision making over the compromise region. If we could collect sets of data over several repetitions of identical projects, then we could model a decision surface that can guide future executions of similar projects. Such typical repetitions of an identical project are most readily apparent in construction projects, for example, residential home development projects.

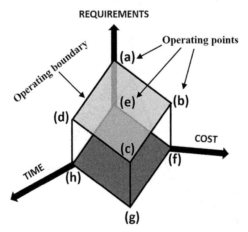

FIGURE 1.2
Operating Boundaries of Systems Constraints.

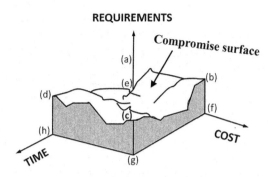

FIGURE 1.3
Compromise surface for cost, time, and requirements trade-off.

Systems Influence Philosophy

Systems influence philosophy suggests the realization that you control the internal environment while only influencing the external environment. In Figure 1.4, the inside (controllable) environment is represented as a black box in the typical input–process–output relationship. The outside (uncontrollable) environment is bounded by a cloud representation. In the comprehensive systems structure, inputs come from the global environment, are moderated by the immediate outside environment, and are delivered to the inside environment. In an unstructured inside environment, functions occur as blobs, as illustrated in Figure 1.5. A "blobby" environment is characterized by intractable activities

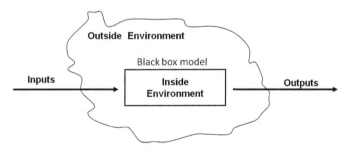

FIGURE 1.4
Outside versus inside environments of a project.

FIGURE 1.5
Blobs in unstructured projects.

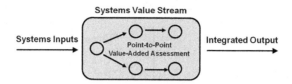

FIGURE 1.6
Systems value-stream structure.

where everyone is busy, but without a cohesive structure of input–output relationships. In such a case, the following disadvantages may be present:

1. Lack of traceability
2. Lack of process control
3. Higher operating cost
4. Inefficient personnel interfaces
5. Unrealized technology potentials

Organizations often inadvertently fall into the blobs structure because it is simple, low cost, and less time consuming until a problem develops. A desired alternative is to model the project system using a systems value-stream structure as shown in Figure 1.6.

This uses a proactive and problem-preempting approach to execute projects. This alternative has the following advantages:

1. Problem diagnosis is easier
2. Accountability is higher
3. Operating waste is minimized
4. Conflict resolution is faster
5. Value points are traceable

Systems Value Modeling

A technique that can be used to assess overall value-added components of a process improvement program is the systems value model (SVM), which is an adaptation of the manufacturing system value model presented by Troxler and Blank (1989). The model provides an analytical decision aid for comparing process alternatives. Value is represented as a p-dimensional vector:

$$V = f\left(A_1, A_2, \ldots, A_p\right)$$

where $A = (A_1, \ldots, A_n)$ is a vector of quantitative measures of tangible and intangible attributes. Examples of process attributes are quality, throughput, capability, productivity, cost, and schedule. Attributes are considered to be a combined function of factors, x_1, expressed as

$$A_k\left(x_1, x_2, \ldots, x_{m_k}\right) = \sum_{i=1}^{m_k} f_i\left(x_i\right)$$

where

$\{x_i\}$ is the set of *m* factors associated with attribute A_k ($k = 1, 2, \ldots, p$)

f_i is the contribution function of factor x_i to attribute A_k

Examples of factors include reliability, flexibility, user acceptance, capacity utilization, safety, and design functionality. Factors are themselves considered to be composed of indicators, v_i, expressed as

$$x_i\left(v_1, v_2, \ldots, v_n\right) = \sum_{j=1}^{n} z_i\left(v_i\right)$$

where

$\{v_j\}$ is the set of *n* indicators associated with factor x_i ($i = 1, 2, \ldots, m$)

z_j is the scaling function for each indicator variable v_j

Examples of indicators are project responsiveness, lead time, learning curve, and work rejects. By combining the aforementioned definitions, a composite measure of the value of a process can be modeled as

Total Value $= f\left(A_1, A_2, \ldots, A_p\right)$

$$= f\left\{\left[\sum_{i=1}^{m_1} f_i\left(\sum_{j=1}^{n} z_j\left(v_j\right)\right)\right]_1, \left[\sum_{i=1}^{m_2} f_i\left(\sum_{j=1}^{n} z_j\left(v_j\right)\right)\right]_2, \ldots, \left[\sum_{i=1}^{m_k} f_i\left(\sum_{j=1}^{n} z_j\left(v_j\right)\right)\right]_p\right\}$$

where *m* and *n* may assume different values for each attribute. A subjective measure to indicate the utility of the decision maker may be included in the model using an attribute weighting factor, w_i, to obtain a weighted total value (TV):

$$\mathrm{TV}_w = f\left(w_1 A_1, w_2 A_2, \ldots, w_p A_p\right)$$

where

$$\sum_{k=1}^{p} w_k = 1 \quad \left(0 \le w_k \le 1\right)$$

With this modeling approach, a set of process options can be compared on the basis of a set of attributes and factors.

Example of Value Vector Modeling

To illustrate the aforementioned model, suppose three Information Technology (IT) options are to be evaluated based on four attribute elements: *capability, suitability, performance,* and *productivity* (see Table 1.1).

For this example, based on the earlier equations, the value vector is defined as

$$V = f \text{ (capability, suitability, performance, productivity)}$$

Capability: The term "capability" refers to the ability of IT equipment to satisfy multiple requirements. For example, a certain piece of IT equipment may only provide computational service. A different piece of equipment may be capable of generating reports in addition to computational analysis, thus increasing the service variety that can be obtained. In Table 1.1, the levels of increase in service variety from the three competing equipment types are 38%, 40%, and 33%, respectively.

Suitability: "Suitability" refers to the appropriateness of the IT equipment for current operations. For example, the respective percentages of operating scope for which the three options are suitable are 12%, 30%, and 53%.

Performance: "Performance," in this context, refers to the ability of the IT equipment to satisfy schedule and cost requirements. In the example, the three options can, respectively, satisfy requirements on 18%, 28%, and 52% of the typical set of jobs.

Productivity: "Productivity" can be measured by an assessment of the performance of the proposed IT equipment to meet workload requirements in relation to the existing equipment. For example, in Table 1.1, the three options show normalized increases of 0.02, −1.0, and −1.1, respectively, on a uniform scale of productivity measurement. A plot of the histograms of the respective "values" of the three IT options is shown in Figure 1.7. Option C is the best "value" alternative in terms of suitability and performance. Option B shows the

TABLE 1.1

Comparison of IT Equipment Value Options

IT Equipment Options	Suitability ($k = 1$)	Capability ($k = 2$)	Performance ($k = 3$)	Productivity ($k = 4$)
Option A	0.12	0.38	0.18	0.02
Option B	0.30	0.40	0.28	−1.00
Option C	0.53	0.33	0.52	−1.10

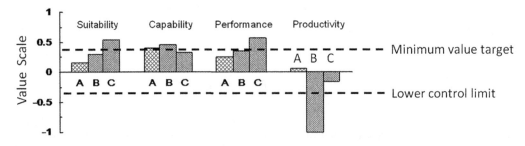

FIGURE 1.7
Relative system value weights of three IT options.

best capability measure, but its productivity is too low to justify the needed investment. Option A offers the best productivity, but its suitability measure is low. The analytical process can incorporate a lower control limit into the quantitative assessment such that any option providing value below that point will not be acceptable. Similarly, a minimum value target can be incorporated into the graphical plot such that each option is expected to exceed the target point on the value scale.

The relative weights used in many justification methodologies are based on subjective propositions of decision makers. Some of those subjective weights can be enhanced by the incorporation of utility models. For example, the weights shown in Table 1.1 could be obtained from utility functions. There is a risk of spending too much time maximizing inputs at "point-of-sale" levels with little time defining and refining outputs at the "wholesale" systems level. Without a systems view, we cannot be sure we are pursuing the right outputs.

Management by Project

Project management continues to grow as an effective means of managing functions in any organization. Project management should be an enterprise-wide systems-based endeavor. Enterprise-wide project management is the application of project management techniques and practices across the full scope of the enterprise. This concept is also referred to as management by project (MBP). MBP is a contemporary concept that employs project management techniques in various functions within an organization. MBP recommends pursuing endeavors as project-oriented activities. It is an effective way to conduct any business activity. It represents a disciplined approach that defines any work assignment as a project. Under MBP, every undertaking is viewed as a project that must be managed just like a traditional project. The characteristics required of each project so defined are

1. An identified scope and a goal
2. A desired completion time
3. Availability of resources
4. A defined performance measure
5. A measurement scale for review of work

An MBP approach to operations helps in identifying unique entities within functional requirements. This identification helps determine where functions overlap and how they are interrelated, thus paving the way for better planning, scheduling, and control. Enterprise-wide project management facilitates a unified view of organizational goals and provides a way for project teams to use information generated by other departments to carry out their functions.

The use of project management continues to grow rapidly. The need to develop effective management tools increases with the increasing complexity of new technologies and processes. The life cycle of a new product to be introduced into a competitive market is a good example of a complex process that must be managed with integrative project management approaches. The product will encounter management functions as it goes from one stage to the next. Project management will be needed throughout the design and production stages of the product. Project management will be needed in developing marketing, transportation, and delivery strategies for the product. When the product finally gets to the customer, project management will be needed to integrate its use with those of other products within the customer's organization.

The need for a project management approach is established by the fact that a project will always tend to increase in size even if its scope is narrowing. The following three literary laws are applicable to any project environment:

1. *Parkinson's law*: Work expands to fill the available time or space.
2. *Peter's principle*: People rise to the level of their incompetence.
3. *Murphy's law*: Whatever can go wrong will.
4. *Badiru's rule*: The grass is always greener where you most need it to be dead.

An integrated systems project management approach can help diminish the adverse impacts of these laws through good project planning, organizing, scheduling, and control.

Integrated Systems Implementation

Project management tools can be classified into three major categories:

1. *Qualitative tools*: There are the managerial tools that aid in the interpersonal and organizational processes required for project management.
2. *Quantitative tools*: These are analytical techniques that aid in the computational aspects of project management.
3. *Computer tools*: These are software and hardware tools that simplify the process of planning, organizing, scheduling, and controlling a project. Software tools can help in both qualitative and quantitative analyses needed for project management.

Although individual books dealing with management principles, optimization models, and computer tools are available, there are few guidelines for the integration of the three areas for project management purposes. In this book, we integrate these three areas for a comprehensive guide to project management. The book introduces the *triad approach* to improve the effectiveness of project management with respect to schedule, cost, and performance constraints within the context of systems modeling. Figure 1.8 illustrates this emphasis. The approach considers not only the management of the project itself but also the management of all the functions that support the project.

It is one thing to have a quantitative model, but it is a different thing to be able to apply the model to real-world problems in a practical form. The systems approach presented in this book illustrates how to make the transition from model to practice.

A systems approach helps increase the intersection of the three categories of project management tools and, hence, improve overall management effectiveness. Crisis should not be the instigator for the use of project management techniques. Project management approaches should be used upfront to prevent avoidable problems rather than to fight them when they develop. What is worth doing is worth doing well, right from the beginning.

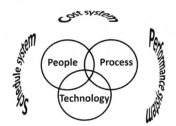

Systems Modeling Environment

Management practices + Computer applications +
Optimization models

FIGURE 1.8
Integration of project management tools.

Critical Factors for Systems Success

The premise of this book is that the critical factors for a system's success revolve around people and the personal commitment and dedication of each person. No matter how good a technology is and no matter how enhanced a process might be, it is ultimately the people involved that determine success. This makes it imperative to take care of people issues first in the overall systems approach to project management. Many organizations recognize this, but only few have been able to actualize the ideals of managing people productively. Execution of operational strategies requires forthrightness, openness, and commitment to get things done. Lip service and arm waving are not sufficient. Tangible programs that cater to the needs of people must be implemented. It is essential to provide incentives, encouragement, and empowerment for people to be self-actuating in determining how best to accomplish their job functions. A summary of critical factors for a system's success encompasses the following:

1. Total system management (hardware, software, and people)
2. Operational effectiveness
3. Operational efficiency
4. System suitability
5. System resilience
6. System affordability
7. System supportability
8. System life cycle cost
9. System performance
10. System schedule
11. System cost

Systems engineering tools, techniques, and processes are essential for project life cycle management to make goals possible within the context of *SMART* principles, which are represented as follows:

1. *Specific*: Pursue specific and explicit outputs.

2. *Measurable*: Design of outputs that can be tracked, measured, and assessed.

3. *Achievable*: Make outputs to be achievable and aligned with organizational goals.

4. *Realistic*: Pursue only the goals that are realistic and result oriented.

5. *Timed*: Make outputs timed to facilitate accountability.

Early Systems Engineering

Systems engineering provides the technical foundation for executing a project success-fully. A systems approach is particularly essential in the early stages of the project to avoid having to reengineer the project at the end of its life cycle. Early systems engineering makes it possible to proactively assess feasibility of meeting user needs, adaptability of new technology, and integration of solutions into regular operations.

DODAF Systems Architecture for Project Management

The military has been a major force in establishing systems engineering platform for executing projects. Through the Department of Defense Architecture Framework (DODAF), the U.S. military executes projects under a consistent platform of an "architecture" perspective borrowed from the conventional physical infrastructure architectural design processes. DODAF is used to standardize the format for architecture descriptions. It seeks to provide a mechanism for operating more efficiently while attending to multiple requirements spread out in multiple and diverse geographical locations. One approach of DODAF adapts traditional architecture to something called *capability architecture*. The reasoning for this is the widespread belief that scores of defense systems are either redundant or do not meet operational needs. As a result, many recent acquisition reform efforts have been aimed at pursuing interoperable and cost-effective *joint* military capabilities.

Traditional architects integrate structure and function with the environment. Their end products, the blueprints, merge various stakeholders' visions and requirements into an acceptable product. They provide sheets, or *views*, that correspond to the homeowner, the plumber, the electrician, the framer, the painter, the residents, and even the neighbors. By contract, the application of systems architecting in the military is not centered on a place of abode (i.e., house), but rather on interoperable weapon systems and diverse spectrum of warfare. This requires a lot of intercomponent coordination. Only a systems view can provide this level of comprehensive appreciation of capability, interdependency, and symbiosis. Systems architecture supports logical interface of capabilities, operations planning, resource requirements, tool development, portfolio management, goal formulation, acquisition, information management, and project phaseout. Some specific requirements for applying systems architecting to program management within the military include the following:

1. The Joint Capabilities Integration and Development System requires that each *capabilities document* contain an annex with a standard Department of Defense (DoD)-formatted architecture. Users and program offices partner to provide those architecture descriptions.

2. The Defense Acquisition System requires architecture to develop systems and manage interoperability of components.

3. Systems that communicate must have *information support plans*, each accompanied by a complete integrated architecture.

4. DOD and U.S. Congress require systems architecture to be used for defense business information systems that cost at least $1 million.

Just as the traditional home architect provides specific *views* to different subcontractors involved in the construction of a house, DODAF prescribes views for various stakeholders involved in a given capability or requirement. There are 26 total views in DODAF organized into three categories:

1. Operational views (OVs)

2. Systems and services views (SVs)

3. Technical standards views (TVs)

The views are a combination of pictures, diagrams, and spreadsheets maintained in an electronic database. The OVs communicate mission-level information and document operational requirements, from a *user* standpoint. The SVs communicate design-level information for use by designers and maintainers. Finally, the TVs document the information technology standards (construction codes) that have been developed for networking compatibility (net centricity). DODAF architecture descriptions are the blueprints for linking key inputs and capabilities for planners, designers, and acquirers. For everyone involved in a large and complex project, a consistent architecture framework can guide the systems-of-systems engineering process. DODAF-integrated architectures provide insight into complex operational relationships, interoperability requirements, and system-related structure.

Grand Challenges for Engineering

The National Academy of Engineering (NAE), in February 2008, released a list of 14 grand challenges for engineering in the coming years. Each area of challenges constitutes a complex project that must be planned and executed strategically. The 14 challenges, which can be viewed as science, technology, engineering, and mathematics (STEM) areas, are listed as follows:

1. Make solar energy affordable

2. Provide energy from fusion

3. Develop carbon sequestration methods

4. Manage the nitrogen cycle

5. Provide access to clean water

6. Restore and improve urban infrastructure

7. Advance health informatics

8. Engineer better medicines
9. Reverse-engineer the brain
10. Prevent nuclear terror
11. Secure cyberspace
12. Enhance virtual reality
13. Advance personalized learning
14. Engineer the tools for scientific discovery

The aforementioned list of existing and forthcoming engineering challenges indicates an urgent need to apply comprehensive systems-based project management to bring about new products, services, and results efficiently within cost and schedule constraints. Project management can effectively be applied to the grand challenges to ensure a realization of objectives.

Systems View of the Grand Challenges

Although the NAE list focuses on engineering challenges, the fact is that every item on the list has the involvement of general areas of STEM, in one form or another. The STEM elements of each area of engineering challenge are contained in the following definitions:

1. *Make solar energy economical*: Solar energy provides less than 1% of the world's total energy, but it has the potential to provide much more.
2. *Provide energy from fusion*: Human-engineered fusion has been demonstrated on a small scale. The challenge is to scale up the process to commercial proportions, in an efficient, economical, and environmentally benign way.
3. *Develop carbon sequestration methods*: Engineers are working on ways to capture and store excess carbon dioxide to prevent global warming.
4. *Manage the nitrogen cycle*: Engineers can help restore balance to the nitrogen cycle with better fertilization technologies and by capturing and recycling waste.
5. *Provide access to clean water*: The world's water supplies are facing new threats; affordable, advanced technologies could make a difference for millions of people around the world.
6. *Restore and improve urban infrastructure*: Good design and advanced materials can improve transportation, energy, water, and waste systems, and also create more sustainable urban environments.
7. *Advance health informatics*: Stronger health information systems not only improve everyday medical visits but also they are essential to counter pandemics and biological or chemical attacks.
8. *Engineer better medicines*: Engineers are developing new systems to use genetic information, sense small changes in the body, assess new drugs, and deliver vaccines.
9. *Reverse-engineer the brain*: The intersection of engineering and neuroscience promises great advances in health care, manufacturing, and communication.

10. *Prevent nuclear terror*: The need for technologies to prevent and respond to a nuclear attack is growing.

11. *Secure cyberspace*: It's more than preventing identity theft. Critical systems in banking, national security, and physical infrastructure may be at risk.

12. *Enhance virtual reality*: True virtual reality creates the illusion of actually being in a difference space. It can be used for training, treatment, and communication.

13. *Advance personalized learning*: Instruction can be individualized based on learning styles, speeds, and interests to make learning more reliable.

14. *Engineer the tools of scientific discovery*: In the century ahead, engineers will continue to be partners with scientists in the great quest for understanding many unanswered questions of nature.

Society will be tackling these grand challenges for the foreseeable decades; and project management is an avenue through which we can ensure that the desired products, services, and results can be achieved. With the positive outcomes these projects achieved, we can improve the quality of life for everyone, and our entire world can benefit positively. In the context of tackling the grand challenges as systems projects, some of the critical issues to address are

1. Strategic implementation plans
2. Strategic communication
3. Knowledge management
4. Evolution of virtual operating environment
5. Structural analysis of projects
6. Analysis of integrative functional areas
7. Project concept mapping
8. Prudent application of technology
9. Scientific control
10. Engineering research and development

Body of Knowledge Methodology

The general Project Management Body of Knowledge (PMBOK®) for project management is published and disseminated by the Project Management Institute (PMI). The body of knowledge comprises specific knowledge areas, which are organized into the following broad areas:

1. Project *integration* management
2. Project *scope* management
3. Project *time* management
4. Project *cost* management
5. Project *quality* management

6. Project *human resource* management

7. Project *communications* management

8. Project *risk* management

9. Project *procurement and subcontract* management

The listed segments of the body of knowledge of project management cover the range of functions associated with any project, particularly complex ones. Multinational projects, particularly, pose unique challenges pertaining to reliable power supply, efficient communication systems, credible government support, dependable procurement processes, consistent availability of technology, progressive industrial climate, trustworthy risk mitigation infrastructure, regular supply of skilled labor, uniform focus on quality of work, global consciousness, hassle-free bureaucratic processes, coherent safety and security system, steady law and order, unflinching focus on customer satisfaction, and fair labor relations. Assessing and resolving concerns about these issues in a step-by-step fashion will create a foundation of success for a large project. While no system can be perfect and satisfactory in all aspects, a tolerable trade-off on factors is essential for a project's success.

Components of the Knowledge Areas

The key components of each element of the body of knowledge are summarized as follows:

1. Integration management
   a. Integrative project charter
   b. Project scope statement
   c. Project management plan
   d. Project execution management
   e. Change control

2. Scope management
   a. Focused scope statements
   b. Cost/benefits analysis
   c. Project constraints
   d. Work breakdown structure
   e. Responsibility breakdown structure
   f. Change control

3. Time management
   a. Schedule planning and control
   b. Program evaluation and review technique (PERT) and Gantt charts
   c. Critical path method
   d. Network models

    e. Resource loading

    f. Reporting

4. Cost management

    a. Financial analysis

    b. Cost estimating

    c. Forecasting

    d. Cost control

    e. Cost reporting

5. Quality management

    a. Total quality management

    b. Quality assurance

    c. Quality control

    d. Cost of quality

    e. Quality conformance

6. Human resource management

    a. Leadership skill development

    b. Team building

    c. Motivation

    d. Conflict management

    e. Compensation

    f. Organizational structures

7. Communications management

    a. Communication matrix

    b. Communication vehicles

    c. Listening and presenting skills

    d. Communication barriers and facilitators

8. Risk management

    a. Risk identification

    b. Risk analysis

    c. Risk mitigation

    d. Contingency planning

9. Procurement and subcontract management

    a. Material selection

    b. Vendor prequalification

    c. Contract types

  d. Contract risk assessment

  e. Contract negotiation

  f. Contract change orders

Step-by-Step and Component-by-Component Implementation

The efficacy of the systems approach is based on a step-by-step and component-by-component implementation of the project management process. The major knowledge areas of project management are administered in a structured outline covering six basic clusters consisting of the following:

1. Initiating
2. Planning
3. Executing
4. Monitoring
5. Controlling
6. Closing

Expanding on the six basic clusters, Figure 1.9 presents a comprehensive representation of project management steps going from problem statement to project closing.

 The implementation clusters represent five process groups that are followed throughout the project life cycle. Each cluster itself consists of several functions and operational steps. When the clusters are overlaid on the nine knowledge areas, we obtain a two-dimensional matrix that spans 44 major process steps. Table 1.2 shows an overlay of the project management knowledge areas and the implementation clusters. The monitoring and controlling clusters are usually administered as one lumped process group (monitoring and controlling). In some cases, it may be helpful to separate them to highlight the essential attributes of each cluster of functions over the project life cycle. In practice, the processes and clusters do overlap. Thus, there is no crisp demarcation of when and where one process ends and where another one begins over the project life cycle. In general, project life cycle defines the following:

1. Resources that will be needed in each phase of the project life cycle
2. Specific work to be accomplished in each phase of the project life cycle

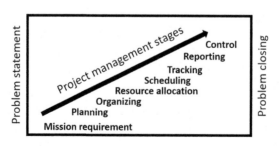

FIGURE 1.9
Expanded project management steps.

TABLE 1.2

Overlay of Project Management Areas and Implementation Clusters

Knowledge Areas	Project Management Process Clusters				
	Initiating	Planning	Executing	Monitoring/ Controlling	Closing
Project integration	Develop project charter Develop preliminary project scope	Develop project management plan	Direct and manage project execution	Monitor and control project work Integrated change control	
Scope		Scope planning Scope definition Create Work Breakdown Structure (WBS)		Scope verification Scope control	
Time		Activity definition Activity sequencing Activity resource estimating Activity duration estimating Schedule development		Schedule control	
Cost		Cost estimating Cost budgeting		Cost control	
Quality		Quality planning	Perform quality assurance	Perform quality control	
Human resources		Human resource planning	Acquire project team Develop project team	Manage project team	
Communication		Communication planning	Information distribution	Performance reporting Manage stakeholders	
Risk		Risk management planning Risk identification Qualitative risk analysis Quantitative risk analysis Risk response planning		Risk monitoring and control	
Procurement		Plan purchases and acquisitions Plan contracting	Request seller responses Select sellers	Contract administration	Contract closure

Figure 1.10 shows the major phases of project life cycle going from the conceptual phase through the close-out phase.

It should be noted that project life cycle is distinguished from the product life cycle. Project life cycle does not explicitly address operational issues, whereas product life cycle is mostly about operational issues starting from the product's delivery to the end of its useful life. Note that, for technical projects, the shape of the life cycle curve may be expedited due to the rapid developments that often occur in science and technology activities. For example, for a high technology project, the entire life cycle may be shortened, with a very

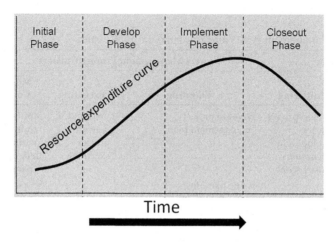

FIGURE 1.10
Project execution phases for systems implementation.

rapid initial phase, even though the conceptualization stage may be very long. Typical characteristics of project life cycle include the following:

1. Cost and staffing requirements are lowest at the beginning of the project and ramp up during the initial and development stages.

2. The probability of successfully completing the project is lowest at the beginning and highest at the end. This is because many unknowns (risks and uncertainties) exist at the beginning of the project. As the project nears its end, there are fewer opportunities for risks and uncertainties.

3. The risks to the project organization (project owner) are lowest at the beginning and highest at the end. This is because not much investment has gone into the project at the beginning, whereas much has been committed by the end of the project. There is a higher sunk cost manifested at the end of the project.

4. The ability of the stakeholders to influence the final project outcome (cost, quality, and schedule) is highest at the beginning and gets progressively lower toward the end of the project. This is intuitive because influence is best exerted at the beginning of an endeavor.

5. Value of scope changes decreases over time during the project life cycle while the cost of scope changes increases over time. The suggestion is to decide and finalize scope as early as possible. If there are to be scope changes, do them as early as possible.

Project Systems Structure

The overall project management systems execution can be outlined as summarized in the following.

Problem Identification

Problem identification is the stage where a need for a proposed project is identified, defined, and justified. A project may be concerned with the development of new products, implementation of new processes, or improvement of existing facilities.

Project Definition

Project definition is the phase at which the purpose of the project is clarified. A *mission statement* is the major output of this stage. For example, a prevailing low level of productivity may indicate a need for a new manufacturing technology. In general, the definition should specify how project management may be used to avoid missed deadlines, poor scheduling, inadequate resource allocation, lack of coordination, poor quality, and conflicting priorities.

Project Planning

A plan represents an outline of a series of actions needed to accomplish a goal. Project planning determines how to initiate a project and execute its objectives. It may be a simple statement of a project goal or it may be a detailed account of procedures to be followed during the project. Project planning is discussed in detail in Chapter 2. Planning can be summarized as

1. Objectives
2. Project definition
3. Team organization
4. Performance criteria (time, cost, quality)

Project Organizing

Project organization specifies how to integrate the functions of the personnel involved in a project. Organizing is usually done concurrently with project planning. Directing is an important aspect of project organization. Directing involves guiding and supervising the project personnel. It is a crucial aspect of management function. Directing requires skillful managers who can interact with subordinates effectively through good communication and motivation techniques. A good project manager will facilitate a project's success by directing his or her staff, through proper task assignments, toward the project goal.

Workers perform better when there are clearly defined expectations. They need to know how their job functions contribute to the overall goals of the project. Workers should be given some flexibility for self-direction in performing their functions. Individual worker needs and limitations should be recognized by the manager when directing project functions. Directing a project requires skills dealing with motivating, supervising, and delegating.

Resource Allocation

Project goals and objectives are accomplished by allocating resources to functional requirements. Resources can consist of money, people, equipment, tools, facilities, information, skills, and so on. These are usually in short supply. The people needed for a particular task may be committed to other ongoing projects. A crucial piece of equipment may be under the control of another team. Chapter 5 addresses resource allocation in detail.

Project Scheduling

Timeliness is the essence of project management. Scheduling is often the major focus in project management. The main purpose of scheduling is to allocate resources so that the overall project objectives are achieved within a reasonable time span. Project objectives are generally conflicting in nature. For example, minimization of the project completion time and minimization of the project cost are conflicting objectives. That is, one objective is improved at the expense of worsening the other objective. Therefore, project scheduling is a multiple-objective decision-making problem.

In general, scheduling involves the assignment of time periods to specific tasks within the work schedule. Resource availability, time limitations, urgency level, required performance level, precedence requirements, work priorities, technical constraints, and other factors complicate the scheduling process. Thus, the assignment of a time slot to a task does not necessarily ensure that the task will be performed satisfactorily in accordance with the schedule. Consequently, careful control must be developed and maintained throughout the project scheduling process.

Project Tracking and Reporting

This phase involves checking whether or not project results conform to project plans and performance specifications. Tracking and reporting are prerequisites for project control. A properly organized report of the project status will help identify any deficiencies in the progress of the project and help pinpoint corrective actions.

Project Control

Project control requires that appropriate actions be taken to correct unacceptable deviations from expected performance. Control is actuated through measurement, evaluation, and corrective action. Measurement is the process of measuring the relationship between planned performance and actual performance with respect to project objectives. The variables to be measured, the measurement scales, and the measuring approaches should be clearly specified during the planning stage. Corrective actions may involve rescheduling, reallocation of resources, or expedition of task performance. Project control is discussed in detail in Chapter 6. Control involves

1. Tracking and reporting
2. Measurement and evaluation
3. Corrective action (plan revision, rescheduling, updating)

Project Termination

Termination is the last stage of a project. The phaseout of a project is as important as its initiation. The termination of a project should be implemented expeditiously. A project should not be allowed to drag on after the expected completion time. A terminal activity should be defined for a project during the planning phase. An example of a terminal activity may be the submission of a final report, the power of new equipment, or the signing of a release order. The conclusion of such an activity should be viewed as the completion of the project. Arrangements may be made for follow-up activities that may improve or extend the outcome of the project. These follow-up or spin-off projects should be managed as new projects but with proper input–output relationships within the sequence of projects.

Project Systems Implementation Outline

While this book is aligned with the main tenets of PMI's PMBOK, the book uses the traditional project management textbook framework encompassing the broad sequence of the following categories:

$$\text{Planning} \rightarrow \text{Organizing} \rightarrow \text{Scheduling} \rightarrow \text{Control} \rightarrow \text{Termination}$$

An outline of the functions to be carried out during a project should be made during the planning stage of the project. A model for such an outline is presented hereafter. It may be necessary to rearrange the contents of the outline to fit the specific needs of a project.

Planning

1. Specify project background
    a. Define current situation and process
        i. Understand the process
        ii. Identify important variables
        iii. Quantify variables
    b. Identify areas for improvement
        i. List and discuss the areas
        ii. Study potential strategy for solution
2. Define unique terminologies relevant to the project
    a. Industry-specific terminologies
    b. Company-specific terminologies
    c. Project-specific terminologies
3. Define project goal and objectives
    a. Write mission statement
    b. Solicit inputs and ideas from personnel
4. Establish performance standards
    a. Schedule

    b. Performance

    c. Cost

  5. Conduct formal project feasibility study

    a. Determine impact on cost

    b. Determine impact on organization

    c. Determine project deliverables

  6. Secure management support

Organizing

  1. Identify project management team

    a. Specify project organization structure

      i. Matrix structure

      ii. Formal and informal structures

      iii. Justify structure

    b. Specify departments involved and key personnel

      i. Purchasing

      ii. Materials management

      iii. Engineering, design, manufacturing, and so on

    c. Define project management responsibilities

      i. Select project manager

      ii. Write project charter

      iii. Establish project policies and procedures

  2. Implement triple C model

    a. Communication

      i. Determine communication interfaces

      ii. Develop communication matrix

    b. Cooperation

      i. Outline cooperation requirements, policies, and procedures

    c. Coordination

      i. Develop work breakdown structure

      ii. Assign task responsibilities

      iii. Develop responsibility chart

Scheduling (Resource Allocation)

  1. Develop master schedule

    a. Estimate task duration

    b. Identify task precedence requirements

      i. Technical precedence

       ii. Resource-imposed precedence

      iii. Procedural precedence

  c. Use analytical models

       i. Critical path method (CPM)

      ii. PERT

     iii. Gantt chart

     iv. Optimization models

Control (Tracking, Reporting, and Correction)

1. Establish guidelines for tracking, reporting, and control
   a. Define data requirements
      i. Data categories
      ii. Data characterization
      iii. Measurement scales
   b. Develop data documentation
      i. Data update requirements
      ii. Data quality control
      iii. Establish data security measures
2. Categorize control points
   a. Schedule audit
      i. Activity network and Gantt charts
      ii. Milestones
      iii. Delivery schedule
   b. Performance audit
      i. Employee performance
      ii. Product quality
   c. Cost audit
      i. Cost containment measures
      ii. Percent completion versus budget depletion
3. Identify implementation process
   a. Comparison with targeted schedules
   b. Corrective course of action
      i. Rescheduling
      ii. Reallocation of resources

Termination (Close, Phaseout)

1. Conduct performance review
2. Develop strategy for follow-up projects
3. Arrange for personnel retention, release, and reassignment

Documentation

1. Document project outcome
2. Submit final report
3. Archive report for future reference

Value of Lean Times

In biblical times, the "seven lean years" were followed by a period of plenty, affluence, and prosperity. Facing a lean period in project management creates value in terms of figuring out how to eliminate or reduce operational waste that is inherent in many human-governed processes. It is a natural fact that having to make do with limited resources creates opportunities for resourcefulness and innovation, which requires an integrated-systems view of what is available and what can be leveraged. The lean principles that are now being embraced by business, industry, and government have been around for a long time. It is just that we are now being forced to implement lean practices due to the escalating shortage of resources. It is unrealistic to expect that problems that have enrooted themselves in different parts of an organization can be solved by a single-point attack. Rather, a systematic probing of all the nooks and corners of the problem must be assessed and tackled in an integrated manner. Like the biblical Joseph, whose life was on the line while interpreting dreams for the Egyptian pharaoh, decision makers cannot afford to misinterpret systems warning signs when managing large and complex projects.

Contrary to the contention in some technocratic circles that budget cuts will stifle innovation, it is a fact that a reduction of resources often forces more creativity in identifying wastes and leveraging opportunities that lie fallow in nooks and crannies of an organization. This is not an issue of wanting more for less. Rather, it is an issue of doing more with less. It is through a systems viewpoint that new opportunities to innovate can be spotted. Necessity does, indeed, spur invention.

Systems Decision Analysis

Systems decision analysis facilitates a proper consideration of the essential elements of decisions in a project systems environment. These essential elements include the problem statement, information, performance measure, decision model, and an implementation of the decision. The recommended steps are enumerated as follows:

Step 1. Problem Statement

A problem involves choosing between competing, and probably conflicting, alternatives. The components of problem solving in project management include

1. Describing the problem (goals, performance measures)
2. Defining a model to represent the problem
3. Solving the model

4. Testing the solution

5. Implementing and maintaining the solution

Problem definition is very crucial. In many cases, *symptoms* of a problem are more readily recognized than its *cause* and *location*. Even after the problem is accurately identified and defined, a benefit/cost analysis may be needed to determine whether the cost of solving the problem is justified.

Step 2. Data and Information Requirements

Information is the driving force for the project decision process. Information clarifies the relative states of past, present, and future events. The collection, storage, retrieval, organization, and processing of raw data are important components for generating information. Without data, there can be no information. Without good information, there cannot be a valid decision. The essential requirements for generating information are

1. Ensuring that an effective data collection procedure is followed

2. Determining the type and the appropriate amount of data to collect

3. Evaluating the data collected with respect to information potential

4. Evaluating the cost of collecting the required data

For example, suppose a manager is presented with a recorded fact that says, "Sales for the last quarter are 10,000 units." This constitutes ordinary data. There are many ways of using the aforementioned data to make a decision, depending on the manager's value system. An analyst, however, can ensure proper use of the data by transforming it into information, such as "Sales of 10,000 units for the last quarter are within x percent of the targeted value." This type of information is more useful to the manager for decision making.

Step 3. Performance Measure

A performance measure for the competing alternatives should be specified. The decision maker assigns a perceived worth or value to available alternatives. Setting measures of performance is crucial to the process of defining and selecting alternatives. Some performance measures, commonly used in project management are project cost, completion time, resource usage, and stability in the workforce.

Step 4. Decision Model

A decision model provides the basis for the analysis and synthesis of information, and is the mechanism by which competing alternatives are compared. To be effective, a decision model must be based on a systematic and logical framework for guiding project decisions. A decision model can be a verbal, graphical, or mathematical representation of the ideas in the decision-making process. A project decision model should have the following characteristics:

1. Simplified representation of the actual situation

2. Explanation and prediction of the actual situation

3. Validity and appropriateness
4. Applicability to similar problems

The formulation of a decision model involves three essential components:

1. *Abstraction*: Determining the relevant factors
2. *Construction*: Combining the factors into a logical model
3. *Validation*: Assuring that the model adequately represents the problem

The basic types of decision models for project management are described next:

1. *Descriptive models*: These models are directed at describing a decision scenario and identifying the associated problem. For example, a project analyst might use a CPM network model to identify bottleneck tasks in a project.
2. *Prescriptive models*: These models furnish procedural guidelines for implementing actions. The triple C approach (Badiru, 2008), for example, is a model that prescribes the procedures for achieving communication, cooperation, and coordination in a project environment.
3. *Predictive models*: These models are used to predict future events in a problem environment. They are typically based on historical data about the problem situation. For example, a regression model based on past data may be used to predict future productivity gains associated with expected levels of resource allocation. Simulation models can be used when uncertainties exist in the task durations or resource requirements.
4. *Satisficing models*: These are models that provide trade-off strategies for achieving a satisfactory solution to a problem, within given constraints. Goal programming and other multicriteria techniques provide good satisficing solutions. For example, these models are helpful in cases where time limitations, resource shortages, and performance requirements constrain the implementation of a project.
5. *Optimization models*: These models are designed to find the best available solution to a problem subject to a certain set of constraints. For example, a linear programming model can be used to determine the optimal product mix in a production environment.

In many situations, two or more of the aforementioned models may be involved in the solution of a problem. For example, a descriptive model might provide insights into the nature of the problem; an optimization model might provide the optimal set of actions to take in solving the problem; a satisficing model might temper the optimal solution with reality; a prescriptive model might suggest the procedures for implementing the selected solution; and a predictive model might forecast a future outcome of the problem scenario.

Step 5. Making the Decision

Using the available data, information, and the decision model, the decision maker will determine the real-world actions that are needed to solve the stated problem. A sensitivity analysis may be useful for determining what changes in parameter values might cause a change in the decision.

Step 6. Implementing the Decision

A decision represents the selection of an alternative that satisfies the objective stated in the problem statement. A good decision is useless until it is implemented. An important aspect of a decision is to specify how it is to be implemented. Selling the decision and the project to management requires a well-organized persuasive presentation. The way a decision is presented can directly influence whether or not it is adopted. The presentation of a decision should include at least the following: an executive summary, technical aspects of the decision, managerial aspects of the decision, resources required to implement the decision, cost of the decision, the time frame for implementing the decision, and the risks associated with the decision.

Group Systems Decision-Making Models

Systems decisions are often complex, diffuse, distributed, and poorly understood. No one has all the information to make all decisions accurately. As a result, crucial decisions are made by a group of people. Some organizations use outside consultants with appropriate expertise to make recommendations for important decisions. Other organizations set up their own internal consulting groups without having to go outside the organization. Decisions can be made through linear responsibility, in which case one person makes the final decision based on inputs from other people. Decisions can also be made through shared responsibility, in which case, a group of people share the responsibility for making joint decisions. The major advantages of group decision making are listed as follows:

1. Facilitation of a systems view of the problem environment.
2. Ability to share experience, knowledge, and resources. Many heads are better than one. A group will possess greater collective ability to solve a given decision problem.
3. Increased credibility. Decisions made by a group of people often carry more weight in an organization.
4. Improved morale. Personnel morale can be positively influenced because many people have the opportunity to participate in the decision-making process.
5. Better rationalization. The opportunity to observe other people's views can lead to an improvement in an individual's reasoning process.
6. Ability to accumulate more knowledge and facts from diverse sources.
7. Access to broader perspectives spanning different problem scenarios.
8. Ability to generate and consider alternatives from different perspectives.
9. Possibility for a broader-base involvement, leading to a higher likelihood of support.
10. Possibility for group leverage for networking, communication, and political clout.

In spite of the much-desired advantages, group decision making does possess the risk of flaws. Some possible disadvantages of group decision making are listed as follows:

1. Difficulty in arriving at a decision.
2. Slow operating time frame.
3. Possibility for individuals' conflicting views and objectives.
4. Reluctance of some individuals in implementing the decision.
5. Potential for power struggle and conflicts among the group.
6. Loss of productive employee time.
7. Too much compromise may lead to a less than optimal group output.
8. Risk of one individual dominating the group.
9. Overreliance on group process may impede agility of management to make decision fast.
10. Risk of dragging feet due to repeated and iterative group meetings.

Brainstorming

Brainstorming is a way of generating many new ideas. In brainstorming, the decision group comes together to discuss alternate ways of solving a problem. The members of the brainstorming group may be from different departments, may have different backgrounds and training, and may not even know one another. The diversity of the participants helps create a stimulating environment for generating different ideas from different viewpoints. The technique encourages free outward expression of new ideas, no matter how farfetched the ideas might appear. No criticism of any new idea is permitted during the brainstorming session. A major concern in brainstorming is that extroverts may take control of the discussions. For this reason, an experienced and respected individual should manage the brainstorming discussion. The group leader establishes the procedure for proposing ideas, keeps the discussions in line with the group's mission, discourages disruptive statements, and encourages the participation of all members.

After the group runs out of ideas, open discussions are held to weed out the unsuitable ones. It is expected that even the rejected ideas may stimulate the generation of other ideas, which may eventually lead to other favored ideas. Guidelines for improving brainstorming sessions are presented as follows:

1. Focus on a specific decision problem.
2. Keep ideas relevant to the intended decision.
3. Be receptive to all new ideas.
4. Evaluate the ideas on a relative basis after exhausting new ideas.
5. Maintain an atmosphere conducive to cooperative discussions.
6. Maintain a record of the ideas generated.

Delphi Method

The traditional approach to group decision making is to obtain the opinion of experienced participants through open discussions. An attempt is made to reach a consensus among the participants. However, open group discussions are often biased because of the influence of subtle intimidation from dominant individuals. Even when the threat of a dominant individual is not present, opinions may still be swayed by group pressure. This is called the "bandwagon effect" of group decision making.

The Delphi method attempts to overcome these difficulties by requiring individuals to present their opinions anonymously through an intermediary. The method differs from the other interactive group methods because it eliminates face-to-face confrontations. It was originally developed for forecasting applications, but it has been modified in various ways for application to different types of decision making. The method can be quite useful for project management decisions. It is particularly effective when decisions must be based on a broad set of factors. The Delphi method is normally implemented as follows:

1. *Problem definition*: A decision problem that is considered significant is identified and clearly described.

2. *Group selection*: An appropriate group of experts or experienced individuals is formed to address the particular decision problem. Both internal and external experts may be involved in the Delphi process. A leading individual is appointed to serve as the administrator of the decision process. The group may operate through the mail or gather together in a room. In either case, all opinions are expressed anonymously on paper. If the group meets in the same room, care should be taken to provide enough room so that each member does not have the feeling that someone may accidentally or deliberately observe their responses.

3. *Initial opinion poll*: The technique is initiated by describing the problem to be addressed in unambiguous terms. The group members are requested to submit a list of major areas of concern in their specialty areas as they relate to the decision problem.

4. *Questionnaire design and distribution*: Questionnaires are prepared to address the areas of concern related to the decision problem. The written responses to the questionnaires are collected and organized by the administrator. The administrator aggregates the responses in a statistical format. For example, the average, mode, and median of the responses may be computed. This analysis is distributed to the decision group. Each member can then see how his or her responses compare with the anonymous views of the other members.

5. *Iterative balloting*: Additional questionnaires based on the previous responses are passed to the members. The members submit their responses again. They may choose to alter or not to alter their previous responses.

6. *Silent discussions and consensus*: The iterative balloting may involve anonymous written discussions of why some responses are correct or incorrect. The process is continued until a consensus is reached. A consensus may be declared after five or six iterations of the balloting or when a specified percentage (e.g., 80%) of the group agrees on the questionnaires. If a consensus cannot be declared on a particular point, it may be displayed to the whole group with a note that it does not represent a consensus.

In addition to its use in technological forecasting, the Delphi method has been widely used in other general decision making. Its major characteristics of anonymity of responses, statistical summary of responses, and controlled procedure make it a reliable mechanism for obtaining numeric data from subjective opinion. The major limitations of the Delphi method are

1. Its effectiveness may be limited in cultures where strict hierarchy, seniority, and age influence decision-making processes.
2. Some experts may not readily accept the contribution of nonexperts to the group decision-making process.
3. Since opinions are expressed anonymously, some members may take the liberty of making ludicrous statements. However, if the group composition is carefully reviewed, this problem may be avoided.

Nominal Group Technique

The nominal group technique is a silent version of brainstorming. It is a method of reaching consensus. Rather than asking people to state their ideas aloud, the team leader asks each member to jot down a minimum number of ideas, for example, five or six. A single list of ideas is then written on a chalkboard for the whole group to see. The group then discusses the ideas and weeds out some iteratively until a final decision is made. The nominal group technique is easier to control. Unlike brainstorming where members may get into shouting matches, the nominal group technique permits members to silently present their views. In addition, it allows introversive members to contribute to the decision without the pressure of having to speak out too often.

In all of the group decision-making techniques, an important aspect that can enhance and expedite the decision-making process is to require that members review all pertinent data before coming to the group meeting. This will ensure that the decision process is not impeded by trivial preliminary discussions. Some disadvantages of group decision making are

1. Peer pressure in a group situation may influence a member's opinion or discussions.
2. In a large group, some members may not get to participate effectively in the discussions.
3. A member's relative reputation in the group may influence how well his or her opinion is rated.
4. A member with a dominant personality may overwhelm the other members in the discussions.
5. The limited time available to the group may create a time pressure that forces some members to present their opinions without fully evaluating the ramifications of the available data.
6. It is often difficult to get all members of a decision group together at the same time.

Despite the noted disadvantages, group decision making definitely has many advantages that may nullify the shortcomings. The advantages as presented earlier will have varying levels of effect from one organization to another. The triple C principle presented in

Chapter 2 may also be used to improve the success of decision teams. Team work can be enhanced in group decision making by adhering to the following guidelines:

1. Get a willing group of people together.
2. Set an achievable goal for the group.
3. Determine the limitations of the group.
4. Develop a set of guiding rules for the group.
5. Create an atmosphere conducive to group synergism.
6. Identify the questions to be addressed in advance.
7. Plan to address only one topic per meeting.

For major decisions and long-term group activities, arrange for team training that allows the group to learn the decision rules and responsibilities together. The steps for the nominal group technique are

1. Silently generate ideas, in writing.
2. Record ideas without discussion.
3. Conduct group discussion for clarification of meaning, not argument.
4. Vote to establish the priority or rank of each item.
5. Discuss vote.
6. Cast final vote.

Interviews, Surveys, and Questionnaires

Interviews, surveys, and questionnaires are important information gathering techniques. They also foster cooperative working relationships. They encourage direct participation and inputs into project decision-making processes. They provide an opportunity for employees at the lower levels of an organization to contribute ideas and inputs for decision making. The greater the number of people involved in the interviews, surveys, and questionnaires, the more valid the final decision. The following guidelines are useful for conducting interviews, surveys, and questionnaires to collect data and information for project decisions:

1. Collect and organize background information and supporting documents on the items to be covered by the interview, survey, or questionnaire.
2. Outline the items to be covered and list the major questions to be asked.
3. Use a suitable medium of interaction and communication: telephone, fax, electronic mail, face-to-face, observation, meeting venue, poster, or memo.
4. Tell the respondent the purpose of the interview, survey, or questionnaire, and indicate how long it will take.
5. Use open-ended questions that stimulate ideas from the respondents.
6. Minimize the use of yes or no type of questions.
7. Encourage expressive statements that indicate the respondent's views.
8. Use the who, what, where, when, why, and how approach to elicit specific information.

9. Thank the respondents for their participation.

10. Let the respondents know the outcome of the exercise.

Multivote

Multivoting is a series of votes used to arrive at a group decision. It can be used to assign priorities to a list of items. It can be used at team meetings after a brainstorming session has generated a long list of items. Multivoting helps reduce such long lists to a few items, usually three to five. The steps for multivoting are

1. Take a first vote. Each person votes as many times as desired, but only once per item.

2. Circle the items receiving a relatively higher number of votes (i.e., majority vote) than the other items.

3. Take a second vote. Each person votes for a number of items equal to one-half the total number of items circled in Step 2. Only one vote per item is permitted.

4. Repeat Steps 2 and 3 until the list is reduced to three to five items depending on the needs of the group. It is not recommended to multivote down to only one item.

5. Perform further analysis of the items selected in Step 4, if needed.

Systems Hierarchy

The traditional concepts of systems analysis are applicable to the project process. The definitions of a project system and its components are presented next:

1. *System*: A project system consists of interrelated elements organized for the purpose of achieving a common goal. The elements are organized to work synergistically to generate a unified output that is greater than the sum of the individual outputs of the components.

2. *Program*: A program is a very large and prolonged undertaking. Such endeavors often span several years. Programs are usually associated with particular systems. For example, we may have a space exploration program within a national defense system.

3. *Project*: A project is a time-phased effort of much smaller scope and duration than a program. Programs are sometimes viewed as consisting of a set of projects. Government projects are often called *programs* because of their broad and comprehensive nature. Industry tends to use the term *project* because of the short-term and focused nature of most industrial efforts.

4. *Task*: A task is a functional element of a project. A project is composed of a sequence of tasks that contribute to the overall project goal.

5. *Activity*: An activity can be defined as a single element of a project. Activities are generally smaller in scope than tasks. In a detailed analysis of a project, an activity

may be viewed as the smallest, practically indivisible work element of the project. For example, we can regard a manufacturing plant as a system. A plantwide endeavor to improve productivity can be viewed as a program. The installation of a flexible manufacturing system is a project within the productivity improvement program. The process of identifying and selecting equipment vendors is a task, and the actual process of placing an order with a preferred vendor is an activity. The systems structure of a project is illustrated in Figure 1.11.

The emergence of systems development has had an extensive effect on project management in recent years. A system can be defined as a collection of interrelated elements brought together to achieve a specified objective. In management context, the purposes of a system are to develop and manage operational procedures and to facilitate an effective decision-making process. Some of the common characteristics of a system include

1. Interaction with the environment
2. Objective
3. Self-regulation
4. Self-adjustment

Representative components of a project system are the organizational subsystem, planning subsystem, scheduling subsystem, information management subsystem, control subsystem, and project delivery subsystem. The primary responsibilities of project analysts involve ensuring the proper flow of information throughout the project system. The classical approach to the decision process follows rigid lines of organizational charts. By contrast, the systems approach considers all the interactions necessary among the various elements of an organization in the decision process.

The various elements (or subsystems) of the organization act simultaneously in a separate but interrelated fashion to achieve a common goal. This synergism helps to expedite the decision process and to enhance the effectiveness of decisions. The supporting commitments from other subsystems of the organization serve to counterbalance the weaknesses of a given subsystem. Thus, the overall effectiveness of the system is greater than the sum of the individual results from the subsystems.

The increasing complexity of organizations and projects makes the systems approach essential in today's management environment. As the number of complex projects increase, there will be an increasing need for project management professionals who can function as

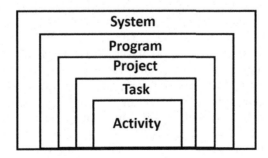

FIGURE 1.11
Hierarchy of a project system.

systems integrators. Project management techniques can be applied to the various stages of implementing a system, as shown in the following guidelines:

1. *Systems definition*: Define the system and associated problems using keywords that signify the importance of the problem to the overall organization. Locate experts in this area who are willing to contribute to the effort. Prepare and announce the development plan.

2. *Personnel assignment*: The project group and the respective tasks should be announced, a qualified project manager should be appointed, and a solid line of command should be established and enforced.

3. *Project initiation*: Arrange an organizational meeting during which a general approach to the problem should be discussed. Prepare a specific development plan and arrange for the installation of needed hardware and tools.

4. *System prototype*: Develop a prototype system, test it, and learn more about the problem from the test results.

5. *Full system development*: Expand the prototype to a full system, evaluate the user interface structure, and incorporate user training facilities and documentation.

6. *System verification*: Get experts and potential users involved, ensure that the system performs as designed, and debug the system as needed.

7. *System validation*: Ensure that the system yields expected outputs. Validate the system by evaluating performance level, such as percentage of success in so many trials, measuring the level of deviation from expected outputs, and measuring the effectiveness of the system output in solving the problem.

8. *System integration*: Implement the full system as planned, ensure the system can coexist with systems already in operation, and arrange for technology transfer to other projects.

9. *System maintenance*: Arrange for continuing maintenance of the system. Update solution procedures as new pieces of information become available. Retain responsibility for system performance or delegate to well-trained and authorized personnel.

10. *Documentation*: Prepare full documentation of the system, prepare a user's guide, and appoint a user consultant.

Systems integration permits sharing of resources. Physical equipment, concepts, information, and skills may be shared as resources. Systems integration is now a major concern of many organizations. Even some of the organizations that traditionally compete and typically shun cooperative efforts are beginning to appreciate the value of integrating their operations. For these reasons, systems integration has emerged as a major interest in business. Systems integration may involve the physical integration of technical components, objective integration of operations, conceptual integration of management processes, or a combination of any of these.

Systems integration involves the linking of components to form subsystems and the linking of subsystems to form composite systems within a single department and/or across departments. It facilitates the coordination of technical and managerial efforts to enhance organizational functions, reduce cost, save energy, improve productivity, and increase the utilization of resources. Systems integration emphasizes the identification and coordination of the interface requirements among the components in an integrated system. The components and subsystems operate synergistically to optimize the performance of the total system. Systems integration ensures that all performance goals are satisfied with a minimum expenditure of time and resources. Integration can be achieved in several forms including the following:

1. *Dual-use integration*: This involves the use of a single component by separate subsystems to reduce both the initial cost and the operating cost during the project life cycle.
2. *Dynamic resource integration*: This involves integrating the resource flows of two normally separate subsystems so that the resource flow from one to or through the other minimizes the total resource requirements in a project.
3. *Restructuring of functions*: This involves the restructuring of functions and reintegration of subsystems to optimize costs when a new subsystem is introduced into the project environment.

Systems integration is particularly important when introducing new technology into an existing system. It involves coordinating new operations to coexist with existing operations. It may require the adjustment of functions to permit the sharing of resources, development of new policies to accommodate product integration, or realignment of managerial responsibilities. It can affect both hardware and software components of an organization. Presented in the following list are guidelines and important questions relevant for systems integration:

1. What are the unique characteristics of each component in the integrated system?
2. How do the characteristics complement one another?
3. What physical interfaces exist among the components?
4. What data/information interfaces exist among the components?
5. What ideological differences exist among the components?
6. What are the data flow requirements for the components?
7. Are there similar integrated systems operating elsewhere?
8. What are the reporting requirements in the integrated system?
9. Are there any hierarchical restrictions on the operations of the components of the integrated system?
10. What internal and external factors are expected to influence the integrated system?
11. How can the performance of the integrated system be measured?
12. What benefit/cost documentations are required for the integrated system?
13. What is the cost of designing and implementing the integrated system?
14. What are the relative priorities assigned to each component of the integrated system?
15. What are the strengths of the integrated system?
16. What are the weaknesses of the integrated system?
17. What resources are needed to keep the integrated system operating satisfactorily?
18. Which section of the organization will have primary responsibility for the operation of an integrated system?
19. What are the quality specifications and requirements for integrated systems?

The integrated approach to project management recommended in this book is represented by the flowchart in Figure 1.12.

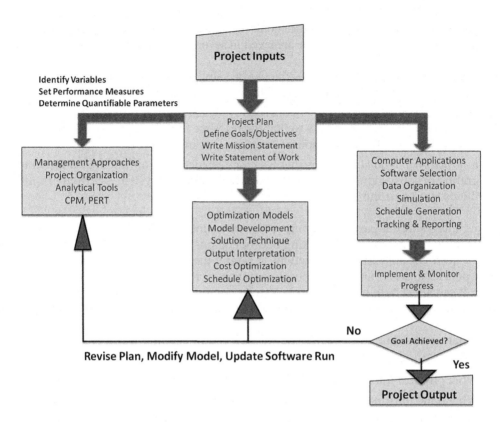

FIGURE 1.12
Flowchart of an integrated project management.

The process starts with a managerial analysis of the project effort. Goals and objectives are defined, a mission statement is written, and a statement of work is developed. After these, traditional project management approaches, such as the selection of an organization structure, are employed. Conventional analytical tools including the CPM and PERT are then mobilized. The use of optimization models is then called upon as appropriate. Some of the parameters to be optimized are cost, resource allocation, and schedule length. It should be understood that not all project parameters will be amenable to optimization. The use of commercial project management software should start only after the managerial functions have been completed. Some project management software have built-in capabilities for the planning and optimization needs.

A frequent mistake in project management is the rush to use a project management software without first completing the planning and analytical studies required by the project. Project management software should be used as a management tool, the same way a word processor is used as a writing tool. It will not be effective to start using the word processor without first organizing the thoughts about what is to be written. Project management is much more than just the project management software. If project management is carried out in accordance with the integration approach presented in the flowchart, the odds of success will be increased. Of course, the structure of the flowchart should not be rigid. Flows and interfaces among the blocks in the flowchart may need to be altered or modified depending on specific project needs.

D-E-J-I Model for Project Execution

The biggest challenge for any project management endeavor is coordinating and integrating the multiple facets that affect the final outputs of a project, where a specific output may be a physical product, a service, or a desired result. Addressing the challenges of project execution from a systems perspective increases the likelihood of success. Badiru (2010) introduced the D-E-J-I (design, evaluate, justify, and integrate) model to facilitate project success through structural implementation. Although originally developed for product development projects, the model is generally applicable to all types of project as every project goes through the stages of process design, evaluation of parameters, justification of the project, and integration of the project into the core business of the organization. The model can be applied across the spectrum of the following elements of an organization:

1. People
2. Process
3. Technology

Figure 1.13 shows the D-E-J-I systems model as a framework of systems design, evaluation, justification, and integration.

Design Stage of D-E-J-I Systems Model

Product or process design should be structured to follow point-to-point transformation. A good technique to accomplish this is the use of state-space transformation, with which we can track the evolution of a project from concept stage to final product stage. For the purpose of project management, we adopt the general definitions and characteristics of state-space

FIGURE 1.13
D-E-J-I model for systems design, evaluation, justification, and integration.

modeling. A state is a set of conditions that describes a process at a specified point in time. A formal definition of *state* in the context of the proposed research is presented as follows:

> The *state* of a project refers to a performance characteristic of the project which relates input to output such that knowledge of the input time function for $t \geq t_0$ and state at time $t = t_0$ determines the expected output for $t \geq t_0$.

A project *state-space* is the set of all possible states of the project life cycle. State-space representation can solve project design problems by moving from an initial state to another state, and eventually to a goal state. The movement from state to state is achieved by means of actions. A goal is a description of an intended state that has not yet been achieved. The process of solving a project problem involves finding a sequence of actions that represents a solution path from the initial state to the goal state. A state-space model consists of state variables that describe the prevailing condition of the project. The state variables are related to inputs by mathematical relationships. Examples of potential project state variables include schedule, output quality, cost, due date, resource, manpower utilization, and productivity level. For a process described by a system of differential equations, the state-space representation is of the form

$$\dot{z} = f(z(t), x(t))$$

$$y(t) = g(z(t), x(t))$$

where f and g are vector-valued functions. For linear systems, the representation is

$$\dot{z} = Az(t) + Bx(t)$$

$$y(t) = Cz(t) + Dx(t)$$

where

$\mathbf{z(t)}$, $\mathbf{x(t)}$, and $\mathbf{y(t)}$ are vectors

A, B, C, and D are matrices

The variable y is the output vector while the variable x denotes the inputs. The state vector $\mathbf{z(t)}$ is an intermediate vector relating $\mathbf{x(t)}$ to $\mathbf{y(t)}$. The state-space representation of a discrete–time linear project design system is represented as

$$z(t+1) = Az(t) + Bx(t)$$

$$y(t) = Cz(t) + Dx(t)$$

In generic terms, a project is transformed from one state to another by a driving function that produces a transitional equation given by

$$S_s = f(x \mid S_p) + \varepsilon$$

where

S_s is the subsequent state

x is the state variable

S_p is the preceding state

ε is the error component

The function f is composed of a given action (or a set of actions) applied to the project. Each intermediate state may represent a significant milestone in the project. Thus, a descriptive state-space model facilitates an analysis of what actions to apply in order to achieve the next desired product state.

State Transformation in Project Systems

Project objectives are achieved by state-to-state transformation of project phases. Figure 1.14 shows a product development example, involving the transformation from one state to another through the application of action. This simple representation can be expanded to cover several components within the product information framework. Hierarchical linking of product elements provides an expanded transformation structure. The product state can be expanded in accordance with implicit requirements. These requirements might include grouping of design elements, precedence linking (both technical and procedural), required communication links, and reporting requirements. The actions to be taken at each state depend on the prevailing product conditions. The nature of subsequent alternate states depends on what actions are implemented. Sometimes, there are multiple paths that can lead to the desired end result. At other times, there exists only one unique path to the desired objective. In conventional

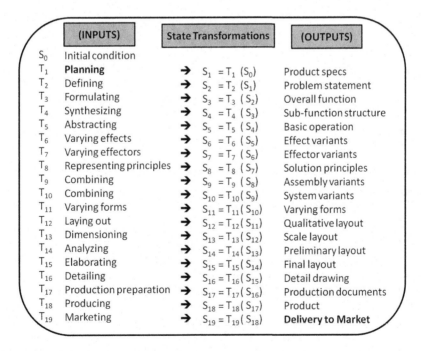

(INPUTS)		State Transformations	(OUTPUTS)
S_0	Initial condition		
T_1	**Planning**	$\rightarrow$ $S_1 = T_1 (S_0)$	Product specs
T_2	Defining	$\rightarrow$ $S_2 = T_2 (S_1)$	Problem statement
T_3	Formulating	$\rightarrow$ $S_3 = T_3 (S_2)$	Overall function
T_4	Synthesizing	$\rightarrow$ $S_4 = T_4 (S_3)$	Sub-function structure
T_5	Abstracting	$\rightarrow$ $S_5 = T_5 (S_4)$	Basic operation
T_6	Varying effects	$\rightarrow$ $S_6 = T_6 (S_5)$	Effect variants
T_7	Varying effectors	$\rightarrow$ $S_7 = T_7 (S_6)$	Effector variants
T_8	Representing principles	$\rightarrow$ $S_8 = T_8 (S_7)$	Solution principles
T_9	Combining	$\rightarrow$ $S_9 = T_9 (S_8)$	Assembly variants
T_{10}	Combining	$\rightarrow$ $S_{10} = T_{10}(S_9)$	System variants
T_{11}	Varying forms	$\rightarrow$ $S_{11} = T_{11}(S_{10})$	Varying forms
T_{12}	Laying out	$\rightarrow$ $S_{12} = T_{12}(S_{11})$	Qualitative layout
T_{13}	Dimensioning	$\rightarrow$ $S_{13} = T_{13}(S_{12})$	Scale layout
T_{14}	Analyzing	$\rightarrow$ $S_{14} = T_{14}(S_{13})$	Preliminary layout
T_{15}	Elaborating	$\rightarrow$ $S_{15} = T_{15}(S_{14})$	Final layout
T_{16}	Detailing	$\rightarrow$ $S_{16} = T_{16}(S_{15})$	Detail drawing
T_{17}	Production preparation	$\rightarrow$ $S_{17} = T_{17}(S_{16})$	Production documents
T_{18}	Producing	$\rightarrow$ $S_{18} = T_{18}(S_{17})$	Product
T_{19}	Marketing	$\rightarrow$ $S_{19} = T_{19}(S_{18})$	**Delivery to Market**

FIGURE 1.14
State-space design flow diagram.

practice, the characteristics of the future states can only be recognized after the fact, thus making it impossible to develop adaptive plans.

In the D-E-J-I implementation, adaptive plans can be achieved because the events occurring within and outside the product state boundaries can be taken into account.

If we describe a product by P state variables s_i, then the composite state of the product at any given time can be represented by a vector S containing P elements. That is,

$$S = \{s_1, s_2, \ldots, s_P\}$$

The components of the state vector could represent either quantitative or qualitative variables (e.g., cost, energy, color, time). We can visualize every state vector as a point in the M-dimensional state-space. The representation is unique since every state vector corresponds to one and only one point in the state-space.

Suppose we have a set of actions (transformation agents) that we can apply to the product information so as to change it from one state to another within the project state-space. The transformation will change a state vector into another state vector. A transformation may be a change in raw material or a change in design approach. Suppose we let T_k be the kth type of transformation. If T_k is applied to the product when it is in state S, the new state vector will be $T_k(S)$, which is another point in the state-space. The number of transformations (or actions) available for a product may be finite or countably infinite. We can construct trajectories that describe the potential states of a product evolution as we apply successive transformations. Each transformation may be repeated as many times as needed. Given an initial state S_0, the sequence of state vectors is represented by the following:

$$S_1 = T_1(S_0)$$

$$S_2 = T_2(S_1)$$

$$S_3 = T_3(S_2)$$

$$\vdots$$

$$S_n = T_n(S_{n-1})$$

The final state, S_n, depends on the initial state S and the effects of actions applied.

Evaluation Stage of D-E-J-I

A project can be evaluated on the basis of cost, quality, and performance. In this section, learning curve modeling is used as the cost evaluation basis of a project with respect to the concept of growth and decay. Here, we use half-life modeling of learning curves to evaluate a technology project. Formal analysis of learning curves first emerged in the mid-1930s in connection with the analysis of the production of airplanes. Learning refers to the improved operational efficiency and cost reduction obtained from repetition of a task. Learning curves have been used for decades to assess improvement achieved over time due the positive impacts of learning. Early analytical modeling of learning curves focused on reduction in cumulative average cost per unit as production level doubled. Several

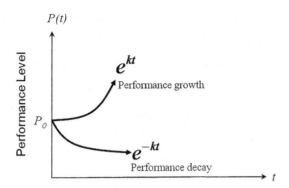

FIGURE 1.15
Concept of learning curve growth and decay.

alternate models of learning curves have been presented in the literature for decades. The classical models have been successfully applied to a variety of problems. In recent years, the deleterious effects of forgetting have also been recognized. It has been shown that workers experience forgetting or decline in performance even while they are making progress along a learning curve. Consequently, contemporary learning curves have attempted to incorporate forgetting components into learning curves. It is of interest to study how fast and how far the forgetting impact can influence overall performance.

Half-life is the amount of time it takes for a quantity to diminish to half of its original size through natural processes. Although the common application of half-life is in natural sciences, the computational analysis lends itself to applications to learning curves. Several research and application studies have confirmed that human performance improves with reinforcement or frequent and consistent repetitions. Reductions in operation processing times achieved through learning curves can directly translate to cost savings. In today's technology-based operations, retention of learning may be threatened by fast-paced shifts in operating requirements. Thus, it is of interest to study the half-life properties of learning curves. Information about the half-life can tell us something about the sustainability of learning-induced performance. This is particularly useful for designing training programs and assessing worker productivity and plant throughput.

Figure 1.15 shows a graphical representation of performance as a function of time under the influence of forgetting (i.e., performance decay). Performance decreases as time progresses. Our interest is to determine when performance has decayed to half of its original level. With half-life computations, a comparative analysis of different learning curves models can be made.

Half-Life Computation for Learning Curves

The basic log-linear model is the most popular learning curve model. It expresses a dependent variable (e.g., production cost) in terms of some independent variable (e.g., cumulative production). The model states that the improvement in productivity is constant (i.e., it has a constant slope) as output increases. That is,

$$C(x) = C_1 x^{-b}$$

where

C(x) is the cumulative average cost of producing x units

C_1 is the cost of the first unit

x is the cumulative production unit

b is the learning curve exponent

The percent productivity gain, p, due the effect of learning is computed as

$$p = 2^{-b}$$

The application of half-life analysis to learning curves can help address questions such as

1. How fast and how far can system performance be improved?
2. What are the limitations to system performance improvement?
3. How resilient is a system to shocks and interruptions to its operation?
4. Are the performance goals that are set for the system achievable?

Figure 1.16 shows a pictorial representation of the basic log-linear model, with the half-life point indicated as $x_{1/2}$. The half-life of the log-linear model is computed as follows.
Let

C_0 be the initial performance level

$C_{1/2}$ be the performance level at half-life

$$C_0 = C_1 x_0^{-b} \text{ and } C_{1/2} = C_1 x_{1/2}^{-b}$$

But $C_{1/2} = 1/2C_0$. Therefore, $C_1 x_{1/2}^{-b} = 1/2C_1 x_0^{-b}$, which leads to $x_{1/2}^{-b} = 1/2x_0^{-b}$, which, by taking the (–1/b)th exponent of both sides, simplifies to yield the following expression as the general expression for the standard log-linear learning curve model

$$x_{1/2} = \left(\frac{1}{2}\right)^{-1/b} x_0, \quad x_0 \geq 1$$

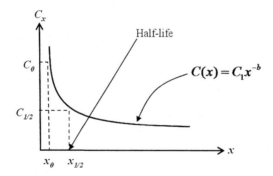

FIGURE 1.16
General profile of the basic learning curve model.

where

$x_{1/2}$ is the half-life

x_0 is the initial point of operation

We refer to $x_{1/2}$ as the *first-order half-life*.

The *second-order half-life* is computed as the time corresponding to half of the preceding half. That is,

$$x_{1/2(2)} = \left(\frac{1}{2}\right)^{-2/b} x_0$$

Similarly, the third-order half-life is

$$x_{1/2(3)} = \left(\frac{1}{2}\right)^{-3/b} x_0$$

In general, the *kth-order half-life* for the log-linear model is represented as

$$x_{1/2(k)} = \left(\frac{1}{2}\right)^{-k/b} x_0$$

Justification Stage of D-E-J-I

Justify on the basis of quantitative value assessment. The total value equation presented earlier in this chapter is a good quantitative technique that can be used here for project justification on the basis of value. The model provides a heuristic decision aid for comparing project alternatives. It is presented here again for the present context. Value is represented as a deterministic vector function that indicates the value of tangible and intangible attributes that characterize the project. It is represented as

$$V = f\left(A_1, A_2, ..., A_p\right)$$

where

V is the value

$A = (A_1, ..., A_n)$ is the vector of quantitative measures or attributes

p is the number of attributes that characterize the project

Examples of project attributes are quality, throughput, capability, productivity, and cost performance. Attributes are considered to be a combined function of factors, x_1, expressed as

$$A_k\left(x_1, x_2, ..., x_{m_k}\right) = \sum_{i=1}^{m_k} f_i\left(x_i\right)$$

where

    $\{x_i\}$ is the set of m factors associated with attribute A_k ($k = 1, 2,..., p$)

    f_i is the contribution function of factor x_i to attribute A_k

Examples of factors are market share, reliability, flexibility, user acceptance, capacity utilization, safety, and design functionality. Factors are themselves considered to be composed of indicators, v_i, expressed as

$$x_i(v_1, v_2,..., v_n) = \sum_{j=1}^{n} z_i(v_i)$$

where

    $\{v_j\}$ is the set of n indicators associated with factor x_i ($i = 1, 2,..., m$)

    z_j is the scaling function for each indicator variable v_j

Examples of indicators are debt ratio, project responsiveness, lead time, learning curve, and scrap volume. By combining the aforementioned definitions, a composite measure of the value of a project is given by the following expression:

$$PV = f\left(A_1, A_2,..., A_p\right)$$

$$= f\left\{ \left[\sum_{i=1}^{m_1} f_i \left(\sum_{j=1}^{n} z_j(v_j) \right) \right]_1, \left[\sum_{i=1}^{m_2} f_i \left(\sum_{j=1}^{n} z_j(v_j) \right) \right]_2,..., \left[\sum_{i=1}^{m_k} f_i \left(\sum_{j=1}^{n} z_j(v_j) \right) \right]_p \right\}$$

where

    PV is the composite project value

    m and n may assume different values for each attribute

A weighting measure to indicate the decision maker's preferences may be included in the model by using an attribute weighting factor, w_i, as follows:

$$PV = f\left(w_1 A_1, w_2 A_2,..., w_p A_p\right)$$

where

$$\sum_{k=1}^{p} w_k = 1 \quad (0 \leq w_k \leq 1)$$

In addition to the quantifiable factors, attributes, and indicators that impinge upon overall PV, human-based subtle factors should also be included in assessing the overall PV. Some of such factors are

1. Project communication
2. Project cooperation
3. Project coordination

Integration Stage of D-E-J-I

We must integrate all the elements of a project on the basis of alignment of functional goals. Systems overlap for integration purposes can conceptually be represented as projection integrals by considering areas bounded by the common elements of subsystems:

$$A = \int\limits_{A_y} \int\limits_{A_x} z(x,y)\,dy\,dx$$

$$B = \int\limits_{B_y} \int\limits_{B_x} z(x,y)\,dy\,dx$$

In Figure 1.17, the projection of a flat plane onto the first quadrant is represented as area *A* while Figure 1.18 shows the projection on an inclined plane as area *B*. The net projection

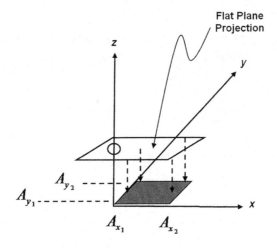

FIGURE 1.17
Flat plane projection for systems integration.

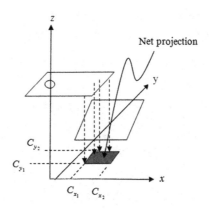

FIGURE 1.18
Inclined plane projection for subsystem alignment and integration.

encompassing the overlap of A and B is represented as area C, shown in Figure 1.19, and computed as

$$C = \int\limits_{C_y} \int\limits_{C_x} z(x,y)\,dy\,dx$$

Notice how each successful net projection area decreases with an increase in the angle of inclination of the project plane. The fact is that in actual project execution, it will be impractical or impossible to model subsystem scenarios as double integrals. But the concept, nonetheless, demonstrates the need to consider where and how project elements overlap for a proper assessment of integration. For mechanical and electrical systems, one can very well develop mathematical representation of systems overlap and integration boundaries. The overall flow diagram for the D-E-J-I model is shown in Figure 1.20, which suggests specific tools and/or techniques for the D-E-J-I stages.

For the purpose of mathematical exposition, double integrals arise in several technical applications. Some examples are

1. Calculation of volumes
2. Calculation of the surface area of a two-dimensional surface (e.g., a plane surface)
3. Calculation of force acting on a two-dimensional surface
4. Calculation of the average of a function
5. Calculation of the mass or moment of inertia of a body
6. Consider the surface area given by the integral

$$A(x) = \int\limits_{c}^{d} f(x,y)\,dy$$

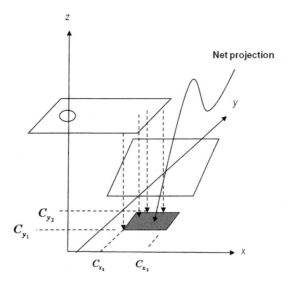

FIGURE 1.19
Reduced net projection area due to steep incline.

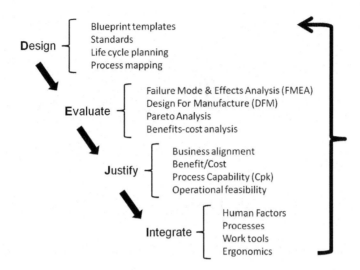

FIGURE 1.20
D-E-J-I model flow diagram.

The variable of integration is y, and x is considered a constant. The cross-sectional area depends on x. Thus, the area is a function of x. That is, $A(x)$. The volume of the slice between x and $x + dx$ is $A(x)dx$. The total volume is the sum of the volumes of all the slices between $x = a$ and $x = b$. That is,

$$V = \int_a^b A(x)\,dx$$

On substituting for $A(x)$, we obtain

$$V = \int_a^b \left[\int_c^d f(x,y)\,dy \right] dx = \int_a^b \int_c^d f(x,y)\,dy\,dx$$

This is an example of an *iterated* integral. One integrates with respect to y first, then x. The integrals with respect to y and x are called the inner and outer integrals, respectively. Alternatively, one can make slices that are parallel to the x-axis. In this case, the volume is given by

$$V = \int_c^d \left[\int_a^b f(x,y)\,dx \right] dy = \int_c^d \int_a^b f(x,y)\,dx\,dy$$

The inner integral corresponds to the cross-sectional area of a slice between y and $y + dy$. The quantities $f(x, y)dy\,dx$ and $f(x, y)dx\,dy$ represent the value of the double integral in the infinitesimally small rectangle between x and $x + dx$ and y and $y + dy$. The length and width of the rectangle are dx and dy, respectively. Hence, $dy\,dx$ (or $dx\,dy$) is the area of the rectangle. Thus, the change in area is $dA = dy\,dx$ or $dA = dx\,dy$.

Computational Example

Consider the double integral

$$V = \iint_R \left(x^2 + xy^3 \right) dA$$

where R is the rectangle $0 \le x \le 1$, $1 \le y \le 2$. Suppose, we integrate with respect to y first. Then

$$V = \int_0^1 \int_1^2 \left(x^2 + xy^3 \right) dy\, dx$$

The inner integral is

$$V = \int_1^2 \left(x^2 + xy^3 \right) dy = \left[x^2 y + x\frac{y^4}{4} \right]_{y=1}^{y=2}$$

Note that we treat x as a constant as we integrate with respect to y. The integral is equal to

$$x^2(2) + x\left(\frac{2^4}{4} \right) - x^2 - \frac{x}{4} = x^2 + \left(\frac{15}{4} \right) x$$

We are now left with the following integral:

$$\int_0^1 \left(x^2 + \frac{15}{4}x \right) dx = \left(\frac{x^3}{3} + \frac{15}{8}x^2 \right)_{x=0}^{x=1} = \frac{1}{3} + \frac{15}{8} = 2.2083$$

Alternatively, we can integrate with respect to x first and then y. We have

$$V = \int_1^2 \int_0^1 \left(x^2 + xy^3 \right) dx\, dy$$

which should yield the same computational result.

In terms of a summary for this chapter, systems integration is the synergistic linking together of various components, elements, and subsystems of a system, where the system may be a complex project, a large endeavor, or an expansive organization. Activities that are resident within the system must be managed both from the technical and managerial standpoints. Any weak link in the system, no matter how small, can be the reason that the overall system fails. In this regard, every component of a project is a critical element that must be nurtured and controlled. Embracing the systems principles for project management will increase the likelihood of success of projects.

Exercises

1.1 Sketch a diagram to illustrate the relationships among a system, a program, a project, a task, and an activity in a project environment.

1.2 Define the functions of project planning, scheduling, and control.

1.3 What project parameters or variables may be important for defining qualitative responsibility for project management?

1.4 What project parameters or variables may be important for defining quantitative accountability for project management?

1.5 Give examples to illustrate the differences between a *project specialist* and a *technical specialist*.

1.6 How can integrated project management tools be used to develop a project implementation strategy?

1.7 Discuss the differences between a leader and a manager.

1.8 Develop a tabulated comparison of the advantages and disadvantages of group decision making.

1.9 How is project management different from traditional management functions?

1.10 Define a project system.

1.11 Expand the flowchart presented in Figure 1.12 to include other important functions within the project management cycle.

1.12 Select your favorite leader in your organization. Discuss how he or she satisfies the characteristics listed in the project leadership loop.

1.13 Select your least favorite leader in your organization. Identify which characteristics in the leadership loop he or she does not have. Discuss how you can counsel this individual on being a better leader.

1.14 Consider the eight operating points in Figure 1.2 (box of scope boundary), analyze and discuss the circumstance of each point with respect to the combination of levels of requirements, cost, and time.

1.15 Project: Consider the compromise surface modeled in Figure 1.3. Develop a table of data (real or hypothetical) to actually model a three-dimensional surface representing a specific type of project of interest.

1.16 List five additional disadvantages of the Blobs structure shown in Figure 1.5 and five additional advantages of value-stream structure shown in Figure 1.6.

1.17 Verify that the system integration double integral here yields the same result of 2.2083 as the computational example presented earlier in this chapter:

$$V = \int\limits_{1}^{2}\int\limits_{0}^{1}\left(x^2 + xy^3\right)dx\,dy$$

References

Badiru, A.B. (2008). *Triple C Model of Project Management*, Taylor & Francis Group, CRC Press, Boca Raton, FL.

Badiru, A.B. (2010). Half-life of learning curves for information technology project management, *International Journal of IT Project Management*, 1(3), 28–45.

Troxler, J.W. and Blank, L. (1989). A comprehensive methodology for manufacturing system evaluation and comparison, *Journal of Manufacturing Systems*, 8(3), 176–183.

2

Systems-Wide Project Planning

Project Planning Objectives

Great discoveries and improvements invariably involve the cooperation of many minds.

Alexander Graham Bell

The key to a successful project is good planning. Project planning provides the basis for the initiation, implementation, and termination of a project. It sets guidelines for specific project objectives, project structure, tasks, milestones, personnel, cost, equipment, performance, and problem resolutions. An analysis of what is needed and what is available should be conducted in the planning phase of new projects. The availability of technical expertise within the organization and outside the organization should be reviewed. If subcontracting is needed, the nature of the contract should undergo a thorough analysis. The question of whether or not the project is needed at all should be addressed. The "make," "buy," "lease," "subcontract," or "do-nothing" alternatives should be compared as part of the project planning process. Here are some guidelines for systems-wide project plans:

1. View a project plan as having tentacles that stretch across the organization.
2. Use project plans to coordinate across functional boundaries.
3. Establish plans as the platform over which project control will be done later on.
4. Leverage the diverse personalities and skills within the project environment.
5. Make room for contingent replanning due to scope changes.
6. Empower workers to manage at the activity level.
7. Identify value-creating tasks and complementing activities.
8. Define specific milestones to facilitate project tracking.
9. Use checklists, tables, charts, and other visual tools project to communicate the plan.
10. Establish project performance metrics.

Although planning is a specific starting step in the project life cycle, it actually stretches over all the other steps of project management. Planning and replanning permeate the project management life cycle. The major knowledge areas of project management, as presented by the Project Management Institute (PMI), are administered in a structured outline

TABLE 2.1

Matrix of Project Management Knowledge Areas and Process Cluster

Knowledge Areas	Project Management Process Clusters				
	Initiating	Planning	Executing	Controlling	Closing
Project Integration	Charter Scope	Project plan	Project execution	Project monitoring Project control	
Scope		Scope planning Create WBS		Scope verification Scope control	
Time		Activities Resources Schedule		Schedule control	
Cost		Cost budget		Cost control	
Quality		Quality	Quality assurance	Quality control	
Human resources		Human resource	Project team	Manage project team	
Communication		Comm. plan	Information distribution	Reporting	
Risk		Risk management Risk response planning		Risk monitoring and control	
Procurement		Purchases Contracting	Vendors	Contract mgmt	Project closure

covering six basic clusters. The implementation clusters represent five process groups that are followed throughout the project life cycle. Each cluster itself consists of several functions and operational steps. When the clusters are overlaid on the nine knowledge areas in the Project Management Book of Knowledge (PMBOK®), we obtain a two-dimensional matrix that spans 44 major process steps shown in Table 2.1. The monitoring and controlling clusters are usually administered as one lumped process group (monitoring and controlling). In some cases, it may be helpful to separate them to highlight the essential attributes of each cluster of functions over the project life cycle. In practice, the processes and clusters do overlap. Thus, there is no crisp demarcation of when and where one process ends and where another one begins over the project life cycle.

In general, project life cycle defines the following:

1. Resources that will be needed in each phase of the project life cycle
2. Specific work to be accomplished in each phase of the project life cycle

Figure 2.1 shows the major phases of project life cycle going from the conceptual phase through the closeout phase. It should be noted that project life cycle is distinguished from product life cycle. Project life cycle does not explicitly address operational issues, whereas product life cycle is mostly about operational issues starting from the product's delivery to the end of its useful life. Note that, for technical projects, the shape of the life cycle curve may be expedited due to the rapid developments that often occur in technology-based activities. For example, for a high-technology project, the entire life cycle may be shortened, with a very rapid initial phase, even though the conceptualization stage may be very long.

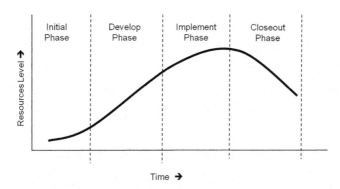

FIGURE 2.1
Resource allocations over phases in project life cycle.

Typical characteristics of project life cycle include the following:

1. Cost and staffing requirements are lowest at the beginning of the project and ramp up during the initial and development stages.
2. The probability of successfully completing the project is lowest at the beginning and highest at the end. This is because many unknowns (risks and uncertainties) exist at the beginning of the project. As the project nears its end, there are fewer opportunities for risks and uncertainties.
3. The risks to the project organization (project owner) are lowest at the beginning and highest at the end. This is because not much investment has gone into the project at the beginning, whereas much has been committed at the end of the project. There is a higher sunk cost manifested at the end of the project.
4. The ability of the stakeholders to influence the final project outcome (cost, quality, and schedule) is highest at the beginning and gets progressively lower toward the end of the project. This is intuitive because influence is best exerted at the beginning of an endeavor.
5. Value of scope changes decreases over time during the project life cycle while the cost of scope changes increases over time. The suggestion is to decide and finalize scope as early as possible. If there are to be scope changes, do them as early as possible.

The specific application context will determine the essential elements contained in the life cycle of the endeavor. Life cycles of business entities, products, and projects have their own nuances that must be understood and managed within the prevailing organizational strategic plan. The components of corporate, product, and project life cycles are summarized as follows:

Corporate (business) life cycle

Policy planning → Needs identification → Business conceptualization
→ Realization → Portfolio management

Product life cycle

       Feasibility studies → Development → Operations → Product obsolescence

Project life cycle

       Initiation → Planning → Execution → Monitoring and control → Closeout

There is no strict sequence for the application of knowledge areas to a specific project. The areas represent a mixed collection of processes that must be followed to achieve a successful project. Thus, some aspects of planning may be found under the knowledge area for communications. In a similar vein, a project may start with the risk management process before proceeding into the integration process. The knowledge areas provide general guidelines. Each project must adapt and tailor the recommended techniques to the specific need and unique circumstances of the project. PMI's PMBOK seeks to standardize project management terms and definitions by presenting a common lexicon for project management activities. It is important to implement the steps of project management in an integrated-systems loop as shown in Figure 2.2.

Specific strategic, operational, and tactical goals and objectives are embedded within each step in the loop. For example, "initiating" may consist of project conceptualization and description. Part of "executing" may include resource allocation and scheduling. "Monitoring" may involve project tracking, data collection, and parameter measurement. "Controlling" implies taking corrective action based on the items that are monitored and evaluated. "Closing" involves phasing out or terminating a project. Closing does not necessarily mean a death sentence for a project as the end of one project may be used as the stepping stone to initiate the next series of endeavors. An overall planning framework is illustrated in Figure 2.3.

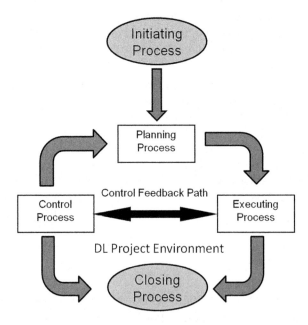

FIGURE 2.2
Integrated loop for project systems implementation.

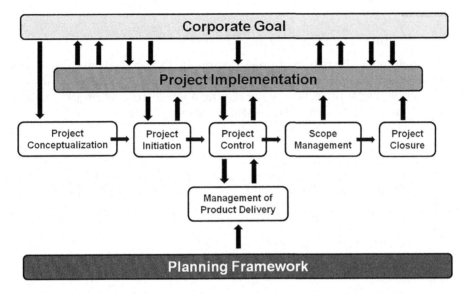

FIGURE 2.3
Planning framework for systems implementation of project management.

In the initial stage of project planning, the internal and external factors that influence the project should be determined and given priority weights. Examples of internal influences on project plans include the following:

1. Infrastructure
2. Project scope
3. Labor relations
4. Project location
5. Project leadership
6. Organizational goal
7. Management approach
8. Technical personnel supply
9. Resource and capital availability

In addition to internal factors, a project plan can be influenced by external factors. An external factor may be the sole instigator of a project or it may manifest itself in combination with other external and internal factors. Such external factors include the following:

1. Public needs
2. Market needs
3. National goals
4. Industry stability
5. State of technology
6. Industrial competition
7. Government regulations

Time–Cost–Performance Criteria for Planning

Project goals determine the nature of project planning. Project goals may be specified in terms of time (schedule), cost (resources), or performance (output). A project can be simple or complex. While simple projects may not require a whole array of project management tools, complex projects may not be successful without all the tools. Project management techniques are applicable to a wide collection of problems ranging from manufacturing to medical services.

The techniques of project management can help achieve goals relating to better product quality, improved resource utilization, better customer relations, higher productivity, and fulfillment of due dates. These can be expressed in terms of the following project constraints:

1. *Performance specifications*
2. *Schedule requirements*
3. *Cost limitations*

Project planning determines the nature of actions and responsibilities needed to achieve the project goal. It entails the development of alternate courses of action and the selection of the best action to achieve the objectives making up the goal. Planning determines what needs to be done, by whom, and when. Whether it is done for long-range (strategic) purposes or for short-range (operational) purposes, planning should be one of the first steps of project management.

Systems Level of Planning

Decisions involving strategic planning lay the foundation for the successful implementation of projects. Planning forms the basis for all actions. Strategic decisions may be divided into three strategy levels: *supralevel planning*, *macrolevel planning*, and *microlevel planning*.

Supralevel planning: Planning at the supralevel deals with the big picture of how the project fits the overall and long-range organizational goals. Questions faced at this level concern potential contributions of the project to the welfare of the organization, its effect on the depletion of company resources, required interfaces with other projects within and outside the organization, risk exposure, management support for the project, concurrent projects, company culture, market share, shareholder expectations, and financial stability.

Macrolevel planning: Planning decisions at the macrolevel address the overall planning within the project boundary. The scope of the project and its operational interfaces should be addressed at this level. Questions faced at the macrolevel include goal definition, project scope, availability of qualified personnel, resource availability, project policies, communication interfaces, budget requirements, goal interactions, deadline, and conflict resolution strategies.

Microlevel planning: The microlevel deals with detailed operational plans at the task levels of the project. Definite and explicit tactics for accomplishing specific project objectives are developed at the microlevel. The concept of management by objective (MBO) may be particularly effective at this level. MBO permits each project member to plan his or her own work at the microlevel. Factors to be considered at the microlevel of project decisions include scheduled time, training requirements, required tools, task procedures, reporting requirements, and quality requirements.

Project decisions at the three levels defined previously will involve numerous personnel within the organization with various types and levels of expertise. In addition to the conventional roles of the project manager, specialized roles may be developed within the project scope. Such roles include the following:

1. *Technical specialist*: This person will have the responsibility for addressing specific technical requirements of the project. In a large project, there will typically be several technical specialists working together to solve project problems.

2. *Operations integrator*: This person will be responsible for making sure that all operational components of the project interface correctly to satisfy project goals. This person should have good technical awareness and excellent interpersonal skills.

3. *Project specialist*: This person has specific expertise related to the specific goals and requirements of the project. Even though a technical specialist may also serve as a project specialist, the two roles should be distinguished. A general electrical engineer may be a technical specialist on the electronic design components of a project. However, if the specific setting of the electronics project is in the medical field, then an electrical engineer with expertise in medical operations may be needed to serve as the project specialist.

Components of a Plan

Planning is an ongoing process that is conducted throughout the project life cycle. Initial planning may relate to overall organizational efforts. This is where specific projects to be undertaken are determined. Subsequent planning may relate to specific objectives of the selected project. In general, a project plan should consist of the following components:

1. *Summary of project plan*: This is a brief description of what is planned. Project scope and objectives should be enumerated. The critical constraints on the project should be outlined. The types of resources required and availability should be specified. The summary should include a statement of how the project complements organizational and national goals, budget size, and milestones.

2. *Objectives*: The objectives should be detailed in outlining what the project is expected to achieve and how the expected achievements will contribute to the overall goal of a project. The performance measures for evaluating the achievement of objectives should be specified.

3. *Approach*: The managerial and technical methodologies of implementing the project should be specified. The managerial approach may relate to project organization, communication network, approval hierarchy, responsibility, and accountability. The technical approach may relate to company experience on previous projects and currently available technology.

4. *Policies and procedures*: Development of a project policy involves specifying the general guidelines for carrying out tasks within the project. Project procedure

involves specifying the detailed method for implementing a given policy relative to the tasks needed to achieve the project goal.

5. *Contractual requirements*: This portion of the project plan should outline reporting requirements, communication links, customer specifications, performance specifications, deadlines, review process, project deliverables, delivery schedules, internal and external contacts, data security, policies, and procedures. This section should be as detailed as practically possible. Any item that has the slightest potential of creating problems later should be documented.

6. *Project schedule*: The project schedule signifies the commitment of resources against time in pursuit of project objectives. A project schedule should specify when the project will be initiated and when it is expected to be completed. The major phases of the project should be identified. The schedule should include reliable time estimates for project tasks. The estimates may come from knowledgeable personnel, past records, or forecasting. Task milestones should be generated on the basis of objective analysis rather than arbitrary stipulations. The schedule in this planning stage constitutes the master project schedule. Detailed activity schedules should be generated under specific project functions.

7. *Resource requirements*: Project resources, budget, and costs are to be documented in this section of the project plan. Capital requirements should be specified by tasks. Resources may include personnel, equipment, and information. Special personnel skills, hiring, and training should be explained. Personnel requirements should be aligned with schedule requirements so as to ensure their availability when needed. Budget size and source should be presented. The basis for estimating budget requirements should be justified and the cost allocation and monitoring approach should be shown.

8. *Performance measures*: Measures of evaluating project progress should be developed. The measures may be based on standard practices or customized needs. The method of monitoring, collecting, and analyzing the measures should also be specified. Corrective actions for specific undesirable events should be outlined.

9. *Contingency plans*: Courses of actions to be taken in the case of undesirable events should be predetermined. Many projects have failed simply because no plans have been developed for emergency situations. In the excitement of getting a project under way, it is often easy to overlook the need for contingency plans.

10. *Tracking, reporting, and auditing*: These involve keeping track of the project plans, evaluating tasks, and scrutinizing the records of the project.

Planning for large projects may include a statement about the feasibility of subcontracting part of the project work. Subcontracting may be needed for various reasons including lower cost, higher efficiency, and logistical convenience.

Motivating the Project Team

Motivation is an essential component of implementing project plans. National leaders, public employees, management staff, producers, and consumers may all need to be motivated about project plans that affect a wide spectrum of society. Those who will play active direct roles in the project must be motivated to ensure productive participation. Direct beneficiaries of the project must be motivated to make good use of the outputs of the project. Other groups must be motivated to play supporting roles to the project.

Motivation may take several forms. For projects that are of a short-term nature, motivation could be either impaired or enhanced by the strategy employed. Impairment may occur if a participant views the project as a mere disruption of regular activities or as a job without long-term benefits. Long-term projects have the advantage of giving participants enough time to readjust to the project efforts. Some of the essential considerations in aligning project plans for motivational purposes include the following elements:

1. Global coordination across functional lines
2. Balancing of task assignments
3. Goal-directed task analysis
4. Human cognitive information flow among the project team
5. Ergonomics and human factors considerations
6. Workload assessment considering fatigue, stress, emotions, sentiments, etc.
7. Interpersonal trust and collegiality
8. Project knowledge transfer lines
9. Harmony of personnel along project lines

Classical concepts of motivation suggest that management involves knowing exactly what workers are expected to do and ensuring that they have the tools and skills to do it well and cost effectively. This means that management requires motivating workers to get things done. Thus, successful management should be able to predict and leverage human behavior. An effective manager should be interested in both results and the people he or she works with. Whatever definition of management is embraced, it ultimately involves some human elements with behavioral and motivational implications. In order to get a worker to work effectively, he or she must be motivated. Some workers are inherently self-motivating, self-directed, and self-actuating. There are other workers for whom motivation is an external force that must be managerially instilled based on the two basic concepts of theory X and theory Y.

Axiom of Theory X

Theory X assumes that the worker is essentially uninterested and unmotivated to perform his or her work. Motivation must be instilled into the worker by the adoption of external motivating agents. A theory X worker is inherently indolent and requires constant supervision and prodding to get him or her to perform. To motivate a theory X worker, a mixture of managerial actions may be needed. The actions must be used judiciously, based on the prevailing circumstances. Examples of motivation approaches under theory X are

1. Rewards to recognize improved effort
2. Strict rules to constrain worker behavior
3. Incentives to encourage better performance
4. Threats to job security associated with performance failure

Axiom of Theory Y

Theory Y assumes that the worker is naturally interested and motivated to perform his or her job. The worker views the job function positively and uses self-control and self-direction to pursue project goals. Under theory Y, management has the task of taking advantage of the worker's positive intuition so that his or her actions coincide with the objectives of the project. Thus, a theory Y manager attempts to use the worker's self-direction as the principal instrument for accomplishing work. In general, theory Y facilitates the following:

1. Worker-designed job methodology
2. Worker participation in decision making
3. Cordial management–worker relationship
4. Worker individualism within acceptable company limits

There are proponents of both theory X and theory Y and managers who operate under each or both can be found in any organization. The important thing to note is that whatever theory one subscribes to, the approach to worker motivation should be conducive to the achievement of the overall goal of the project.

Maslow's Hierarchy of Needs

The needs of project participants must be taken into consideration in any project planning in accordance with the prevailing personal and behavior landscape of the project. A common tool for accomplishing this is *Maslow's hierarchy of needs*, which stresses that human needs are ordered in a hierarchical fashion consisting of five categories:

1. *Physiological needs*: The needs for the basic things of life, such as food, water, housing, and clothing. This is the level where access to money is most critical.
2. *Safety needs*: The needs for security, stability, and freedom from threat of physical harm. The fear of adverse environmental impact may inhibit project efforts.
3. *Social needs*: The needs for social approval, friends, love, affection, and association. For example, public service projects may bring about a better economic outlook that may enable individuals to be in a better position to meet their social needs.
4. *Esteem needs*: The needs for accomplishment, respect, recognition, attention, and appreciation. These needs are important not only at the individual level but also at the national level.
5. *Self-actualization needs*: These are the needs for self-fulfillment and self-improvement. They also involve the availability of opportunity to grow professionally. Work improvement projects may lead to self-actualization opportunities for individuals to assert themselves socially and economically. Job achievement and professional recognition are two of the most important factors that lead to employee satisfaction and better motivation.

Hierarchical motivation implies that the particular motivation technique utilized for a given person should depend on where the person stands in the hierarchy of needs. For example, the need for esteem takes precedence over physiological needs when the latter are relatively well satisfied. Money, for example, cannot be expected to be a very successful motivational factor for an individual who is already on the fourth level of the hierarchy of needs. The hierarchy of needs emphasizes the fact that things that are highly craved in youth tend to assume less importance later in life.

Hygiene Factors and Motivators

There are two motivational factors classified as the *hygiene factors* and *motivators*. Hygiene factors are necessary but not sufficient conditions for a contented worker. The negative aspects of the factors may lead to a disgruntled worker, whereas their positive aspects do not necessarily enhance the satisfaction of the worker. Examples include the following:

1. *Administrative policies*: Bad policies can lead to the discontent of workers, while good policies are viewed as routine with no specific contribution to improving worker satisfaction.
2. *Supervision*: A bad supervisor can make a worker unhappy and less productive, while a good supervisor cannot necessarily improve worker performance.
3. *Worker conditions*: Bad working conditions can enrage workers, but good working conditions do not automatically generate improved productivity.
4. *Salary*: Low salaries can make a worker unhappy, disruptive, and uncooperative, but a raise will not necessarily provoke him to perform better. While a raise in salary will not necessarily increase professionalism, a reduction in salary will most certainly have an adverse effect on morale.
5. *Personal life*: A miserable personal life can adversely affect worker performance, but a happy life does not imply that he or she will be a better worker.
6. *Interpersonal relationships*: Good peer, superior, and subordinate relationships are important to keep a worker happy and productive, but extraordinarily good relations do not guarantee that he or she will be more productive.
7. *Social and professional status*: Low status can force a worker to perform at *his* or *her* level, whereas high status does not imply performance at a higher level.
8. *Security*: A safe environment may not motivate a worker to perform better, but an unsafe condition will certainly impede productivity.

Motivators are motivating agents that should be inherent in the work itself. If necessary, work should be redesigned to include inherent motivating factors. Some guidelines for incorporating motivators into jobs are as follows:

1. *Achievement*: The job design should give consideration to opportunities for worker achievement and avenues to set personal goals to excel.
2. *Recognition*: The mechanism for recognizing superior performance should be incorporated into the job design. Opportunities for recognizing innovation should be built into the job.

3. *Work content*: The work content should be interesting enough to motivate and stimulate the creativity of the worker. The amount of work and the organization of the work should be designed to fit a worker's needs.

4. *Responsibility*: The worker should have some measure of responsibility for how his or her job is performed. Personal responsibility leads to accountability, which invariably yields better work performance.

5. *Professional growth*: The work should offer an opportunity for advancement so that the worker can set his or her own achievement level for professional growth within a project plan.

The aforementioned examples may be described as job enrichment approaches with the basic philosophy that work can be made more interesting in order to induce an individual to perform better. Normally, work is regarded as an unpleasant necessity (a necessary evil). A proper design of work will encourage workers to become anxious to go to work to perform their jobs.

Management by Objective

MBO is a management concept whereby a worker is allowed to take responsibility for the design and performance of a task under controlled conditions. It gives workers a chance to set their own objectives in achieving project goals. Workers can monitor their own progress and take corrective actions when needed without management intervention. Workers under the concept of theory Y appear to be the best suited for the MBO concept. MBO has some disadvantages which include the possible abuse of the freedom to self-direct and possible disruption of overall project coordination. The advantages of MBO include the following:

1. It encourages workers to find better ways of performing their jobs.
2. It avoids oversupervision of professionals.
3. It helps workers become better aware of what is expected of them.
4. It permits timely feedback on worker performance.

Management by Exception

Management by exception (MBE) is an after-the-fact management approach to control. Contingency plans are not made, and there is no rigid monitoring. Deviations from expectations are viewed as exceptions to the normal course of events. When intolerable deviations from plans occur, they are investigated, and then an action is taken. The major advantage of MBE is that it lessens the management workload and reduces the cost of management. However, it is a dangerous concept to follow, especially for high-risk technology-based projects. Many of the problems that can develop in complex projects are such that after-the-fact corrections are expensive or even impossible. As a result, MBE should be carefully evaluated before adopting it. The previously described motivational concepts can be implemented successfully for specific large projects. They may be used as single approaches or in a combined strategy. The motivation approaches may be directed at individuals or groups of individuals, locally, or at the national level.

Project Feasibility Study

The feasibility of a project can be ascertained in terms of technical factors, economic factors, or both. A feasibility study is documented with a report showing all the ramifications of the project.

1. *Technical feasibility*: Technical feasibility refers to the ability of the process to take advantage of the current state of the technology in pursuing further improvement. The technical capability of the personnel as well as the capability of the available technology should be considered.

2. *Managerial feasibility*: Managerial feasibility involves the capability of the infrastructure of a process to achieve and sustain process improvement. Management support, employee involvement, and commitment are key elements required to ascertain managerial feasibility.

3. *Economic feasibility*: This involves the feasibility of the proposed project to generate economic benefits. A benefit–cost analysis and a break-even analysis are important aspects of evaluating the economic feasibility of new industrial projects. The tangible and intangible aspects of a project should be translated into economic terms to facilitate a consistent basis for evaluation.

4. *Financial feasibility*: Financial feasibility should be distinguished from economic feasibility. Financial feasibility involves the capability of the project organization to raise the appropriate funds needed to implement the proposed project. Project financing can be a major obstacle in large multiparty projects because of the level of capital required. Loan availability, credit worthiness, equity, and loan schedule are important aspects of financial feasibility analysis.

5. *Cultural feasibility*: Cultural feasibility deals with the compatibility of the proposed project with the cultural setup of the project environment. In labor-intensive projects, planned functions must be integrated with local cultural practices and beliefs. For example, religious beliefs may influence what an individual is willing to do or not to do.

6. *Social feasibility*: Social feasibility addresses the influences that a proposed project may have on the social system in the project environment. The ambient social structure may be such that certain categories of workers may be in short supply or nonexistent. The effect of the project on the social status of the project participants must be assessed to ensure compatibility. It should be recognized that workers in certain industries may have certain status symbols within the society.

7. *Safety feasibility*: Safety feasibility is another important aspect that should be considered in project planning. Safety feasibility refers to an analysis of whether the project is capable of being implemented and operated safely with minimal adverse effects on the environment. Unfortunately, environmental impact assessment is often not adequately addressed in complex projects.

8. *Political feasibility*: A politically feasible project may be referred to as a "politically correct project." Political considerations often dictate the direction for a proposed project. This is particularly true for large projects with national visibility that may have significant government inputs and political implications. For example, political necessity may be a source of support for a project regardless of the project's merits. On the other hand, worthy projects may face insurmountable opposition simply because of political factors. Political feasibility analysis requires an evaluation of the compatibility of project goals with the prevailing goals of the political system.

Scope of Feasibility Analysis

In general terms, the elements of a feasibility analysis for a project should cover the following items:

1. *Need analysis*: This indicates recognition of a need for the project. The need may affect the organization itself, another organization, the public, or the government. A preliminary study is then conducted to confirm and evaluate the need. A proposal of how the need may be satisfied is then made. Pertinent questions that should be asked include the following:
   a. Is the need significant enough to justify the proposed project?
   b. Will the need still exist by the time the project is completed?
   c. What are the alternate means of satisfying the need?
   d. What are the economic, social, environmental, and political impacts of the need?
2. *Process work*: This is the preliminary analysis done to determine what will be required to satisfy the need. The work may be performed by a consultant who is an expert in the project field. The preliminary study often involves system models or prototypes. For technology-oriented projects, artist's conceptions and scaled-down models may be used for illustrating the general characteristics of a process. A simulation of the proposed system can be carried out to predict the outcome before the actual project starts.
3. *Engineering and design*: This involves a detailed technical study of the proposed project. Written quotations are obtained from suppliers and subcontractors as needed. Technological capabilities are evaluated as needed. Product design, if needed, should be done at this stage.
4. *Cost estimate*: This involves estimating project cost to an acceptable level of accuracy. Levels of around −5% to +15% are common at this stage of a project plan. Both the initial and operating costs are included in the cost estimation. Estimates of capital investment and recurring and nonrecurring costs should also be contained in the cost estimate document. Sensitivity analysis can be carried out on the estimated cost values to see how sensitive the project plan is to the estimated cost values.
5. *Financial analysis*: This involves an analysis of the cash flow profile of the project. The analysis should consider rates of return, inflation, sources of capital, payback periods, break-even point, residual values, and sensitivity. This is a critical analysis, since it determines whether or not and when funds will be available to the

project. The project cash flow profile helps support the economic and financial feasibility of the project.

6. *Project impacts*: This portion of the feasibility study provides an assessment of the impact of the proposed project. Environmental, social, cultural, political, and economic impacts may be some of the factors that will determine how a project is perceived by the public. The value-added potential of the project should also be assessed. A value-added tax may be assessed based on the price of a product and the cost of the raw material used in making the product. The tax so collected may be viewed as a contribution to government coffers.

7. *Conclusions and recommendations*: The feasibility study should end with the overall outcome of the project analysis. This may indicate an endorsement or disapproval of the project. Recommendations on what should be done should be included in this section of the feasibility study.

Contents of Project Proposals

Once a project is shown to be feasible, the next step is to issue a *request for proposal* (RFP) depending on the funding sources involved. Proposals are classified as either "solicited" or "unsolicited." Solicited proposals are those written in response to a request for a proposal, while unsolicited ones are those written without a formal invitation from the funding source. Many companies prepare proposals in response to inquiries received from potential clients. Many proposals are written under competitive bids. If an RFP is issued, it should include statements about project scope, funding level, performance criteria, and deadlines.

The purpose of an RFP is to identify companies that are qualified to successfully conduct the project in a cost-effective manner. Formal RFPs are sometimes issued to only a selected list of bidders who have been preliminarily evaluated as being qualified. These may be referred to as *targeted* RFPs. In some cases, general or open RFPs are issued, and whoever is interested may bid for the project. This, however, has been found to be inefficient in many respects. Ambitious, but unqualified, organizations waste valuable time preparing losing proposals. The receiving agency, on the other hand, spends much time reviewing and rejecting worthless proposals. Open proposals do have proponents who praise their "equal opportunity" approach.

In industry, each organization has its own RFP format, content, and procedures. The request is called by different names including PI (procurement invitation), PR (procurement request), RFB (request for bid), or IFB (invitation for bids). In some countries, it is sometimes referred to as request for tender. Irrespective of the format used, an RFP should request information on bidder's costs, technical capability, management, and other characteristics. It should, in turn, furnish sufficient information on the expected work. A typical detailed RFP should include

1. *Project background*: Need, scope, preliminary studies, and results.

2. *Project deliverables and deadlines*: What products are expected from the project, when the products are expected, and how the products will be delivered should be contained in this document.

3. *Project performance specifications*: Sometimes, it may be more advisable to specify system requirements rather than rigid specifications. This gives the system or

project analysts the flexibility to utilize the most updated and most cost-effective technology in meeting the requirements. If rigid specifications are given, what is specified is what will be provided regardless of the cost and the level of efficiency.

4. *Funding level*: This is sometimes not specified because of nondisclosure policies because of budget uncertainties. However, whenever possible, the funding level should be indicated in the RFP.

5. *Reporting requirements*: Project reviews, format, number and frequency of written reports, oral communication, financial disclosure, and other requirements should be specified.

6. *Contract administration*: Guidelines for data management, proprietary work, progress monitoring, proposal evaluation procedure, requirements for inventions, trade secrets, copyrights, and so on should be included in the RFP.

7. *Special requirements (as applicable)*: Facility access restrictions, equal opportunity/ affirmative actions, small business support, access facilities for the handicapped, false statement penalties, cost sharing, compliance with government regulations, and so on should be included if applicable.

8. *Boilerplates (as applicable)*: There are special requirements that specify the specific ways certain project items are handled. Boilerplates are usually written based on organizational policy and are not normally subject to conditional changes. For example, an organization may have a policy that requires that no more than 50% of a contract award will be paid prior to the completion of the contract. Boilerplates are quite common in government-related projects. Thus, large projects may need boilerplates dealing with environmental impacts, social contributions, and financial requirements.

Proposal Preparation

Whether responding to an RFP or preparing an unsolicited proposal, a proposing organization must take care to provide enough detail to permit an accurate assessment of a project proposal. The proposing organization will need to find out the following:

1. Project time frame
2. Level of competition
3. Organization's available budget
4. Organization of the agency
5. Person to contact within the agency
6. Previous contracts awarded by the agency
7. Exact procedures used in awarding contracts
8. Nature of the work done by the funding agency

The proposal should present a detailed plan for executing the proposed project. The proposal may be directed to a management team within the same organization or to an external organization. However, the same level of professional preparation should be practiced for both internal and external proposals. The proposal contents may be written in two parts: technical section and management section.

1. Technical section of project proposal
   a. Project background
      i. Expertise in the project area
      ii. Project scope
      iii. Primary objectives
      iv. Secondary objectives
   b. Technical approach
      i. Required technology
      ii. Available technology
      iii. Problems and their resolutions
      iv. Work breakdown structure
   c. Work statement
      i. Task definitions and list
      ii. Expectations
   d. Schedule
      i. Gantt charts
      ii. Milestones
      iii. Deadlines
   e. Project deliverables
   f. Value of the project
      i. Significance
      ii. Benefit
      iii. Impact
2. Management section of project proposal
   a. Project staff and experience
      i. Staff vita
   b. Organization
      i. Task assignment
      ii. Project manager, liaison, assistants, consultants, and so on
   c. Cost analysis
      i. Personnel cost
      ii. Equipment and materials
      iii. Computing cost
      iv. Travel
      v. Documentation preparation
      vi. Cost sharing
      vii. Facilities cost
   d. Delivery dates
      i. Specified deliverables

  e. Quality control measures
      i. Rework policy
  f. Progress and performance monitoring
      i. Productivity measurement
  g. Cost control measures

An executive summary or cover letter may accompany the proposal. The summary should briefly state the capability of the proposing organization in terms of previous experience on similar projects, unique qualification of the project personnel, advantages of the organization over other potential bidders, and reasons why the project should be awarded to the bidder.

Proposal Incentives

In some cases, it may be possible to include an incentive clause in a proposal in an attempt to entice the funding organization. An example is the use of cost sharing arrangements. Other frequently used project proposal incentives include bonus and penalty clauses, employment of minorities, public service, and contribution to charity. If incentives are allowed in project proposals, their nature should be critically reviewed. If not controlled, a project incentive arrangement may turn out to be an opportunity for an organization to buy itself into a project contract.

Budget Planning

After the planning for a project has been completed, the next step is the allocation of resources required to implement the project plan. This is referred to as budgeting or capital rationing. Budgeting is the process of allocating scarce resources to the various endeavors of an organization. It involves the selection of a preferred subset of a set of acceptable projects due to overall budget constraints. Budget constraints may result from restrictions on capital expenditures, shortage of skilled personnel, shortage of materials, or mutually exclusive projects. The budgeting approach can be used to express the overall organizational policy. The budget serves many useful purposes including the following:

1. Performance measure
2. Incentive for efficiency
3. Project selection criterion
4. Expression of organizational policy
5. Plan of resource expenditure
6. Catalyst for productivity improvement
7. Control basis for managers and administrators
8. Standardization of operations within a given horizon

The preliminary effort in the preparation of a budget is the collection and proper organization of relevant data. The preparation of a budget for a project is more difficult than the preparation of budgets for regular and permanent organizational endeavors. Recurring endeavors usually generate historical data that serve as inputs to subsequent estimating functions. Projects, on the other hand, are often onetime undertakings without the benefits of prior data. The input data for the budgeting process may include inflationary trends, cost of capital, standard cost guides, past records, and forecast projections. Budget data collection may be accomplished by one of several available approaches, including top-down budgeting and bottom-up budgeting.

Top-Down Budgeting

Top-down budgeting involves collecting data from upper-level sources such as top and middle managers. The cost estimates supplied by the managers may come from their judgments, past experiences, or past data on similar project activities. The cost estimates are passed to lower-level managers, who then break the estimates down into specific work components within the project. These estimates may, in turn, be given to line managers, supervisors, and so on to continue the process. At the end, individual activity costs are developed. The top management issues the global budget while the line worker generates specific activity budget requirements.

One advantage of the top-down budgeting approach is that individual work elements need not be identified before approving the overall project budget. Another advantage of the approach is that the aggregate or overall project budget can be reasonably accurate even though specific activity costs may contain substantial errors. There is, consequently, a keen competition among lower-level managers to get the biggest slice of the budget pie.

Bottom-Up Budgeting

Bottom-up budgeting is the reverse of top-down budgeting. In this method, elemental activities, their schedules, descriptions, and labor skill requirements are used to construct detailed budget requests. The line workers who are actually performing the activities are requested to furnish cost estimates. Estimates are made for each activity in terms of labor time, materials, and machine time. The estimates are then converted to dollar values. The dollar estimates are combined into composite budgets at each successive level up the budgeting hierarchy. If estimate discrepancies develop, they can be resolved through intervention to senior management, junior management, functional managers, project managers, accountants, or financial consultants. Analytical tools such as learning curve analysis, work sampling, and statistical estimation may be used in the budgeting process as appropriate to improve the quality of cost estimates.

All component costs and departmental budgets are combined into an overall budget and sent to top management for approval. A common problem with bottom-up budgeting is that individuals tend to overstate their needs with the notion that top management may cut the budget by some percentage. It should be noted, however, that sending erroneous and misleading estimates will only lead to a loss of credibility. Properly documented and justified budget requests are often spared the budget ax. Honesty and accuracy are invariably the best policies for budgeting.

Zero-Base Budgeting

Zero-base budgeting is a budgeting approach that bases the level of project funding on previous performance. It is normally applicable to recurring programs especially in the public sector. Accomplishments in past funding cycles are weighed against the level of resource expenditure. Programs that are stagnant in terms of their accomplishments relative to budget size do not receive additional budgets. Programs that have suffered decreasing yields are subjected to budget cuts or even elimination. On the other hand, programs that experience increments in accomplishments are rewarded with larger budgets.

A major problem with zero-base budgeting is that it puts participants under tremendous data collection, organization, and program justification pressures. Too much time may be spent documenting program accomplishments to the extent that productivity improvement on current projects may be sacrificed. For this reason, the approach has received only limited use in practice. However, proponents believe it is a good means of making managers and administrators more conscious of their management responsibilities. In a project management context, the zero-base budgeting approach may be used to eliminate specific activities that have not contributed to project goals in the past.

Project Work Breakdown Structure

Work Breakdown Structure (WBS) represents a family tree hierarchy of project operations required to accomplish project objectives. It is particularly useful for the purpose of planning, scheduling, and control. Tasks that are contained in the WBS collectively describe the overall project. The tasks may involve physical products (e.g., steam generators), services (e.g., testing), and data (e.g., reports, sales data). The WBS serves to describe the link between the end objective and the operations required to reach that objective. It shows work elements in the conceptual framework for planning and controlling. The objective of developing a WBS is to study the elemental components of a project in detail. It permits the implementation of the "divide and conquer" concepts. Overall project planning and control can be improved by using a WBS approach. A large project may be broken down into smaller subprojects, which may, in turn, be systematically broken down into task groups.

Individual components in a WBS are referred to as WBS elements, and the hierarchy of each is designated by a level identifier. Elements at the same level of subdivision are said to be of the same WBS level. Descending levels provide increasingly detailed definition of project tasks. The complexity of a project and the degree of control desired determine the number of levels in the WBS. An example of a WBS is shown in Figure 2.4.

Each WBS component is successively broken down into smaller details at lower levels. The process may continue until specific project activities are reached. The basic approach for preparing a WBS is as follows:

1. *Level 1*: It contains only the final project purpose. This item should be identifiable directly as an organizational budget item.
2. *Level 2*: It contains the major subsections of the project. These subsections are usually identified by their contiguous location or by their related purposes.
3. *Level 3*: It contains definable components of the level 2 subsections.

Subsequent levels are constructed in more specific detail depending on the level of control desired. If a complete WBS becomes too crowded, separate WBSs may be drawn for level 2 components. A *specification of work* or WBS summary should normally accompany

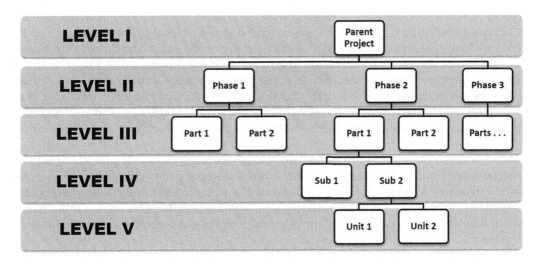

FIGURE 2.4
Project work breakdown structure.

the WBS. A statement of work is a narrative of the work to be done. It should include the objectives of the work, its nature, resource requirements, and tentative schedule. Each WBS element is assigned a code that is used for its identification throughout the project life cycle. Alphanumeric codes may be used to indicate element level as well as component groups.

Legal Systems Considerations

In a comprehensive systems environment, managing a project is tougher and more complicated than traditional projects. The workforce is more diverse and multigenerational, technology is more dynamic, and the customer is less predictable. Governmental changes, institutional changes, and personnel changes are some of the factors that complicate project management with respect to legal requirements. The number of legal issues that arise is increasing at an alarming rate. The job of project management has, consequently, become more strenuous. Any prudent manager of today should give serious considerations to the legal implications of project operations. Many organizations that have failed to recognize legal consequences have paid dearly for their mistakes.

There are several examples of project errors and legal problems. Some of the families of the astronauts killed in the space shuttle *Challenger* sued the National Aeronautics and Space Administration (NASA) for millions of dollars. Some of the managerial staff involved in the Chernobyl accident in the Soviet Union lost their jobs and were legally convicted for dereliction of duty and criminal negligence, and several were sentenced to stiff jail terms. The Love Canal incident near Niagara Falls is still haunting residents and those responsible. The U.S. Justice Department filed a lawsuit against Hooker Chemicals & Plastics Company, the company responsible for dumping the Love Canal industrial waste. In 1980, New York State sued the same company for $635 million for negligence. The Three Mile Island accident of March 28, 1979 in Pennsylvania caused the loss of $2 billion and an erosion of public confidence in the nuclear industry. The Bhopal disaster in India on December 3, 1984 left over 3,000 dead, some 250,000 disabled, and is still costing the Union Carbide Company a great number of legal problems. The oil spill in Alaska had legal repercussions that stretched over several years. The British Petroleum (BP) oil spill in the Gulf of Mexico on April 20, 2010 caused devastating losses, and its adverse impacts will last for decades. It is the largest accidental marine oil spill in the history of the petroleum industry. The nuclear disaster in Japan in 2011, following the devastating earthquake and Tsunami will have lingering legal issues and professional project accountability queries for many years. All of these disasters have project planning failure implications.

With the emergence of new technology and complex systems, it is only prudent to anticipate dangerous and unmanageable events from a systems perspective. The key to preventing disasters is thoughtful planning and cautious preparation. Industrial projects are particularly prone to legal problems, the most pronounced of which are related to environmental damage. Industrial project planning should include a comprehensive evaluation of the potential legal aspects of the project.

Systems Information Flow

Information flow is crucial in project planning. Information is the driving force for project decisions. The value of information is measured in terms of the quality of decisions that can be generated from the information. What appears to be valuable information to one user may be useless to another. Similarly, the timing of information can significantly affect its decision-making value. The same information that is useful in one instance may be useless in another. Some of the crucial factors affecting the value of information include accuracy, timeliness, relevance, reliability, validity, completeness, clearness, and comprehensibility. Proper information flow in project management ensures that tasks are accomplished when, where, and how they are needed. Figure 2.5 illustrates the flow of information for decision making in project management.

Information starts with raw data (e.g., numbers, facts, specifications). The data may pertain to raw material, technical skills, or other factors relevant to the project goal. The data is processed to generate information in the desired form. The information feedback model acts as a management control process that monitors project status and generates

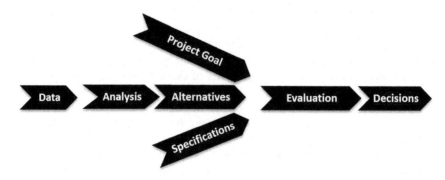

FIGURE 2.5
Information flow for project decision making.

appropriate control actions. The contribution of the information to the project goal is fed back through an information feedback loop. The feedback information is used to enhance future processing of input data to generate additional information. The final information output provides the basis for improved management decisions. The key questions to ask when requesting, generating, or evaluating information for project management are as follows:

1. What data are needed to generate the information?
2. Where are the data going to come from?
3. When will the data be available?
4. Is the data source reliable?
5. Are there enough data?
6. Who needs the information?
7. When is the information needed?
8. In what form is the information needed?
9. Is the information relevant to project goals?
10. Is the information accurate, clear, and timely?

As an example, the information flow model described before may be implemented to facilitate the inflow and outflow of information linking several functional areas of an organization, such as the design department, manufacturing department, marketing department, and customer relations department. The lack of communication among functional departments has been blamed for many of the organizational problems in industry. The use of a standard information flow model can help alleviate many of the communication problems. The information flow model can be expanded to take into account the uncertainties that may occur in the project environment.

Cost and Value of Information

Information is the basis for planning. However, too much information is as bad as too little information. Too much information can impede the progress of a project. The marginal benefit of information decreases as its size increases. However, the marginal cost of obtaining additional information may increase as the size of the information increases. Figure 2.6 shows the potential behaviors of the value and cost curves with respect to the size of information.

The optimum size of information is determined by the point that represents the widest positive difference between the value of information and its cost. The costs associated with information can often be measured accurately. However, the benefits may be more difficult to document. The size of information may be measured in terms of a number of variables, including number of pages of documentation, number of lines of code, and the size of the computer storage requirement. The amount of information presented for project management purposes should be condensed to cover only what is needed to make effective decisions. Information condensation may involve pruning the information that is already available or limiting what needs to be acquired.

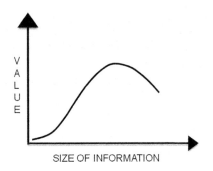
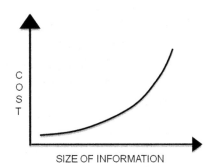

FIGURE 2.6
Value and cost curves for project information.

The cost of information is composed of the cost of the resources required to generate the information. The required resources may include computer time, personnel hours, software, and so on. Unlike the value of information, which may be difficult to evaluate, the cost of information is somewhat more tractable. However, the development of accurate cost estimates prior to actual project execution is not trivial. The degree of accuracy required in estimating the cost of information depends on the intended use of the estimates. Cost estimate may be used as general information for control purposes. Cost estimates may also be used as general guides for planning purposes or for developing standards. The bottom-up cost estimation approach is a popular method used in practice. This method proceeds by breaking the cost structure down into its component parts. The cost of each element is then established. The sum of these individual cost elements gives an estimate of the total cost of information.

It is important to assess the value of project information relative to its cost before deciding to acquire it. Investments for information acquisition should be evaluated just like any other capital investment. The value of information is determined by how the information is used. In project management, information has two major uses. The first use relates to the need for information to run the daily operations of a project. Resource allocation, material procurement, replanning, rescheduling, hiring, and training are just a few of the daily functions for which information is needed. The second major use of information in project management relates to the need for information to make long-range project decisions. The value of information for such long-range decision making is even more difficult to estimate since the future cost of not having the information today is unknown.

The classical approach for determining the value of information is based on the value of perfect information. The expected value of perfect information is the maximum expected loss due to imperfect information. Using probability analysis or other appropriate quantitative methods, the project analyst can predict what a project outcome might be if certain information is available or not available. For example, if it is known for sure that it will rain on a certain day, a project manager might decide to alter the project schedule so that only nonweather-sensitive tasks are planned on that particular day. The value of the perfect information about the weather would then be measured in terms of what loss could have been incurred if that information were not available. The loss may be in terms of lateness penalty, labor idle time, equipment damage, or ruined work.

An experienced project manager can accurately estimate the expected losses, and hence, the value of the perfect information about the weather. The cost of the same information may be estimated in terms of what it would cost to consult with a weather forecaster or the cost of buying a subscription to a special weather forecast channel on cable television.

Triple C Model

The Triple C model is an effective project planning tool. The model states that project management can be enhanced by implementing it within the integrated functions of

1. Communication
2. Cooperation
3. Coordination

The model facilitates a systematic approach to project planning, organizing, scheduling, and control. The Triple C model is distinguished from the 3C approach commonly used in military operations. The military approach emphasizes personnel management in the hierarchy of command, control, and communication. This places communication as the last function. The Triple C, by contrast, suggests communication as the first and foremost function. The Triple C model can be implemented for project planning, scheduling, and control purposes. The model is shown graphically in Figure 2.7.

It highlights what must be done and when. It can also help to identify the resources (personnel, equipment, facilities, etc.) required for each effort. It points out important questions such as

1. Does each project participant know what the objective is?
2. Does each project participant know his or her role in achieving the objective?
3. What obstacles may prevent a participant from playing his or her role effectively?

Communication

What we have here is a failure to communicate.

Common regret in project execution

FIGURE 2.7
Triple C model for project planning.

Communication makes working together possible. The communication function of project management involves making all those concerned become aware of project requirements and progress. Those who will be affected by the project directly or indirectly, as direct participants or as beneficiaries, should be informed as appropriate regarding the following:

1. Scope of the project
2. Personnel contribution required
3. Expected cost and merits of the project
4. Project organization and implementation plan
5. Potential adverse effects if the project should fail
6. Alternatives, if any, for achieving the project goal
7. Potential direct and indirect benefits of the project

The communication channel must be kept open throughout the project life cycle. In addition to internal communication, appropriate external sources should also be consulted. The project manager must

1. Exude commitment to the project
2. Utilize the communication responsibility matrix
3. Facilitate multichannel communication interfaces
4. Identify internal and external communication needs
5. Resolve organizational and communication hierarchies
6. Encourage both formal and informal communication links

Table 2.2 presents an example of a possible responsibility matrix layout for organizing a systems broadcast program for a distance learning (DL) educational project. The responsibility codes in the table are R (responsible), I (inform), S (support), and C (consult) while the task tracking codes are D (done), O (on Track), and L (late). In order for the matrix to be effective, it must be disseminated appropriately so that all who need to know, indeed, are aware of the project and their respective roles to remove ambiguity.

When clear communication is maintained between management and employees and among peers, many project problems can be averted. Project communication may be carried out in one or more of the following formats:

1. One-to-many
2. One-to-one
3. Many-to-one
4. Written and formal
5. Written and informal
6. Oral and formal
7. Oral and informal
8. Nonverbal gesture

Good communication is effected when what is implied is perceived as intended. Effective communications are vital to the success of any project. Despite the awareness that proper

TABLE 2.2

Example of Responsibility Matrix for DL Project Coordination

Tasks	Person Responsible				Status of Task			
	Person A	Person B	Person C	Mgr	28 Feb	15 Mar	30 Apr	31 May
Planning meeting	R	R	R	R	D			
Identify instructors				R		O		
Select classroom	I	R	R			O		
Select delivery mode	R	R				D		
Do publicity		C	R	I	O	O	D	
Lecture materials		C	R					D
Schedule lectures			R			L	L	
Select visual aids			R		L	L	L	
Coordinate activities			R				L	
Periodic review	R	R	R	S				D
Monitor progress	C	R	R			O	L	
Review progress	R				O	O	L	L
Closure	R							L
Final review	R	R	R	R			D	

communications form the blueprint for project success, many organizations still fail in their communication functions. The study of communication is complex. Factors that influence the effectiveness of communication within a project organization structure include the following:

1. *Personal perception*: Each person perceives events on the basis of personal, psychological, social, cultural, and experiential background. As a result, no two people can interpret a given event the same way. The nature of events is not always the critical aspect of a problem situation. Rather, the problem is often the different perceptions of the different people involved.

2. *Psychological profile*: The psychological makeup of each person determines personal reactions to events or words. Thus, individual needs and level of thinking will dictate how a message is interpreted.

3. *Social environment*: Communication problems sometimes arise because people have been conditioned by their prevailing social environment to interpret certain things in unique ways. Vocabulary, idioms, organizational status, social stereotypes, and economic situation are among the social factors that can thwart effective communication.

4. *Cultural background*: Cultural differences are among the most pervasive barriers to project communications, especially in today's multinational organizations. Language and cultural idiosyncrasies often determine how communication is approached and interpreted.

5. *Semantic and syntactic factors*: Semantic and syntactic barriers to communications usually occur in written documents. Semantic factors are those that relate to the intrinsic knowledge of the subject of the communication. Syntactic factors are those that relate to the form in which the communication is presented. The problems

created by these factors become acute in situations where response, feedback, or reaction to the communication cannot be observed.

6. *Organizational structure*: Frequently, the organization structure in which a project is conducted has a direct influence on the flow of information and, consequently, on the effectiveness of communication. Organization hierarchy may determine how different personnel levels perceive a given communication.

7. *Communication media*: The method of transmitting a message may also affect the value ascribed to the message and, consequently, how it is interpreted or used. The common barriers to project communications are listed as follows:

   a. Inattentiveness
   b. Lack of organization
   c. Outstanding grudges
   d. Preconceived notions
   e. Ambiguous presentation
   f. Emotions and sentiments
   g. Lack of communication feedback
   h. Sloppy and unprofessional presentation
   i. Lack of confidence in the communicator
   j. Lack of confidence by the communicator
   k. Low credibility of communicator
   l. Unnecessary technical jargon
   m. Too many people involved
   n. Untimely communication
   o. Arrogance or imposition
   p. Lack of focus

Some suggestions on improving the effectiveness of communication are presented next. The recommendations may be implemented as appropriate for any of the forms of communication listed earlier. The recommendations are for both the communicator and the audience.

1. Never assume that the integrity of the information sent will be preserved, as the information passes through several communication channels. Information is generally filtered, condensed, or expanded by the receivers before relaying it to the next destination. When preparing a communication that needs to pass through several organization structures, one safeguard is to compose the original information in a concise form to minimize the need for recomposition.

2. Give the audience a central role in the discussion. A leading role can help make a person feel a part of the project effort and responsible for the project's success. He or she can then have a more constructive view of project communication.

3. Do homework and think through the intended accomplishment of the communication. This helps eliminate trivial and inconsequential communication efforts.

4. Carefully plan the organization of the ideas embodied in the communication. Use indexing or points of reference whenever possible. Grouping ideas into related chunks of information can be particularly effective. Present the short message first. Short messages help create focus, maintain interest, and prepare the mind for the longer messages to follow.

5. Highlight why the communication is of interest and how it is intended to be used. Full attention should be given to the content of the message with regard to the prevailing project situation.

6. Elicit the support of those around you by integrating their ideas into the communication. The more people feel they have contributed to the issue, the more expeditious they are in soliciting the cooperation of others. The effect of the multiplicative rule can quickly garner support for the communication purpose.

7. Be responsive to the feelings of others. It takes two to communicate. Anticipate and appreciate the reactions of members of the audience. Recognize their operational circumstances and present your message in a form they can relate to.

8. Accept constructive criticism. Nobody is infallible. Use criticism as a springboard to higher communication performance.

9. Exhibit interest in the issue in order to arouse the interest of your audience. Avoid delivering your message as a matter of a routine organizational requirement.

10. Obtain and furnish feedback promptly. Clarify vague points with examples.

11. Communicate at the appropriate time, at the right place, to the right people.

12. Reinforce words with positive action. Never promise what cannot be delivered. Value your credibility.

13. Maintain eye contact in oral communication and read the facial expressions of your audience to obtain real-time feedback.

14. Concentrate on listening as much as speaking. Evaluate both the implicit and explicit meanings of statements.

15. Document communication transactions for future references.

16. Avoid asking questions that can be answered yes or no. Use relevant questions to focus the attention of the audience. Use questions that make people reflect upon their words, such as "How do you think this will work?" compared to "Do you think this will work?"

17. Avoid patronizing the audience. Respect their judgment and knowledge.

18. Speak and write in a controlled tempo. Avoid emotionally charged voice inflections.

19. Create an atmosphere for formal and informal exchanges of ideas.

20. Summarize the objectives of communication and how they will be achieved.

Figure 2.8 shows an example of a design of communication responsibility matrix. A communication responsibility matrix shows the linking of sources of communication and targets of communication. Cells within the matrix indicate the subject of the desired communication. There should be at least one filled cell in each row and each column of the matrix. This assures that each individual of a department has at least one communication source or target associated with him or her. With a communication responsibility matrix, a clear understanding of what needs to be communicated to whom can be developed.

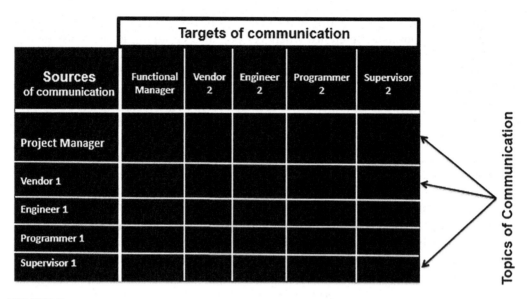

FIGURE 2.8
Triple C communication matrix.

Communication in a project environment can take any of several forms. The specific needs of a project may dictate the most appropriate mode. Three popular computer communication modes are discussed in the context of communicating data and information for project management.

1. *Simplex communication*: This is a unidirectional communication arrangement in which one project entity initiates communication to another entity or individual within the project environment. The entity addressed in the communication does not have a mechanism or capability for responding to the communication. An extreme example of this is a one-way, top-down communication from top management to the project personnel. In this case, the personnel have no communication access or input to top management. A budget-related example is a case where top management allocates budget to a project without requesting and reviewing the actual needs of the project. Simplex communication is common in authoritarian organizations.

2. *Half-duplex communication*: This is a bidirectional communication arrangement whereby one project entity can communicate with another entity and receive a response within a certain time lag. Both entities can communicate with each other but not at the same time. An example of half-duplex communication is a project organization that permits communication with top management without a direct meeting. Each communicator must wait for a response from the target of the communication. Request and allocation without a budget meeting is another example of half-duplex data communication in project management.

3. *Full-duplex communication*: This involves a communication arrangement that permits dialog between the communicating entities. Both individuals and entities can communicate with each other at the same time or face to face. As long as there is no clash of words, this appears to be the most receptive communication mode. It allows participative project planning in which each project personnel has an opportunity to contribute to the planning process.

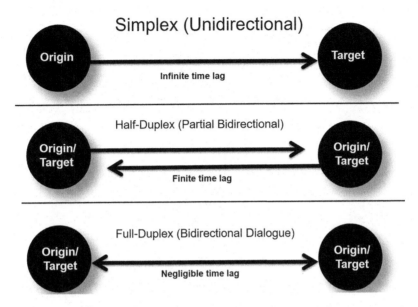

FIGURE 2.9
Project communication modes.

Figure 2.9 presents a graphical representation of the communication modes discussed earlier. Each member of a project team needs to recognize the nature of the prevailing communication mode in the project. Management must evaluate the prevailing communication structure and attempt to modify it if necessary to enhance project functions. An evaluation of who is to communicate with whom about what may help improve the project data/information communication process. A communication matrix may include notations about the desired modes of communication between individuals and groups in the project environment. Table 2.3 summarizes the types of communication, cooperation, and coordination that can be in effect in a project environment.

Complexity of Multiperson Communication

Communication complexity increases with an increase in the number of communication channels. It is one thing to wish to communicate freely, but it is another thing to contend with the increased complexity when more people are involved. The statistical formula of a combination can be used to estimate the complexity of communication as a function of the number of communication channels or number of participants. The combination formula is used to calculate the number of possible combinations of r objects from a set of n objects. This is written as

$$_nC_r = \frac{n!}{r![n-r]!}$$

In the case of communication, for illustration purposes, we assume communication is between two members of a team at a time. That is, combination of two from n team members. That is, number of possible combinations of two members out of a team of n people.

TABLE 2.3

Types of Communication, Cooperation, and Coordination

Types of communication	Verbal
	Written
	Body language
	Visual tools (i.e., graphical tools)
	One-to-one
	One-to-many
	Many-to-one
Types of cooperation	Proximity
	Functional
	Professional
	Social
	Power influence
	Social
	Authority influence
	Hierarchical
	Lateral
	Cooperation by intimidation
	Cooperation by enticement
Types of coordination	Teaming
	Delegation
	Supervision
	Partnership
	Token passing
	Baton handoff

Thus, the formula for communication complexity reduces to the expression as follows, after some of the computation factors cancel out:

$$_nC_2 = \frac{n(n-1)}{2}$$

In a similar vein, Badiru (2008) introduced a formula for cooperation complexity based on the statistical concept of permutation. Permutation is the number of possible arrangements of k objects taken from a set of n objects. The permutation formula is written as

$$_nP_k = \frac{n!}{(n-k)!}$$

Thus, for the number of possible permutations of two members out of a team of n members is estimated as

$$_nP_2 = n(n-1)$$

Permutation formula is used for cooperation because cooperation is bidirectional. Full cooperation requires that if A cooperates with B, then B must cooperate with A. However,

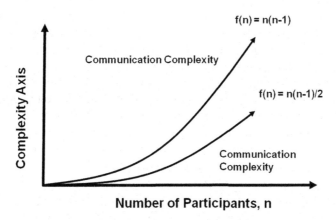

FIGURE 2.10
Plots of communication and cooperation complexities.

A cooperating with B does not necessarily imply B cooperating with A. In notational form, that is,

$$A \rightarrow B \text{ does not necessarily imply } B \rightarrow A$$

Figure 2.10 shows the relative plots of communication complexity and cooperation complexity as a function of project team size *n*.

Cooperation

> If you want to be incrementally better, be competitive. If you want to be exponentially better, be cooperative
>
> *Unknown source*

The cooperation of the project personnel must be explicitly elicited. Merely voicing a consent for a project is not enough assurance for full cooperation. The participants and beneficiaries of the project must be convinced of the merits of the project. Some of the factors that influence cooperation in a project environment include personnel requirements, resource requirements, budget limitations, past experiences, conflicting priorities, and lack of uniform organizational support. A structured approach to seeking cooperation should clarify the following:

1. Cooperative efforts required
2. Precedents for future projects
3. Implication of lack of cooperation
4. Criticality of cooperation to a project's success
5. Organizational impact of cooperation
6. Time frame involved in the project
7. Rewards of good cooperation

Cooperation is a basic virtue of human interaction. More projects fail due to a lack of cooperation and commitment than any other project factors. To secure and retain the cooperation of project participants, you must elicit a positive first reaction to the project. The most positive aspects of a project should be the first item of project communication. For project management, there are different types of cooperation that should be understood.

1. *Functional cooperation*: This is cooperation induced by the nature of functional relationship between two groups. The two groups may be required to perform related functions that can only be accomplished through mutual cooperation.

2. *Social cooperation*: This is the type of cooperation effected by the social relationship between two groups. The prevailing social relationship motivates cooperation that may be useful in getting the project work done.

3. *Legal cooperation*: Legal cooperation is the type of cooperation that is imposed through some authoritative requirement. In this case, the participants may have no choice other than to cooperate.

4. *Administrative cooperation*: This is cooperation brought on by administrative requirements that make it imperative that two groups work together on a common goal.

5. *Associative cooperation*: This type of cooperation may also be referred to as collegiality. The level of cooperation is determined by the association that exists between two groups.

6. *Proximity cooperation*: Cooperation due to the fact that two groups are geographically close is referred to as proximity cooperation. Being close makes it imperative that the two groups work together.

7. *Dependency cooperation*: This is cooperation caused by the fact that one group depends on another group for some important aspect. Such dependency is usually of a mutual two-way nature. One group depends on the other for one thing while the latter group depends on the former for some other thing.

8. *Imposed cooperation*: In this type of cooperation, external agents must be employed to induce cooperation between two groups. This is applicable for cases where the two groups have no natural reason to cooperate. This is where the approaches presented earlier for seeking cooperation can become very useful.

9. *Lateral cooperation*: Lateral cooperation involves cooperation with peers and immediate associates. Lateral cooperation is often easy to achieve because existing lateral relationships create a conducive environment for project cooperation.

10. *Vertical cooperation*: Vertical or hierarchical cooperation refers to cooperation that is implied by the hierarchical structure of the project. For example, subordinates are expected to cooperate with their vertical superiors.

Whichever type of cooperation is available in a project environment, the cooperative forces should be channeled toward achieving project goals. Documentation of the prevailing level of cooperation is useful for winning further support for a project. Clarification of project priorities will facilitate personnel cooperation. Relative priorities of multiple projects should be specified so that a project that is of high priority to one segment of an organization is also of high priority to all groups within the organization. Some guidelines for securing cooperation for most projects are as follows:

1. Establish achievable goals for the project.
2. Clearly outline the individual commitments required.

3. Integrate project priorities with existing priorities.

4. Eliminate the fear of job loss due to industrialization.

5. Anticipate and eliminate potential sources of conflict.

6. Use an open-door policy to address project grievances.

7. Remove skepticism by documenting the merits of the project.

Commitment: Cooperation must be supported with commitment. To cooperate is to support the ideas of a project. To commit is to willingly and actively participate in project efforts again and again through the thick and thin of the project. Provision of resources is one way that management can express commitment to a project.

$$\text{Triple C} + \text{Commitment} = \text{Project success}$$

Coordination

After the communication and cooperation functions have been successfully initiated, the efforts of the project personnel must be coordinated. Coordination facilitates harmonious organization of project efforts. The construction of a responsibility chart can be very helpful at this stage. A responsibility chart is a matrix consisting of columns of individuals or functional departments and rows of required actions. Cells within the matrix are filled with relationship codes that indicate who is responsible for what. Table 2.4 illustrates an example of a responsibility matrix for the planning of a seminar program.

TABLE 2.4

Case Example of Responsibility Matrix for Project Coordination and Tracking

Tasks	Person Responsible				Status of Task			
	Staff A	Staff B	Staff C	Mgr	31 Jan	15 Feb	28 Mar	21 Apr
Brainstorming meeting	R	R	R	R	D			
Identify speakers				R		O		
Select seminar location	I	R	R			O		
Select banquet location	R	R				D		
Prepare publicity materials		C	R	I	O	O	D	
Draft brochures		C	R					D
Develop schedule			R			L	L	
Arrange for visual aids			R		L	L	L	
Coordinate activities			R				L	
Periodic review of tasks	R	R	R	S				D
Monitor progress of program	C	R	R			O	L	
Review program progress	R				O	O	L	L
Closing arrangements	R							L
Postprogram review and evaluation	R	R	R	R			D	

Responsibility codes: R (responsible); I (inform); S (support); C (consult).
Task codes: D (done); O (on track); L (late).

The matrix helps avoid neglecting crucial communication requirements and obligations. It can help resolve questions such as

1. Who is to do what?
2. How long will it take?
3. Who is to inform whom of what?
4. Whose approval is needed for what?
5. Who is responsible for which results?
6. What personnel interfaces are required?
7. What support is needed from whom and when?

Resolving Project Conflicts with Triple C

When implemented as an integrated process, the Triple C model can help avoid conflicts in a project. When conflicts do develop, it can help in resolving the conflicts. Several sources of conflicts can exist in large projects. Some of these are discussed next.

1. *Schedule conflict*: Conflicts can develop because of improper timing or sequencing of project tasks. This is particularly common in large multiple projects. Procrastination can lead to having too much to do at once, thereby creating a clash of project functions and discord among project team members. Inaccurate estimates of time requirements may lead to infeasible activity schedules. Project coordination can help avoid schedule conflicts.

2. *Cost conflict*: Project cost may not be generally acceptable to the clients of a project. This will lead to project conflict. Even if the initial cost of the project is acceptable, a lack of cost control during project implementation can lead to conflicts. Poor budget allocation approaches and the lack of a financial feasibility study will cause cost conflicts later on in a project. Communication and coordination can help prevent most of the adverse effects of cost conflicts.

3. *Performance conflict*: If clear performance requirements are not established, performance conflicts will develop. Lack of clearly defined performance standards can lead each person to evaluate his or her own performance based on personal value judgments. In order to uniformly evaluate quality of work and monitor project progress, performance standards should be established by using the Triple C approach.

4. *Management conflict*: There must be a two-way alliance between management and the project team. The views of management should be understood by the team. The views of the team should be appreciated by management. If this does not happen, management conflicts will develop. A lack of a two-way interaction can lead to strikes and industrial actions that can be detrimental to project objectives. The Triple C approach can help create a conducive dialog environment between management and the project team.

5. *Technical conflict*: If the technical basis of a project is not sound, technical conflicts will develop. New industrial projects are particularly prone to technical conflicts because of their significant dependence on technology. Lack of a comprehensive

technical feasibility study will lead to technical conflicts. Performance requirements and systems specifications can be integrated through the Triple C approach to avoid technical conflicts.

6. *Priority conflict*: Priority conflicts can develop if project objectives are not defined properly and applied uniformly across a project. Lack of a direct project definition can lead each project member to define his or her own goals, which may be in conflict with the intended goal of a project. Lack of consistency of the project mission is another potential source of priority conflict. Overassignment of responsibilities with no guidelines for relative significance levels can also lead to priority conflicts. Communication can help defuse priority conflicts.

7. *Resource conflict*: Resource allocation problems are a major source of conflict in project management. Competition for resources, including personnel, tools, hardware, software, and so on, can lead to disruptive clashes among project members. The Triple C approach can help secure resource cooperation.

8. *Power conflict*: Project politics lead to a power play that can adversely affect the progress of a project. Project authority and project power should be clearly delineated. Project authority is the control that a person has by virtue of his or her functional post. Project power relates to the clout and influence a person can exercise due to connections within the administrative structure. People with popular personalities can often wield a lot of project power in spite of low or nonexistent project authority. The Triple C model can facilitate a positive marriage of project authority and power to the benefit of project goals. This will help define clear leadership for a project.

9. *Personality conflict*: Personality conflict is a common problem in projects involving a large group of people. The larger a project, the larger the size of the management team needed to keep things running. Unfortunately, the larger management team creates an opportunity for personality conflicts. Communication and cooperation can help defuse personality conflicts.

In summary, conflict resolution through Triple C can be achieved by observing the following guidelines:

1. Confront the conflict and identify the underlying causes.

2. Be cooperative and receptive to negotiation as a mechanism for resolving conflicts.

3. Distinguish between proactive, inactive, and reactive behaviors in a conflict situation.

4. Use communication to defuse internal strife and competition.

5. Recognize that short-term compromise can lead to long-term gains.

6. Use coordination to work toward a unified goal.

7. Use communication and cooperation to turn a competitor into a collaborator.

Classical Abilene Paradox

A classic example of conflict in project planning is illustrated by the *Abilene Paradox*, which is narrated as follows. It was a July afternoon in Coleman, a tiny Texas town. It was a hot afternoon. The wind was blowing fine-grained West Texas topsoil through the house. Despite the harsh weather, the afternoon was still tolerable and potentially enjoyable. There was a fan blowing on the back porch; there was cold lemonade; and finally, there was entertainment: dominoes. Perfect for the conditions. The game required little more physical exertion than an occasional mumbled comment, "Shuffle them," and an unhurried movement of the arm to place the spots in the appropriate position on the table. All in all, it had the makings of an agreeable Sunday afternoon in Coleman, until Jerry's father-in-law suddenly said, "Let's get in the car and go to Abilene and have dinner at the cafeteria."

Jerry thought, "What, go to Abilene? Fifty-three miles? In this dust storm and heat? And in a nonairconditioned 1958 Buick?" But Jerry's wife chimed in with, "Sounds like a great idea. I'd like to go. How about you, Jerry?" Since Jerry's own preferences were obviously out of step with the rest, he replied, "Sounds good to me," and added, "I just hope your mother wants to go."

"Of course I want to go," said Jerry's mother-in-law. "I haven't been to Abilene in a long time." So into the car and off to Abilene they went. Jerry's predictions were fulfilled. The heat was brutal. The group was coated with a fine layer of dust that was cemented with perspiration by the time they arrived. The food at the cafeteria provided a first-rate testimonial material for antacid commercials.

Some 4 hours and 106 miles later, they returned to Coleman, hot and exhausted. They sat in front of the fan for a long time in silence. Then, both to be sociable and to break the silence, Jerry said, "It was a great trip, wasn't it?" No one spoke. Finally, his father-in-law said, with some irritation, "Well, to tell the truth, I really didn't enjoy it much and would rather have stayed here. I just went along because the three of you were so enthusiastic about going. I wouldn't have gone if you all hadn't pressured me into it."

Jerry couldn't believe what he just heard. "What do you mean, 'you all'?" he said. "Don't put me in the 'you all' group. I was delighted to be doing what we were doing. I didn't want to go. I only went to satisfy the rest of you. You're the culprits." Jerry's wife looked shocked. "Don't call me a culprit. You and Daddy and Mama were the ones who wanted to go. I just went along to be sociable and to keep you happy. I would have had to be crazy to want to go out in heat like that."

Her father entered the conversation abruptly. "Hell!" he said. He proceeded to expand on what was already absolutely clear. "Listen, I never wanted to go to Abilene. I just thought you might be bored. You visit so seldom, I wanted to be sure you enjoyed it. I would have preferred to play another game of dominoes and eat the leftovers in the icebox."

After the outburst of recrimination, they all sat back in silence. There they were, four reasonably sensible people who, of their own volition, had just taken a 106-mile trip across a godforsaken desert in a furnace-like temperature through a cloud-like dust storm to eat unpalatable food at a hole-in-the-wall cafeteria in Abilene, when one of them had really wanted to go. In fact, to be more accurate, they'd done just the opposite of what they wanted to do. The whole situation simply didn't make sense. It was a paradox of agreement.

This example illustrates a problem that can be found in many organizations or project environments. Organizations often take actions that totally contradict their stated goals and objectives. They do the opposite of what they really want to do. For most organizations, the adverse effects of such diversion, measured in terms of human distress and economic loss, can be immense. A family group that experiences the Abilene paradox would soon get over the distress, but for an organization engaged in a competitive market, the distress may last a very long time. Six specific symptoms of the paradox are identified as follows:

1. Organization members agree privately, as individuals, as to the nature of the situation or problem facing the organization.

2. Organization members agree privately, as individuals, as to the steps that would be required to cope with the situation or solve the problem they face.

3. Organization members fail to accurately communicate their desires and/or beliefs to one another. In fact, they do just the opposite and, thereby, lead one another into misinterpreting the intentions of others. They misperceive the collective reality. Members often communicate inaccurate data (e.g., "Yes, I agree"; "I see no problem with that"; "I support it") to other members of the organization. No one wants to be the lone dissenting voice in the group.

4. With such invalid and inaccurate information, organization members make collective decisions that lead them to take actions contrary to what they want to do and, thereby, arrive at results that are counterproductive to the organization's intent and purpose. For example, the Abilene group went to Abilene when it preferred to do something else.

5. As a result of taking actions that are counterproductive, organization members experience frustration, anger, irritation, and dissatisfaction with their organization. They form subgroups with supposedly trusted individuals and blame other subgroups for the organization's problems.

6. The cycle of the Abilene paradox repeats itself with increasing intensity if the organization members do not learn to manage their agreement.

This author has witnessed many project situations where, in private conversations, individuals express their discontent about a project and yet fail to repeat their statements in a group setting. Consequently, other members are never aware of the dissenting opinions. In large organizations, the Triple C model, considering the individual needs of all subsystems, can help in managing communication, cooperation, and coordination functions to avoid the Abilene paradox. The lessons to be learned from proper approaches to project planning can help avoid unwilling trips to Abilene.

Exercises

2.1 Develop a definition for the Triple C model of project management.

2.2 List and discuss five major functions in project planning.

2.3 Describe the role of the project manager in project planning.

2.4 Develop a project planning model for the construction of a new manufacturing facility. The facility will produce products to be sold in the international market.

2.5 List some of the factors that can impede the flow of information for project planning purposes. How can these factors be controlled?

2.6 Discuss the impact of hygiene factors and motivators in project planning.

2.7 Discuss how the concepts of MBO and MBE can be integrated for project planning purposes.

2.8   Discuss some of the criteria that can be used to assess the technical feasibility of a project.

2.9   Prepare the technical section of a project proposal dealing with environmental impact assessment for a new industry. Select a specific industry type for your response.

2.10 Prepare the management section of a project proposal dealing with environmental impact assessment for a new industry. Use the same industry type selected in Exercise 2.9.

2.11 Use the combination and permutation formulas to compute and plot the complexity of project communication against varying values of n, the number of discussants. Discuss what you observe in the characteristics of the plots.

Reference

Badiru, A.B. (2008). *Triple C Model of Project Management*, Taylor & Francis Group, CRC Press, Boca Raton, FL.

3

Project Systems Organization

Environmental Factors in Project Organization

We trained hard, but it seemed that every time we were beginning to form into teams we would be *reorganized*. I was to learn later in life that we tend to meet any new situation by *reorganizing*; and what a wonderful method it can be for creating the illusion of progress while producing confusion, inefficiency and demoralization.

Petronius Arbiter, 210 B.C.

If the benefits of systems structure are to be realized, a comprehensive change is needed across the project environment in terms of recognizing and responding to all the requirements of personnel interfaces, technology, resource allocation, and work process. First, change must occur in the physical and information domains. New communications, operating systems (OS), and other supporting infrastructure and information flow must be developed. Information age technologies developed in specific application domains must be adapted for the current project environment. Second, change must occur in the social and cognitive domains. This entails changing the ways people, organizations, and their processes interact. It requires new ways of selecting, training, and assigning project personnel, and it depends on the ability to overcome traditional obstacles to information sharing and functional collaboration. It requires that trust and confidence be fostered amongst a variety of people from diverse backgrounds (functions and services). Breaking down existing stovepipes becomes an imperative within any project for systems that limit information shareability between functions. Breaking down barriers that limit collaboration and interoperability between one department and another is equally essential. The specific organization structure that is selected and utilized for a project can facilitate success with the challenges cited earlier.

Issues in Social and Cognitive Domains

Leadership Development

Develop adaptive and innovative leaders who are comfortable with broad functions across organizational units.

Develop project leaders who are comfortable in the information environment and do not feel the need to micromanage.

Require that project leaders be familiar with and operate information systems throughout their career to develop the level of comfort and knowledge required to leverage the most from the available information systems. This requires formal education, organizational training, and continuing education programs.

Develop project leaders who are comfortable operating in a continuous, concurrent planning process versus the continuous, sequential planning process. This requires a type of "benevolent hierarchy," where leaders operating at varying levels and subordinate personnel are freely sharing information and working more in a collegial systems type of environment to accomplish project goals.

Preparing the Project Personnel

Project personnel management systems need to recognize and codify the new skill sets required for multidimensional projects. Develop and codify the multifunctional project personnel. Individuals working with project information systems should be identified by skill and utilized in that type of assignments leveraging the resident skills.

The common arrangement of having stove-piped assignments should be modified to permit cross-training and collaborative execution of tasks.

Cultural change is required to break down information sharing barriers and create an environment that better encourages resource sharing. Project personnel at all levels must be quicker and more flexible to provide an adequate response to the rapid evolution of the project scope based on the speed of operations and the improvement in shared situational awareness. A hierarchical and directive model will not always work. The goal should be a professional and collegial atmosphere that emphasizes the rapid interchange of information rather than a send–receive–respond method of implementing project requirements.

Training is at the core of every successful operation. Success of the project system is a reflection of direct and realistic training of project personnel in the specific requirements of the project. A systems approach requires familiarity with and confidence in the project team itself as well as the available tools, tactics, techniques, policies, procedures, and work process. Most projects are information rich. The challenge is how to extract and leverage the inherent information to accomplish a project.

Project Office

Establish and operate a project management office (PMO) to serve as the command post for project operations.

The PMO must develop a plan to train and integrate project personnel across the project system.

Operating a project system requires flexibility and must support a degree of ad hoc responsiveness to quickly adjust processes and structural design to maximize the sharing of information and increase collaboration in varying situations.

The PMO can be the command-and-control center with on-the-move capabilities from the management level down to the secretarial support staff.

Organizational Breakdown Structure

There are three major categories of breakdown structures. The *project breakdown structure (PBS)* is a project-specific structure that refers to the breakdown of tasks to be performed in a specific project. This is often referred to as the work breakdown structure (WBS). The OS refers to the organization of the environment within which a project is to be carried out. The *organizational breakdown structure (OBS)* refers to the identification and organization of the resources required to carry out the activities associated with a project. OBS is a model that provides a way of organizing resources into groups for better management within an organization, which may be a company, a project, or a division of a large enterprise. OBS can be used to keep track of resource allocation and specific work assignments. There is a strong interdependency between OBS and WBS. A good breakdown of work helps in estimating resource requirements more accurately. Good team organization is essential for the success of OBS. For example, a mixed team composed of individuals with different technical backgrounds is required for technology-based projects. OBS team requirements are

1. Background of the team leader: experience, education, technical knowledge
2. Scope of work to be accomplished: hours, skills, locations
3. Estimate the number and type of personnel required
4. Equipment and workspace requirements
5. Reporting procedures

Selecting an Organization Structure

After project planning, the next step is the selection of OS for the project. This involves the selection of an OS that shows the management line and responsibilities of the project personnel. Any of the several approaches may be utilized. Before selecting an OS, the project team should assess the nature of job to be performed and its requirements. The structure may be defined in terms of functional specializations, departmental proximity, standard management boundaries, operational relationships, or product requirements.

Large and complex projects should be based on well-designed structures that permit effective information and decision processes. The primary function of an organizational design is to facilitate effective information flow. Traditional organization models consist of decision making, bureaucracy, social, and system structures. The decision-making structure handles the policies and general directions of the overall organization. The bureaucracy is concerned with the administrative process. Some of the administrative functions may not be directly relevant to the main goal of the organization, but they are, nonetheless, deemed necessary. Bureaucratic processes are potential sources of delay for public projects because of government involvement. The social structure facilitates amiable interactions among the personnel. Such interpersonal relationships are essential for the group effort needed to achieve company objectives. The systems structure can better be described as the link among the various synergistic segments of the organization.

Many organizations still use the traditional or classical organization structure. A traditional organization is often utilized in service-oriented companies. The structure is sometimes referred to as a pure functional structure because groups with similar functional

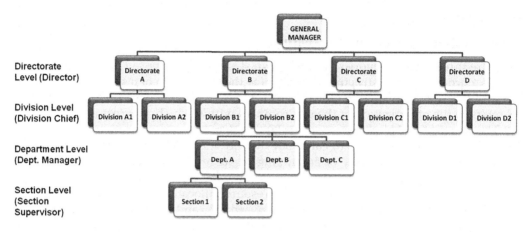

FIGURE 3.1
Traditional organization structure.

responsibilities are clustered at the same level of the structure. Figure 3.1 illustrates the traditional organization structure.

The positive characteristics of the structure are

1. Availability of broad personnel base
2. Identifiable technical line of control
3. Grouping of specialists to share technical knowledge
4. Collective line of responsibility
5. Possible assignment of personnel to several different projects
6. Clear hierarchy for supervision
7. Continuity and consistency of functional disciplines
8. Possibility for departmental policies, procedures, and missions

However, the traditional structure does have some negative characteristics:

1. No one individual is directly responsible for the total project.
2. Project-oriented planning may be impeded.
3. There may not be a clear line of reporting up from the lower levels.
4. Coordination is complex.
5. A higher level of cooperation is required between adjacent levels.
6. The strongest functional group may wrongfully claim project authority.

Formal and Informal Structures

The formal organization structure represents the officially sanctioned structure of a functional area. The informal organization, on the other hand, develops when people organize themselves in an unofficial way to accomplish an objective that is in line with the overall project goals. The informal organization is often subtle in that not everyone in the organization is aware of its existence. Both formal and informal organizations are practiced in every project environment. Even organizations with strict hierarchical structures, such as the military, still have some elements of informal organization.

Span of Control

The functional organization calls attention to the form and span of management that are suitable for company goals. The span of management (also known as the span of control) can be wide or narrow. In a narrow span, the functional relationships are streamlined with fewer subordinate units reporting to a single manager. Wide and narrow spans of control are illustrated in Figure 3.2. The wide span of management permits several subordinate units to report to the same boss. The span of control required for a project is influenced by a combination of the following:

1. Level of planning required
2. Level of communication desired
3. Effectiveness of delegating authority
4. Dynamism and nature of the subordinate's job
5. Competence of the subordinate in performing his or her job

Given a project environment that is conducive, the wide span of management can be very effective. From a motivational point of view, workers tend to have a better identification with upper management since there are fewer hierarchical steps to go through to reach the top. More professional growth is possible because workers assume more responsibilities. In addition, the wide span of management is more economical because of the absence of extra layers of supervision. However, the narrow span of management does have its own appeal in situations where there are several mutually exclusive skill levels in the organization.

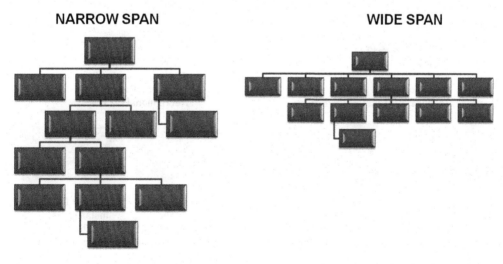

FIGURE 3.2
Wide and narrow spans of control.

Functional Organization

The most common type of formal organization is known as the functional organization, whereby people are organized into groups dedicated to particular functions. Depending on the size and type of auxiliary activities involved in a project, several minor, but supporting, functional units can be developed for a project.

Projects that are organized along functional lines are normally resident in a specific department or area of specialization. The project home office or headquarters is located in the specific functional department. For example, projects that involve manufacturing operations may be under the control of the vice president of manufacturing, while a project involving new technology may be assigned to the vice president for advanced systems. Figure 3.3 shows examples of projects that are functionally organized.

The advantages of a functional organization structure are

1. Improved accountability
2. Discernible line of control
3. Flexibility in manpower utilization
4. Enhanced comradeship of technical staff
5. Improved productivity of specially skilled personnel
6. Potential for staff advancement along functional path
7. Use of home office can serve as a refuge for project problems

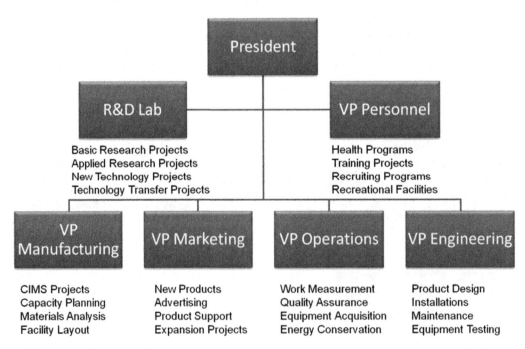

FIGURE 3.3
Functional organization structure.

TABLE 3.1

Functional versus Project Views

Project concerns	Functional concerns
What is the project?	How will the task be done?
When will the project be done?	Where will the task be done?
Why is the project needed?	Who will do the task?
What resources are available?	How do functional inputs affect the project?
What is the project status?	How does the project affect the organization?

The disadvantages of a functional organization structure are

1. Divided attention between project goals and regular functions
2. Conflict between project objectives and regular functions
3. Poor coordination of similar project responsibilities
4. Unreceptive attitude by the surrogate department
5. Multiple layers of management
6. Lack of concentrated effort

It is difficult to separate the project environment from the traditional functional environment. There must be integration. The project management approach affects the functional management approach and vice versa. The questions in Table 3.1 illustrate a comparison of the views of the functional and project environments. Since tasks are the basic components of a project and tasks are the major focus of functional endeavors, they form the basis for the integration of the project and functional environments.

Product Organization

Another approach to organizing a project is to use the end product or goal of the project as the determining factor for personnel structure. This is often referred to as the pure project organization or, simply, project organization. The project is set up as a unique entity within the parent organization. It has its own dedicated technical staff and administration. It is linked to the rest of the system through progress reports, organizational policies, procedures, and funding. The interface between product-organized projects and other elements of the organization may be strict or liberal depending on the organization. An example of a pure project organization is shown in Figure 3.4. The project staff is assembled by assigning personnel from different functional areas.

The product organization is common in large project-oriented organizations or organizations that have multiple product lines. Unlike the functional structure, the product organization decentralizes functions. It creates a unit consisting of specialized skills around a given project or product. Sometimes referred to as team, task force, or product group, the product organization is common in public, research, and manufacturing organizations where specially organized and designated groups are assigned specific functions. A major advantage of the product organization is that it gives the project members a feeling of dedication to the project as well as identification with a particular goal.

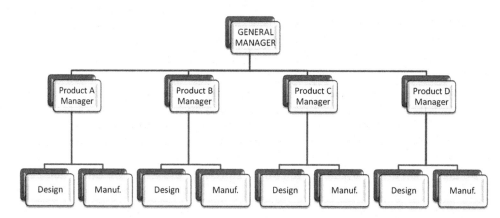

FIGURE 3.4
Product organization structure.

A possible shortcoming of the product organization is the requirement that the product group be sufficiently funded to be able to stand alone without sharing resources or personnel with other functional groups or programs. The product group may be viewed as an ad hoc unit that is formed for the purpose of a specific goal. The personnel involved in the project are dedicated to the particular mission at hand. At the conclusion of the mission, they may be reassigned to other projects. The product organization can facilitate the most diverse and flexible grouping of project participants and permits highly dedicated attention to the project at hand. The advantages of the product organization structure are

1. Simplicity of structure
2. Unity of project purpose
3. Localization of project failures
4. Condensed and focused communication lines
5. Full authority on the project manager
6. Quicker decisions due to centralized authority
7. Skill development due to project specialization
8. Improved motivation, commitment, and concentration
9. Flexibility in determining time, cost, performance trade-offs
10. Accountability of project team to one boss (project manager)
11. Individual acquisition and maintenance of expertise on a given project

The disadvantages are

1. Narrow view of project personnel (as opposed to global organization view)
2. Mutually exclusive allocation of resources (one man to one project)
3. Duplication of efforts on different but similar projects
4. Monopoly of organizational resources
5. Concern about life after the project
6. Reduced skill diversification

Matrix Organization Structure

The matrix organization is a popular choice of management professionals. A matrix organization exists where there are multiple managerial accountability and responsibility for a job function. It attempts to combine the advantages of the traditional structure and the product organization structure. In pure product organization, technology utilization and resource sharing are limited because there is no single group responsible for overall project planning. In the traditional organization structure, time and schedule efficiency are sacrificed. Matrix organization can be defined as follows:

> Matrix organization is a structure of management that facilitates maximum resource utilization and increased performance within time, cost, and performance constraints.

There are usually two chains of command: horizontal and vertical. The horizontal line deals with the functional line of responsibility while the vertical line deals with the project line of responsibility. The project manager has total responsibility and accountability for a project's success. The functional managers have the responsibility to achieve and maintain high technical performance of the project. An example of a project organized under the matrix model is given in Figure 3.5.

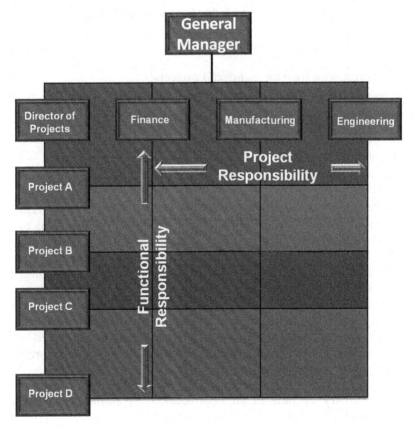

FIGURE 3.5
Matrix organization structure.

The project that is organized under a matrix structure may relate to specific problems, marketing issues, product quality improvement, and so on. The project line in the matrix is usually of a temporary nature while the functional line is more permanent. The matrix organization is dynamic. Its actual structure is determined by the prevailing activities of the project.

The matrix organization has the following advantages:

1. Good team interaction
2. Consolidation of objectives
3. Multilateral flow of information
4. Lateral mobility for job advancement
5. Opportunity to work on a variety of projects
6. Efficient sharing and utilization of resources
7. Reduced project cost due to sharing of personnel
8. Continuity of functions after project completion
9. Stimulating interactions with other functional teams
10. Cooperation of functional lines to support project efforts
11. Home office for personnel after project completion
12. Equal availability of company knowledge base to all projects

The disadvantages of matrix organization are

1. Slow matrix response time for fast-paced projects
2. Independent operation of each project organization
3. High overhead cost due to additional lines of command
4. Potential conflict of project priorities
5. Problems of multiple bosses
6. Complexity of the structure

Despite its disadvantages, the matrix organization is widely used in practice. Its numerous advantages seem to outweigh the disadvantages. In addition, the problems of the matrix organization structure can be overcome with good project planning, which can set the tone for a smooth organization structure. Matrix organization is a collaborative effort between product and functional organization structures. It permits both vertical and horizontal flow of information. The matrix model is sometimes called a multiple-boss organization. It is a model that is becoming increasingly popular as the need for information sharing increases. For example, large technical projects require the integration of specialties from different functional areas. Under matrix organization, projects are permitted to share critical resources as well as management expertise. Several project situations are suitable for implementing a matrix organization.

1. When the primary outputs of an organization are numerous, complex, and resource critical
2. When a complicated design calls for innovation and widespread expertise

3. When expensive, sophisticated, and scarce technologies are needed in designing, building, and testing products

4. When emergency response and flexibility are required for a project

Traditionally, industrial projects are conducted in serial functional implementations such as R&D, engineering, manufacturing, and marketing. At each stage, unique specifications and work patterns may be used without consulting the preceding and succeeding phases. The consequence is that the end product may not possess the original intended characteristics. For example, the first project in the series might involve the production of one component while the subsequent projects might involve the production of other components. The composite product may not achieve the desired performance because the components were not designed and produced from a unified point of view. In today's interdependent market-oriented projects, such lack of a unified design will lead to overall project failure.

The major appeal of matrix organization is that it attempts to provide synergy within groups in an organization. This synergy can be realized if certain ground rules are observed when implementing a matrix organization. Some of those rules are as follows:

1. There must be an individual who is devoted full time to the project.

2. There must be both horizontal and vertical channels for communication.

3. There must be a quick access and conflict resolution strategy between managers.

4. All managers must have an input into the planning process.

5. Functional and project managers must be willing to negotiate and commit resources.

Mixed Organization Structure

Another approach for organizing a project is to adopt a combined implementation of the functional, product, and matrix structures. This permits the different structures to coexist simultaneously in the same project. In an industrial project, for example, the project of designing a new product may be organized using a matrix structure while the subproject of designing the production line may be organized along functional lines. The mixed model facilitates flexibility in meeting special problem situations. The structure can adapt to the prevailing needs of the project or the needs of the overall organization. However, a disadvantage is the difficulty in identifying the lines of responsibility within a given project. There is a wide array of mixed organizations based on the matrix and other structures. These range from a single product/project manager with strong dependence on the functional organization, to a large product/ project organization with little dependence on the functional organization. The functional personnel may be located in the PMO or in a separate geographical location. They may be fully dedicated to a single project manager or they may serve many project managers. In the next section, we present new ideas for unique organization structures that cater to specific project situations.

Alternate Organization Structures

In addition to the traditional OSs, new structures are often needed to address unique organizational or project needs. Some of the structures presented hereafter will, no doubt, be of a temporary nature, and new ones will always be needed to accommodate unique needs that develop within the project system.

Bubble Organization Structure

The bubble organization, which can also be called the *blob organization*, is a structure that allows functional teams to rally around a central project goal. This may be suitable for grass-roots movement among society groups canvassing for a national need. The bubble structure will, most often, be temporary in nature. It disorganizes as soon as its goal is accomplished or it is deemed no longer worthwhile. Figure 3.6 illustrates the bubble organization structure.

Market Organization Structure

As world markets expand, it is important to be more responsive to the changing market forms. The market organization structure permits a project to adapt to market conditions. The market organization structure is illustrated in Figure 3.7.

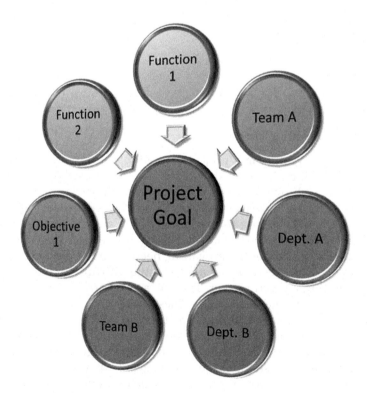

FIGURE 3.6
Bubble organization structure.

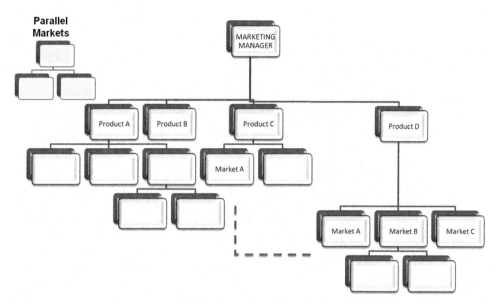

FIGURE 3.7
Market organization structure.

Chronological Organization Structure

The chronological organization is suitable for projects where time sequence is very essential in organizing tasks. This is for time-critical sets of tasks. A training program is suitable for the use of a chronological organization structure. Figure 3.8 presents an illustration of the chronological organization structure.

Sequential Organization Structure

The sequential organization is similar to the chronological structure, except that magnitude or quality of output rather than time is the basis for organizing the project. The quality

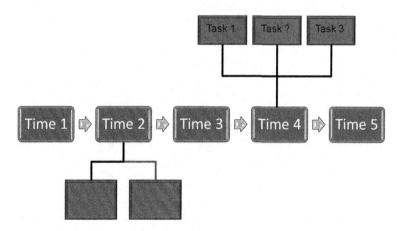

FIGURE 3.8
Chronological organization structure.

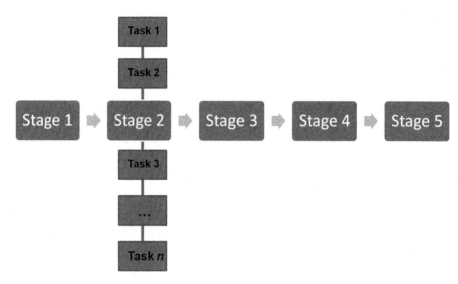

FIGURE 3.9
Sequential organization structure.

of the output at each stage of the organization structure is needed to carry out the functions of the organization sequentially. A value-added production facility is an example of a system that is suitable for a sequential organization structure. Figure 3.9 presents a sequential organization structure.

Military Organization Structure

The military organization follows a strict hierarchical structure and chain of command. It discourages informal lines of communication or responsibility. The name does not necessarily refer to the traditional military structure or configuration, but as in the military command structure, the block at the top of the military organization structure is notably more powerful than the lower blocks. A military organization structure is presented in Figure 3.10.

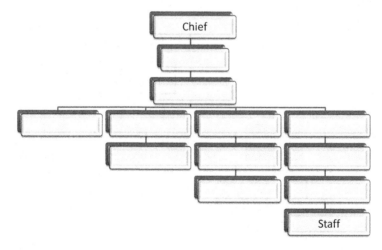

FIGURE 3.10
Military organization structure.

Political Organization Structure

A political organization structure can be viewed as a rotary type of structure that is dynamic with respect to a time cycle. It has a very large base. The large base is collectively more powerful than the few blocks at the top. This may also be referred to as a democratic organization structure. The political organization structure is shown in Figure 3.11.

Autocratic Organization Structure

An autocratic organization may be viewed as the reverse of the political organization structure. There is a single block at the top that is infinitely more powerful than the rest of the organization. Despite the large number of blocks at the lower levels of the structure, the top block remains almightily powerful. A major difference between the military and the autocratic organization structures is that there is a higher prospect that the block at the top of the military structure can be replaced. An autocratic organization structure is presented in Figure 3.12.

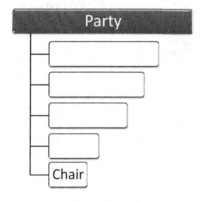

FIGURE 3.11
Political organization structure.

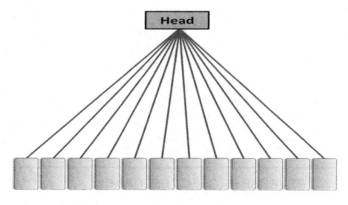

FIGURE 3.12
Autocratic organization structure.

Project Transfer Organization

Project and organizational transfers are important aspects of project management. The OS that is in effect during a project can influence the transfer of the final product of a project. The organic (internal) OS and the external OS must be linked by a discernible transfer path shown in Figure 3.13.

Figure 3.14 shows how products, ideas, concepts, and decisions move from one project environment to another. The receiving organization (referred to as the *transfer target*) uses the transferred elements to generate new products, ideas, concepts, and decisions, which follow a reverse transfer path to the *transfer source*. Thereby, both project environments operate on a symbiotic basis, with each contributing something to the other. Figure 3.15 shows the local adaptation of the transferred project.

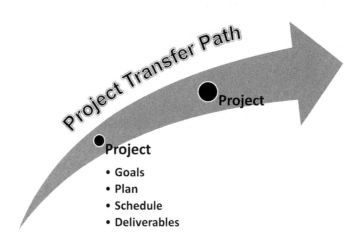

FIGURE 3.13
Internal and external organizational linkage.

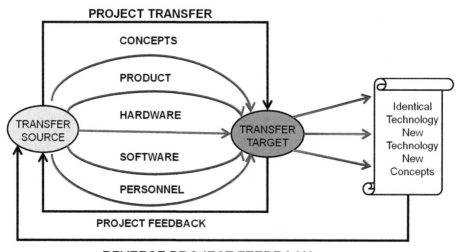

FIGURE 3.14
Project transfer and feedback model.

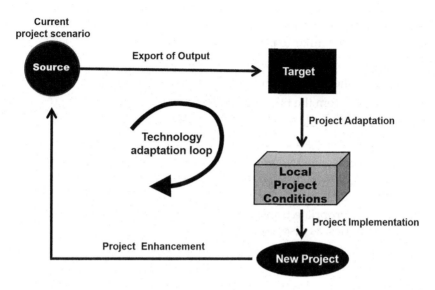

FIGURE 3.15
Local adaptation loop for project transfer.

Project transfer can be achieved in various forms, as outlined in the following text.

1. Transfer of complete project products. In this case, a fully developed product is transferred from one project to another. Very little product development effort is carried out at this receiving point. However, information about the operations of the product is fed back to the source so that necessary product enhancements can be pursued. Hence, the recipient generates product information that serves as a resource for future work of the transfer source.

2. Transfer of procedures and guidelines. In this transfer mode, procedures and guidelines are transferred from one project to another. The project blueprints are implemented locally to generate the desired services and products. The use of local raw materials and personnel is encouraged for local operations. Under this mode, the implementation of the transferred procedures can generate new operating procedures that can be fed back to enhance the original project. With this symbiotic arrangement, a loop system is created whereby both the transferring and receiving organizations obtain useful benefits.

3. Transfer of project concepts, theories, and ideas. This strategy involves the transfer of basic concepts, theories, and ideas associated with a project. The transferred elements can then be enhanced, modified, or customized within local constraints to generate new project outputs. The local modifications and enhancements have the potential to initiate an identical project; a new related project; or a new set of project concepts, theories, and ideas. These derived products may then be transferred back to the transfer source.

The important questions to ask about project transfer include the following:

1. What exactly is being transferred?
2. Who is receiving the transferred elements?

3. What is the cost of the transfer?
4. How is this project similar to previous projects?
5. How is this project different from previous projects?
6. Are the goals of the projects similar?
7. What is expected from the transferred project?
8. Is in-house skill adequate to use the transferred project?
9. Is the prevailing management culture receptive to the new project?
10. Is the current infrastructure capable of supporting the project?
11. What modifications to the project will be necessary?

The selection of an appropriate transfer mode is particularly important for projects that cross national boundaries, as discussed in the next section.

Organizing Multinational Projects

Projects that cross national boundaries either in concept or implementation have unique characteristics that create project management problems. In multinational projects, individual organizational policies are not enough to govern operations. Factors that normally influence these projects include

1. Territorial laws and regulations
2. Geographical segregation and restricted access
3. Time differences
4. Different scientific standards of measure
5. Trade agreements
6. Different government and political ideologies
7. Different social, cultural, and labor practices
8. Different stages of industrialization
9. National security concerns
10. Protection of proprietary technology information
11. Strategic military implications
12. Traditional national allies and adversaries
13. Taxes, duties, and other import/export charges
14. Foreign currency exchange rates
15. National extradition/protection agreements
16. Paperwork, permits, and restrictions
17. Health, weather, and environmental considerations
18. Poor, slow, or incompatible communication links
19. Different native languages

International communication is perhaps one of the most difficult to deal with among all the factors listed previously. The task of international transfer of technology and mutual project support takes on critical dimensions because of differences in the structures, objectives, and interests of the different countries involved. One common communication problem is that information destined for another country may have to pass through several levels of approval before reaching the point of use. The information is subject to all types of distortions and perils in its arduous journey. The integrity of the information may not be preserved as it is passed from one point to another. When implementing international projects, the following considerations should be reviewed:

1. Product
   a. Type of product expected
   b. Portability of the product
   c. Product maintenance
   d. Required training
   e. Availability of spare parts
   f. Feasibility for intended use
   g. Local versus overseas productions
2. Technology
   a. Local availability of required technology
   b. Import/export restrictions on the technology
   c. Implementation requirements
   d. Supporting technologies
   e. Adaptation to local situations
   f. Lag between development and applications
   g. Operational approvals required
3. Political and social environment
   a. Leadership and consistency of national policies
   b. Political and social stability
   c. Management views
   d. Cultural adaptations
   e. Bureaucracies
   f. Structures of formal and informal organizations
   g. Acceptance of foreigners and formation of relationships
   h. Immigration laws
   i. Ethnicity
   j. General economic situation
   k. Religious situations
   l. Population pressures
   m. Local development plans
   n. Decision-making bureaucracies

4. Labor
   a. Union regulations
   b. Wage structures
   c. Personnel dedication, loyalty, and motivation
   d. Educational background and opportunities
   e. Previous experience
   f. Management relationships
   g. Economic condition and level of contentment
   h. Interests, attitudes, personalities, and leisure activities
   i. Taxation policies
   j. Logistics of employee relocation
   k. Productivity consciousness
   l. Demarcation of private and business activities
   m. Local communication practices and facilities
   n. Local customs
   o. Language barriers

5. Market
   a. Market needs
   b. Inflation
   c. Stability
   d. Variety and availability of products
   e. Cash, credit, and billing requirements
   f. Exchange rates
   g. National budget and gross national product
   h. Competition and size of market
   i. Transportation facilities

6. Plant and residential amenities
   a. Location
   b. Structural condition
   c. Accessibility
   d. Facilities available
   e. Proximity to business centers
   f. Topography
   g. Basic amenities (water, light, sewage, etc.)

7. Financial services
   a. Banking
   b. International money transfer
   c. Currency strength and stability
   d. Local sources of capital

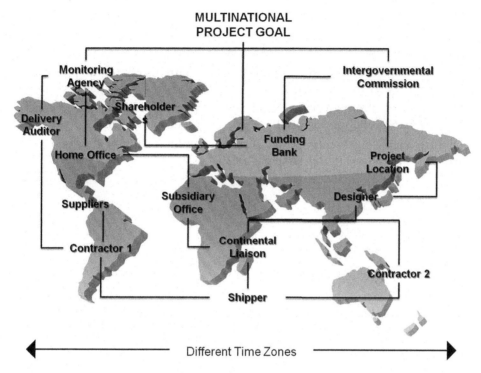

FIGURE 3.16
Multinational intercontinental organizational structure.

  e. Interest rates
  f. Efficiency in conducting transactions
  g. Investment laws

Because of all these various factors, international project managers must undergo more extensive training than their conventional counterparts. A foreign manager in an international project must be open-minded, flexible, adaptive, and able to learn quickly. Since he or she will be working in an unfamiliar combination of social, cultural, political, and religious settings, he or she must have a keen sense of awareness and should be unassuming and responsive to local practices. An evaluation of the factors presented earlier in the context of the specific countries involved should help the international project manager to be better prepared for his or her expanded role. The complexity of an OS for a multinational intercontinental project is illustrated by the example in Figure 3.16.

Exercises

3.1 Suggest a suitable organization structure for an industrial development project. Justify your recommendation with a discussion.

3.2 Discuss how a poor organization structure can lead to project failure.

3.3   Which organization structure was used on the last project that you worked on?

3.4   Discuss how the matrix structure can be implemented in an organization that traditionally adheres to strict hierarchical structure.

3.5   Develop a model for a mixed organization structure using matrix, product, and functional structures.

3.6   Draw a market organization structure for a new product to be marketed worldwide. Select a specific product type to answer this exercise.

3.7   Develop a table of advantages and disadvantages of the different organizational structures presented in this chapter.

4

Project Systems Scheduling

Fundamentals of Network Analysis

Project scheduling is distinguished from job shop, flow shop, and other production sequencing problems because of the unique nature of many of the activities that make up a project. In production scheduling, the scheduling problem follows a standard procedure that determines the characteristics of production operations. A scheduling technique that works for one production run may be expected to work equally effectively for succeeding and identical production runs. By contrast, projects usually involve onetime endeavors that may not be duplicated in identical circumstances. In some cases, it may be possible to duplicate the concepts of the whole project or a portion of it.

Several techniques have been developed for the purpose of planning, scheduling, and controlling projects. The available scheduling models and solution approaches can be categorized as follows:

1. Project scheduling
   a. Resources unconstrained
      i. Critical path analysis
      ii. Time–cost trade-off problem
   b. Resources constrained
      i. Heuristic techniques
      ii. Mathematical programming techniques

Project schedules may be complex, unpredictable, and dynamic. Complexity may be due to interdependencies of activities, multiple resource requirements, multiple concurrent events, conflicting objectives, technical constraints, and schedule conflicts. Unpredictability may be due to equipment breakdowns, raw material inconsistency (delivery and quality), operator performance, labor absenteeism, and unexpected events. Dynamism may be due to resource variability, changes in work orders, and resource substitutions. We define *predictive scheduling* as a scheduling approach that attempts to anticipate the potential causes of schedule problems. These problems are corrected by contingency plans. We define *reactive scheduling* as a scheduling approach that reacts to problems that develop in the scheduling environment.

The most widely used scheduling aids involve network techniques, two of which are Critical Path Method (CPM) and Project Evaluation and Review Technique (PERT). Network analysis procedures originated from the traditional Gantt chart, or bar chart, developed by Henry L. Gantt during World War I. There have been several mathematical techniques for scheduling activities, especially where resource constraints are a major factor. Unfortunately, the mathematical formulations are not generally practical due to the complexity involved in implementing them realistically for large projects. Even computer implementations of the mathematical techniques sometimes become too cumbersome for real-time managerial applications. It should be recalled that the people responsible for project schedules are the managers who, justifiably, prefer simple and quick decision aids. To a project scheduler, a complex mathematical procedure constitutes an impediment rather than an aid in the scheduling routine. Nonetheless, the premise of the mathematical formulations rests on their applicability to small projects consisting of very few activities. Many of the techniques have been evaluated, applied, and reported in the literature.

A more practical approach to scheduling is the use of heuristics. If the circumstances of a problem satisfy the underlying assumptions, a good heuristic will yield schedules that are feasible enough to work with. A major factor in heuristic scheduling is to select a heuristic whose assumptions are widely applicable. A wide variety of scheduling heuristics exists for a wide variety of special cases. The procedure for using heuristics to schedule projects involves prioritizing activities in the assignment of resources and time slots in the project schedule. Many of the available priority rules consider activity durations and resource requirements in the scheduling process.

If all activities are assigned priorities at the beginning and then scheduled, the scheduling heuristic is referred to as a *serial method*. If priorities are assigned to the set of activities eligible for scheduling at a given instant and the schedule is developed concurrently, then the scheduling heuristic is referred to as a *parallel method*. In the serial method, the relative priorities of activities remain fixed. In the parallel method, the priorities change with the current composition of activities.

The network of activities contained in a project provides the basis for scheduling the project. CPM and PERT are the two most popular techniques for project network analysis. The Precedence Diagramming Method (PDM) has gained in popularity in recent years because of the move toward concurrent engineering in manufacturing operations. A project network is the graphical representation of the contents and objectives of the project. The basic project network analysis is typically implemented in three phases: network planning phase, network scheduling phase, and network control phase.

Network planning is sometimes referred to as activity planning. This involves the identification of the relevant activities for the project. The required activities and their precedence relationships are determined. Precedence requirements may be determined on the basis of technological, procedural, or imposed constraints. The activities are then represented in the form of a network diagram. The two popular models for network drawing are the *activity-on-arrow* (*AOA*) and the *activity-on-node* (*AON*) conventions. In the AOA approach, arrows are used to represent activities, while nodes represent starting and ending points of activities. In the AON approach, nodes represent activities, while arrows represent precedence relationships. Time, cost, and resource requirement estimates are developed for each activity during the network planning phase. The estimates may be based on historical records, time standards, forecasting, regression functions, or other quantitative models.

Network scheduling is performed by using forward and backward pass computational procedures. These computations give the earliest and latest starting (LS) and finishing times for each activity. The amount of slack or float associated with each activity is determined. The activity path with the minimum slack in the network is used to determine the critical activities. This path also determines the duration of the project. Resource allocation and time–cost trade-offs are other functions performed during network scheduling.

Network control involves tracking the progress of a project on the basis of the network schedule and taking corrective actions when needed. An evaluation of actual performance versus expected performance determines deficiencies in the project progress. The advantages of project network analysis are as follows:

1. Advantages for communication
   a. Clarifies project objectives
   b. Establishes the specifications for project performance
   c. Provides a starting point for more detailed task analysis
   d. Presents a documentation of the project plan
   e. Serves as a visual communication tool

2. Advantages for control
   a. Presents a measure for evaluating project performance
   b. Helps determine what corrective actions are needed
   c. Gives a clear message of what is expected
   d. Encourages team interactions

3. Advantages for team interaction
   a. Offers a mechanism for a quick introduction to the project
   b. Specifies functional interfaces on the project
   c. Facilitates ease of application

Figure 4.1 shows the graphical representation for an AON network.
The components of the network are explained next.

1. *Node*: A node is a circular representation of an activity.
2. *Arrow*: An arrow is a line connecting two nodes and having an arrowhead at one end. The arrow implies that the activity at the tail of the arrow precedes the one at the head of the arrow.
3. *Activity*: An activity is a time-consuming effort required to perform a part of the overall project. An activity is represented by a node in the AON system or by an arrow in the AOA system. The job the activity represents may be indicated by a short phrase or symbol inside the node or along the arrow.
4. *Restriction*: A restriction is a precedence relationship that establishes the sequence of activities. When one activity must be completed before another activity can begin, the first is said to be a predecessor of the second.
5. *Dummy*: A dummy is used to indicate one event of a significant nature (e.g., milestone). It is denoted by a dashed circle and treated as an activity with zero time duration. A dummy is not required in the AON method. However, it may be included for convenience, network clarification, or to represent a milestone in the progress of the project.

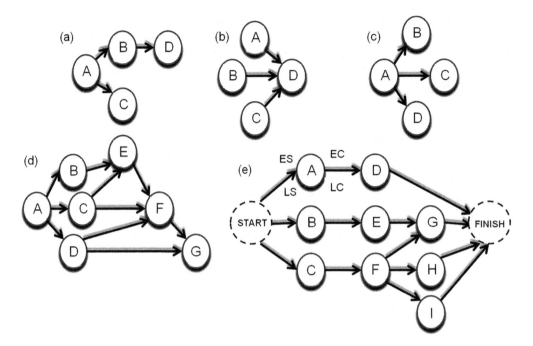

FIGURE 4.1
Graphical representation of an AON network.

6. *Predecessor activity*: A predecessor activity is one that immediately precedes the one being considered. In Figure 4.1a, A is a predecessor of B and C.

7. *Successor activity*: A successor activity is one that immediately follows the one being considered. In Figure 4.1a, activities B and C are successors to A.

8. *Descendent activity*: A descendent activity is any activity restricted by the one under consideration. In Figure 4.1a, activities B, C, and D are all descendants of activity A.

9. *Antecedent activity*: An antecedent activity is any activity that must precede the one being considered. Activities A and B are antecedents of D. Activity A is an antecedent of B, and A has no antecedent.

10. *Merge point*: A merge point (see Figure 4.1b) exists when two or more activities are predecessors to a single activity. All activities preceding the merge point must be completed before the merge activity can commence.

11. *Burst point*: A burst point (see Figure 4.1c) exists when two or more activities have a common predecessor. None of the activities emanating from the same predecessor activity can be started until the burst point activity is completed.

12. *Precedence diagram*: A precedence diagram (see Figure 4.1d) is a graphical representation of the activities making up a project and the precedence requirements needed to complete the project. Time is conventionally shown to be from left to right, but no attempt is made to make the size of the nodes or arrows proportional to time.

Critical Path Method

Precedence relationships in a CPM network fall into three major categories:

1. Technical precedence
2. Procedural precedence
3. Imposed precedence

Technical precedence requirements are caused by the technical relationships among activities in a project. For example, in conventional construction, walls must be erected before the roof can be installed. Procedural precedence requirements are determined by policies and procedures. Such policies and procedures are often subjective, with no concrete justification. Imposed precedence requirements can be classified as resource-imposed, state-imposed, or environment-imposed. For example, resource shortages may require that one task be before another. The current status of a project (e.g., percent completion) may determine that one activity be performed before another. The environment of a project, for example, weather changes or the effects of concurrent projects, may determine the precedence relationships of the activities in a project.

The primary goal of a CPM analysis of a project is the determination of the *critical path*. The critical path determines the minimum completion time for a project. The computational analysis involves *forward* and *backward pass* procedures. The forward pass determines the earliest start (ES) time and the earliest completion (EC) time for each activity in the network. The backward pass determines the LS time and the latest completion (LC) time for each activity. Conventional network logic is always drawn from left to right. If this convention is followed, there is no need to use arrows to indicate the directional flow in the activity network. The notations used for activity A in the network are explained as follows:

1. *A*: Activity identification
2. ES: Earliest starting time
3. EC: Earliest completion time
4. LS: Latest starting time
5. LC: Latest completion time
6. *t*: Activity duration

During the forward pass analysis of the network, it is assumed that each activity will begin at its ES time. An activity can begin as soon as the last of its predecessor is finished. The completion of the forward pass determines the EC time of the project. The backward pass analysis is the reverse of the forward pass analysis. The project begins at its LC time and ends at the LS time of the first activity in the project network. The rules for implementing the forward pass and backward pass analyses in CPM are presented in the following. These rules are implemented iteratively until the ES, EC, LS, and LC have been calculated for all nodes in the activity network.

Rule 1

Unless otherwise stated, the starting time of a project is set equal to time 0. That is, the first node, *node 1*, in the network diagram has an ES time of 0. Thus,

$$ES(1) = 0$$

If a desired starting time, t_0, is specified, then

$$ES(1) = t_0$$

Rule 2

The ES time for any node (activity j) is equal to the maximum of the EC times of the immediate predecessors of the node. That is,

$$ES(i) = Max\{EC(j)\}$$

$$j \in P\{i\}$$

where $P\{i\}$ = {set of immediate predecessors of activity i}.

Rule 3

The EC time of activity i is the activity's ES time plus its estimated time, t_i. That is,

$$EC(i) = ES(i) + t_i$$

Rule 4

The EC time of a project is equal to the EC time of the very last node, *node n*, in the project network. That is,

$$EC(\text{Project}) = EC(n)$$

Rule 5

Unless the LC time of a project is explicitly specified, it is set equal to the EC time of the project. This is called the *zero project slack convention*. That is,

$$LC(\text{Project}) = EC(\text{Project})$$

Rule 6

If a desired deadline, T_p, is specified for the project, then

$$LC(\text{Project}) = T_p$$

It should be noted that an LC time or deadline may sometimes be specified for a project on the basis of contractual agreements.

Rule 7

The LC time for activity j is the smallest of the LS times of the activity's immediate successors. That is,

$$LC(i) = Min\{LS(j)\}$$

$$j \in S\{i\}$$

where $S\{i\}$ = {immediate successors of activity i}.

Rule 8

The LS time for activity j is the LC time minus the activity time. That is,

$$LS(i) = LC(i) - t_i$$

CPM Calculation Example

Table 4.1 presents the data for a simple project network. This network and extensions of it will be used for computational examples in this chapter and subsequent chapters. The AON network for the example is given in Figure 4.2. Dummy activities are included in the network to designate single starting and ending points for the network.

Forward Pass

The forward pass calculations are shown in Figure 4.3. Zero is entered as the ES for the initial node. Since the initial node for the example is a dummy node, its duration is 0. Thus, the EC for the starting node is equal to its ES. The ES values for the immediate successors of the starting node are set equal to the EC of the start node, and the resulting EC values are computed.

Each node is treated as the start node for its successors. However, if an activity has more than one predecessor, the maximum of the EC times of the preceding activities is used as

TABLE 4.1

Data for Sample Project for CPM Analysis

Activity	Predecessor	Duration (Days)
A	—	2
B	—	6
C	—	4
D	A	3
E	C	5
F	A	4
G	B, D, E	2

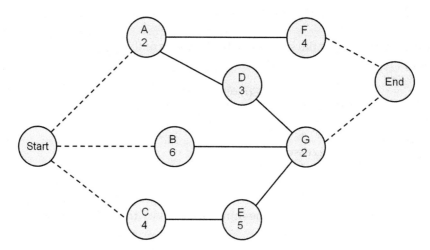

FIGURE 4.2
Example of activity network.

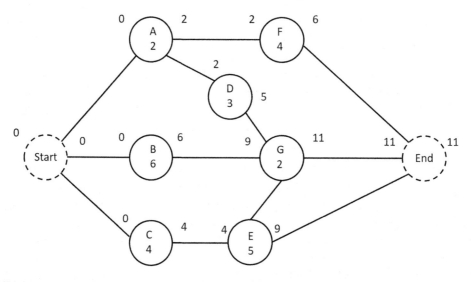

FIGURE 4.3
Forward pass analysis for CPM example.

the activity's starting time. This happens in the case of activity G, whose ES is determined as Max{6, 5, 9} = 9. The earliest project completion time for the example is 11 days. Note that this is the maximum of the immediately preceding EC time: Max{6, 11} = 11. Since the dummy ending node has no duration, its EC time is set equal to its ES time of 11 days.

Backward Pass

The backward pass computations establish the LS and LC times for each node in the network. The results of the backward pass computations are shown in Figure 4.4. Since no deadline is specified, the LC time of the project is set equal to the EC time. By back-tracking and using the network analysis rules presented earlier, the LC and LS times are

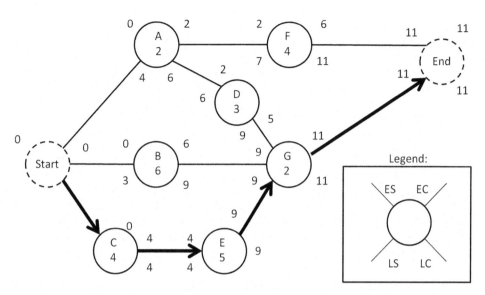

FIGURE 4.4
Backward pass analysis for CPM example.

determined for each node. Note that, in the case of activity A with two immediate successors, the LC time is determined as the minimum of the immediately succeeding LS time.
That is, Min{6, 7} = 6. A similar situation occurs for the dummy starting node. In that case,
the LC time of the dummy start node is Min{0, 3, 4} = 0. Since this dummy node has no
duration, the LS time of the project is set equal to the node's LC time. Thus, the project
starts at time 0 and is expected to be completed by time 11.

Within a project network, there are usually several possible paths and a number of
activities that must be performed sequentially and some activities that may be performed
concurrently. If an activity has ES and EC times that are not equal, then the actual start
and completion times of that activity may be flexible. The amount of flexibility an activity
possesses is called slack time. The slack time is used to determine the critical activities in
the network as discussed next.

Determination of Critical Activities

The critical path is defined as the path with the least slack in the network diagram. All
the activities on the critical path are said to be critical activities. These activities can create bottlenecks in the network if they are delayed. The critical path is also the longest
path in the network diagram. In some networks, particularly large ones, it is possible to
have multiple critical paths or a critical path subnetwork. If there are a large number of
paths in the network, it may be very difficult to visually identify all the critical paths.
The slack time of an activity is also referred to as its *float*. There are four basic types of
activity slack:

Total slack (TS): Total slack is defined as the amount of time an activity may be delayed
from its ES time without delaying the LC time of the project.

The TS of activity i is the difference between the LC time and the EC time of the activity, or the difference between the LS time and the ES time of the activity.

$$TS_i = LC_i - EC_i$$
$$\text{or}$$
$$TS_i = LS_i - ES_i$$

TS is the measure that is used to determine the critical activities in a project network. The critical activities are identified as those having the minimum TS in the network diagram. If there is only one critical path in the network, then all the critical activities will be on that one path.

Free slack: Free slack (FS) is the amount of time an activity may be delayed from its ES time without delaying the starting time of any of its immediate successors. Activity FS is calculated as the difference between the minimum ES time of the activity's successors and the EC time of the activity.

$$FS_i = \text{Min}\{ES_j\} - EC_i$$

$$j \in S(i)$$

where
  FS_i is the FS for activity i
  ES_j is the earliest starting time of a succeeding activity, j, from the set of successors of
      activity i
  $S(i)$ is the set of successor of activity i
  EC_i is the earliest completion of activity i

Interfering slack: Interfering slack (IS) or interfering float is the amount of time by which an activity interferes with (or obstructs) its successors when its TS is fully used. This is rarely used in practice. The interfering float is computed as the difference between TS and FS.

$$IS_i = TS_i - FS_i$$

Independent float: Independent float (IF) or independent slack is the amount of float that an activity will always have regardless of the completion time of its predecessors or the starting times of its successors. IF is computed as

$$IF_i = \text{Max}\{0, (\text{Min } ES_j - \text{Max } LC_k - t_i)\}$$

$$j \in S(i) \quad k \in P(i)$$

where
  ES_j is the earliest starting time of succeeding activity j
  LC_k is the latest completion time of preceding activity k
  t_i is the duration of the activity i, whose IF is being calculated

IF takes a pessimistic view of the situation of an activity. It evaluates the situation whereby the activity is pressured from either side, that is, when its predecessors are delayed as late as possible, while its successors are to be started as early as possible. IF is useful for conservative planning purposes, but it is not used much in practice. Despite its low level of use, IF does have practical implications for better project management. Activities can be buffered with *IFs* as a way to handle contingencies.

For Figure 4.4, the TS and FS for activity A are calculated, respectively, as

$$TS = 6 - 2 = 4 \text{ days}$$

$$FS = Min\{2, 2\} - 2 = 2 - 2 = 0$$

Similarly, the TS and FS for activity F are

$$TS = 11 - 6 = 5 \text{ days}$$

$$FS = Min\{11\} - 6 = 11 - 6 = 5 \text{ days}$$

Table 4.2 presents a tabulation of the results of the CPM example.

The table contains the earliest and latest times for each activity as well as the total and *FSs*. The results indicate that the minimum TS in the network is 0. Thus, activities C, E, and G are identified as critical activities. The critical path is highlighted in Figure 4.4 and consists of the following sequence of activities:

$$START \rightarrow C \rightarrow E \rightarrow G \rightarrow END$$

The TS for the overall project itself is equal to the TS observed on the critical path. The minimum slack in most networks will be zero since the ending LC is set equal to the ending EC. If a deadline is specified for a project, then we would set the project's LC time to the specified deadline. In that case, the minimum TS in the network would be given by

$$TS_{min} = \text{Project deadline} - \text{EC of the last node in the network}$$

This minimum TS will then appear as the TS for each activity on the critical path. If a specified deadline is lower than the EC at the finish node, then the project will start out with a negative slack. That means that it will be behind schedule before it even starts. It may then become necessary to expedite some activities (i.e., crashing) in order to overcome the negative slack. Figure 4.5 shows an example with a specified project deadline. In this case, the deadline of 18 days comes after the EC time of the last node in the network.

TABLE 4.2

Result of CPM Analysis for Sample Project

Activity	Duration	ES	EC	LS	LC	TS	FS	Critical
A	2	0	2	4	6	4	0	—
B	6	0	6	3	9	3	3	—
C	4	0	4	0	4	0	0	Critical
D	3	2	5	6	9	4	4	—
E	5	4	9	4	9	0	0	Critical
F	4	2	6	7	11	5	5	—
G	2	9	11	9	11	0	0	Critical

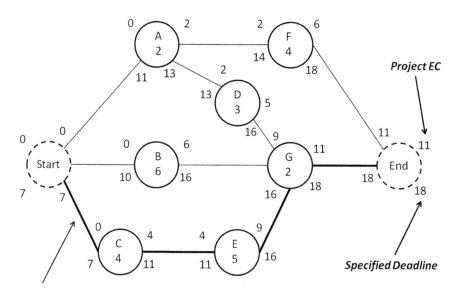

Critical Path with Float of 7 Days

FIGURE 4.5
CPM network with deadline.

Using Forward Pass to Determine the Critical Path

The critical path in CPM analysis can be determined from the forward pass only. This can be helpful in cases where it is desired to quickly identify the critical activities without performing all the other calculations needed to obtain LS times, LC times, and TSs. The steps for determining the critical path from the forward pass only are as follows:

1. Complete the forward pass in the usual manner.
2. Identify the last node in the network as a critical activity.
3. If activity i is an immediate predecessor of activity j, which is determined as a critical activity, then check EC_i and ES_j. If $EC_i = ES_j$, then label activity i as a critical activity. When all immediate predecessors of activity j are considered, mark activity j.
4. Continue the backtracking from each unmarked critical activity until the project starting node is reached. Note that if there is a single starting node or a single ending node in the network, then that node will always be on the critical path.

Subcritical Paths

In a large project network, there may be paths that are near critical. Such paths require almost as much attention as the critical path since they have a high potential of becoming critical when changes occur in the network. Analysis of subcritical paths may help in the classification of tasks into A, B, and C categories on the basis of Pareto analysis. Pareto analysis separates the "vital" few activities from the "trivial" many activities. This permits a more efficient allocation of resources. The principle of Pareto analysis originated from the work of Italian economist Vilfredo Pareto (1848–1923). In his studies, Pareto discovered that most of the wealth in his country was held by a few individuals.

For project control purposes, the Pareto principle states that 80% of the bottlenecks are caused by only 20% of the tasks. This principle is applicable to many management processes. For example, in cost analysis, one can infer that 80% of the total cost is associated with only 20% of cost items. Similarly, 20% of an automobile's parts cause 80% of maintenance problems. In personnel management, about 20% of employees account for about 80% of absenteeism. For critical path analysis, 20% of the network activities will take up 80% of our control efforts. The ABC classification based on Pareto analysis divides items into three priority categories: A (most important), B (moderately important), and C (least important). Appropriate percentages (e.g., 20%, 25%, 55%) are assigned to the categories.

With Pareto analysis, attention can be shifted from focusing only on the critical path to managing critical and near-critical tasks. The level of criticality of each path may be assessed by the following procedure:

1. *Step 1*: Sort activities in increasing order of TS.
2. *Step 2*: Partition the sorted activities into groups based on the magnitude of their TSs.
3. *Step 3*: Sort the activities within each group in increasing order of their ES times.
4. *Step 4*: Assign the highest level of criticality to the first group of activities (e.g., 100%). This first group represents the usual critical path.
5. *Step 5*: Calculate the relative criticality indices for the other groups in decreasing order of criticality.

Define the following variables:

1. α_1 is the minimum TS in the network
2. α_2 is the maximum TS in the network
3. β is the TS for the path whose criticality is to be calculated

Compute the path's criticality as

$$\lambda = \frac{\alpha_2 - \beta}{\alpha_2 - \alpha_1}(100\%)$$

This procedure yields relative criticality levels between 0% and 100%. Table 4.3 presents a hypothetical example of path criticality indices. The criticality level may be converted to a scale between 1 (least critical) and 10 (most critical) by the following expression:

$$\lambda' = 1 + 0.09\lambda$$

Gantt Charts

When the results of CPM analysis are fitted to a calendar time, the project plan becomes a schedule. The Gantt chart is one of the most widely used tools for presenting a project schedule. A Gantt chart can show the planned and actual progress of activities. The timescale is indicated along the horizontal axis, while horizontal bars or lines representing

TABLE 4.3

Analysis of Subcritical Paths

Path Number	Activities on Path	Total Slack	λ	λ'
1	A, C, G, H	0	100%	10
2	B, D, E	1	97.56	9.78
3	F, I	5	87.81	8.90
4	J, K, L	9	78.05	8.03
5	O, P, Q, R	10	75.61	7.81
6	M, S, T	25	39.02	4.51
7	N, AA, BB, U	30	26.83	3.42
8	V, W, X	32	21.95	2.98
9	Y, CC, EE	35	17.14	2.54
10	DD, Z, FF	41	0	1.00

activities are ordered along the vertical axis. As the project progresses, markers are made on the activity bars to indicate actual work accomplished. Gantt charts must be updated periodically to indicate project status. Figure 4.6 presents the Gantt chart for our illustrative example using the ES times from Table 4.2. Figure 4.7 presents the Gantt chart for the example based on the LS time. Critical activities are indicated by the shaded bars.

Figure 4.6 shows that the starting time of activity F can be delayed from day 2 until day 7 (i.e., TS = 5) without delaying the overall project. Likewise, A, D, or both may be delayed by a combined total of 4 days (TS = 4) without delaying the overall project. If all the 4 days of slack are used up by A, then D cannot be delayed. If A is delayed by 1 day, then D can be delayed by up to 3 days without causing a delay of G, which determines the project

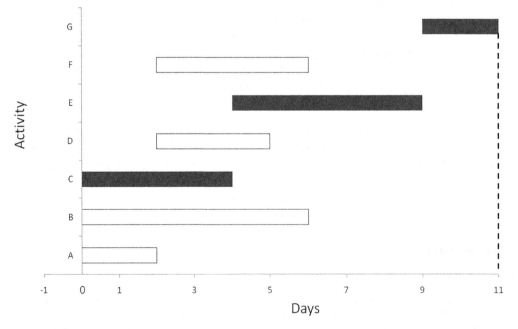

FIGURE 4.6
Gantt chart based on ES times.

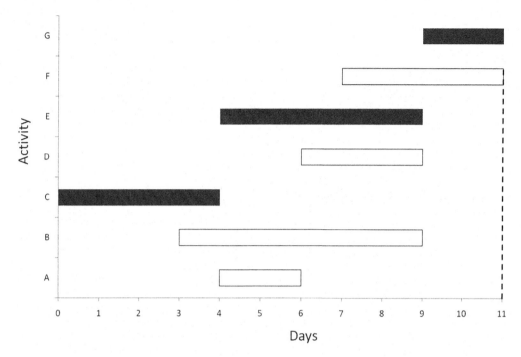

FIGURE 4.7
Gantt chart based on LS times.

completion. The Gantt chart also indicates that activity B may be delayed by up to 3 days without affecting the project completion time.

In Figure 4.7, the activities are scheduled by their LC times. This represents the extreme case where activity slack times are fully used. No activity in this schedule can be delayed without delaying the project. In Figure 4.7, only one activity is scheduled over the first 3 days. This may be compared to the schedule in Figure 4.6, which has three starting activities. The schedule in Figure 4.7 may be useful if there is a situation that permits only a few activities to be scheduled in the early stages of the project. Such situations may involve shortage of project personnel, lack of initial budget, time for project initiation, time for personnel training, allowance for learning period, or general resource constraints. Scheduling of activities based on ES times indicates an optimistic view. Scheduling on the basis of LS times represents a pessimistic approach.

Gantt Chart Variations

The basic Gantt chart does not show the precedence relationships among activities. The chart can be modified to show these relationships by linking appropriate bars, as shown in Figure 4.8. However, the linked bars become cluttered and confusing for large networks.

Figure 4.9 shows a Gantt chart that presents a comparison of planned and actual schedules. Note that two tasks are in progress at the current time indicated in the figure. One of the ongoing tasks is an unplanned task. Figure 4.10 shows a Gantt chart on which important milestones have been indicated. Figure 4.11 shows a Gantt chart in which bars represent a combination of related tasks.

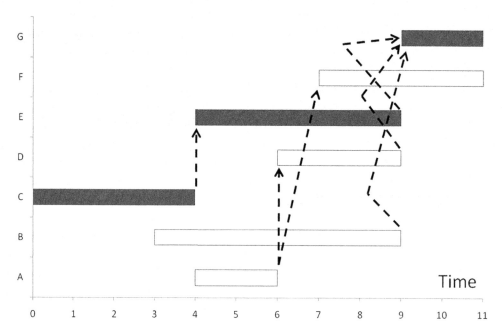

FIGURE 4.8
Linked bars in Gantt chart.

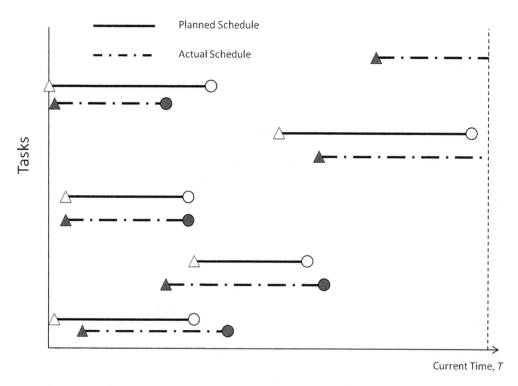

FIGURE 4.9
Progress-monitoring Gantt chart.

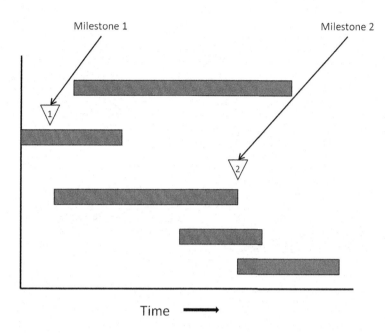

FIGURE 4.10
Milestone Gantt chart.

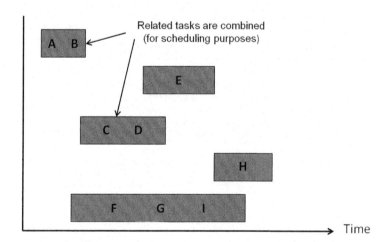

FIGURE 4.11
Task-combination Gantt chart.

Tasks may be combined for scheduling purposes or for conveying functional relationships required on a project. Figure 4.12 presents a Gantt chart of project phases. Each phase is further divided into parts. Figure 4.13 shows a multiple-project Gantt chart. Multiple-project charts are useful for evaluating resource allocation strategies. Resource loading over multiple projects may be needed for capital budgeting and cash flow analysis decisions. Figure 4.14 shows a project slippage chart that is useful for project tracking and control. Other variations of the basic Gantt chart may be developed for specific needs.

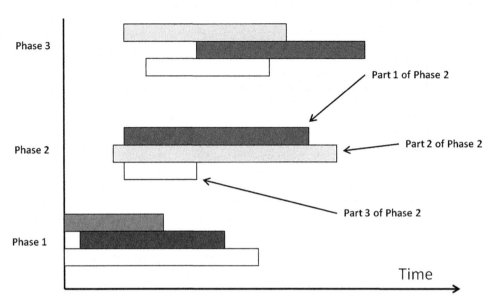

FIGURE 4.12
Phase-based Gantt chart.

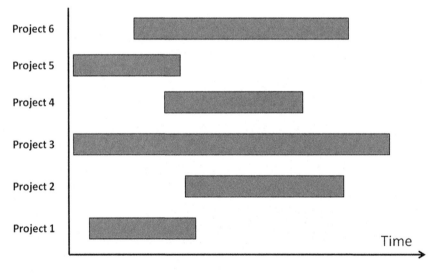

FIGURE 4.13
Multiple-projects Gantt chart.

Activity Crashing and Schedule Compression

Schedule compression refers to reducing the length of a project network. This is often accomplished by crashing activities. *Crashing,* sometimes referred to as expediting, reduces activity durations, thereby reducing project duration. Crashing is done as a trade-off between shorter task duration and higher task cost. It must be determined whether the total cost savings realized from reducing the project duration is enough to justify the higher costs

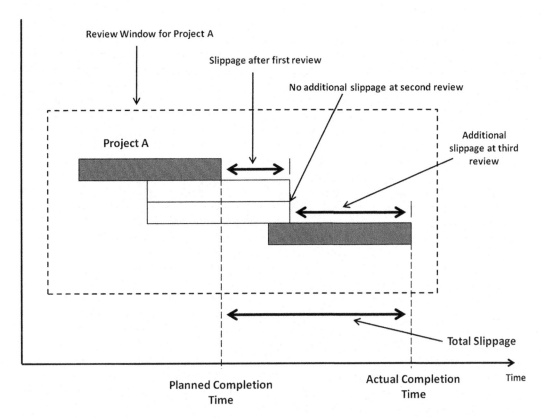

FIGURE 4.14
Project-slippage-tracking Gantt chart.

associated with reducing individual task durations. If there is a delay penalty associated with a project, it may be possible to reduce the total project cost even though individual task costs are increased by crashing. If the cost savings on a delay penalty are higher than the incremental cost of reducing the project duration, then crashing is justified. Under conventional crashing, the further the duration of a project is compressed, the higher the total cost of the project. The objective is to determine at what point to terminate further crashing in a network. *Normal task duration* refers to the time required to perform a task under normal circumstances. *Crash task duration* refers to the reduced time required to perform a task when additional resources are allocated to it.

If each activity is assigned a range of time and cost estimates, then several combinations of time and cost values will be associated with the overall project. Iterative procedures are used to determine the best time or cost combination for a project. Time–cost trade-off analysis may be conducted, for example, to determine the marginal cost of reducing the duration of the project by one time unit. Table 4.4 presents an extension of the data for the example problem to include normal and crash times as well as normal and crash costs for each activity. The normal duration of the project is 11 days, as seen earlier, and the normal cost is $2,775.

If all the activities are reduced to their respective crash durations, the total crash cost of the project will be $3,545. In that case, the crash time is found by CPM analysis to be 7 days. The CPM network for the fully crashed project is shown in Figure 4.15. Note that activities

TABLE 4.4

Normal and Crash Time and Cost Data

Activity	Normal Duration	Normal Cost	Crash Duration	Crash Cost	Crashing Ratio
A	2	$210	2	$210	0
B	6	400	4	600	100
C	4	500	3	750	250
D	3	540	2	600	60
E	5	750	3	950	100
F	4	275	3	310	35
G	2	100	1	125	25
		$2,775		$3,545	

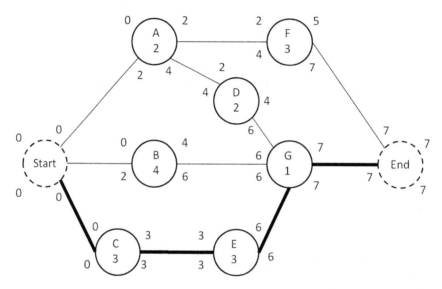

FIGURE 4.15
Example of a fully crashed CPM network.

C, E, and G remain critical. Sometimes, the crashing of activities may result in additional critical paths. The Gantt chart in Figure 4.16 shows a schedule of the crashed project using the ES times.

In practice, one would not crash all activities in a network. Rather, some heuristic would be used to determine which activity should be crashed and by how much. One approach is to crash only the critical activities or those activities with the best ratios of incremental cost versus time reduction. The last column in Table 4.4 presents the respective ratios for the activities in our example. The crashing ratio is computed as

$$r = \frac{\text{Crash cost} - \text{Normal cost}}{\text{Normal duration} - \text{Crash duration}}$$

This method of computing the crashing ratio gives crashing priority to the activity with the lowest cost slope. It is a commonly used approach of expediting in CPM networks.

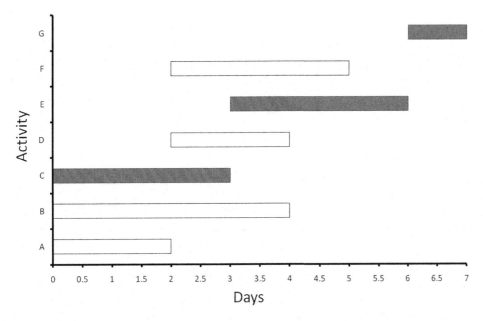

FIGURE 4.16
Gantt chart of a fully crashed CPM network.

Activity G offers the lowest cost per unit time reduction of $25. If our approach is to crash only one activity at a time, we may decide to crash activity G first and evaluate the increase in project cost versus the reduction in project duration. The process can then be repeated for the next best candidate for crashing, which in this case is activity F. The project completion time is not reduced any further since activity F is not a critical activity. After F has been crashed, activity D can then be crashed. This approach is repeated iteratively in order of activity preference, until no further reduction in project duration can be achieved or until the total project cost exceeds a specified limit.

A more comprehensive analysis is to evaluate all possible combinations of activities that can be crashed. However, such a complete enumeration would be prohibitive, since there would be a total of 2^c crashed networks to evaluate, where c is the number of activities that can be crashed out of the n activities in the network ($c \leq n$). For our example, only six out of seven activities in the network can be crashed. Thus, a complete enumeration will involve $2^6 = 64$ alternate networks. Table 4.5 shows 7 of the 64 crashing options. Activity G, which offers the best crashing ratio, reduces the project duration by only 1 day. Even though activities F, D, and B are crashed by a total of 4 days at an incremental cost of $295, they do not generate any reduction in project duration. Activity E is crashed by 2 days and it generates a reduction of 2 days in project duration. Activity C, which is crashed by 1 day, generates a further reduction of 1 day in the project duration. It should be noted that the activities that generate reductions in project duration are the ones that were identified earlier as the critical activities.

Figure 4.17 shows the crashed project duration versus the crashing options and a plot of the total project cost after crashing. As more activities are crashed, the project duration decreases while the total project cost increases. If full enumeration was performed, the plot would contain additional points between the minimum possible project duration of 7 days (fully crashed) and the normal project duration of 11 days (no crashing). Similarly,

TABLE 4.5

Selected Crashing Options for CPM Example

Option Number	Activities Crashed	Network Duration	Time Reduction	Incremental Cost	Total Cost
1	None	$210	2	$210	0
2	G	400	4	600	100
3	G, F	500	3	750	250
4	G, F, D	540	2	600	60
5	G, F, D, B	750	3	950	100
6	G, F, D, B, E	275	3	310	35
7	G, F, D, B, E, C	100	1	125	25

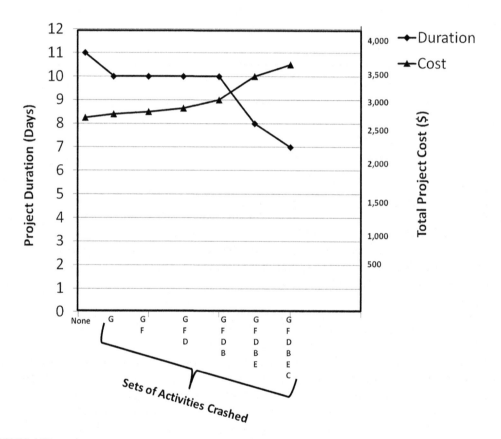

FIGURE 4.17
Plot of duration and cost as a function of crashing option.

the plot for total project cost would contain additional points between the normal cost of $2,775 and the crash cost of $3,545.

In general, there may be more than one critical path, so one needs to check for the set of critical activities with the least total crashing ratio in order to minimize the total crashing cost. Also, one needs to update the critical paths every time a set of activities is crashed, because new activities may become critical in the meantime. For the network given in

Figure 4.15, the path C–E–G is the only critical path throughout $7 \leq T \leq 11$. Therefore, one need not consider crashing other jobs since the incurred cost will not affect the project completion time. There are 12 possible ways one can crash activities C, G, and E to reduce the project time.

Table 4.6 defines possible strategies and crashing costs for durations of $7 \leq T \leq 11$. Again, the strategies involve only critical arcs (activities), since crashing a noncritical arc is clearly fruitless. Figure 4.18 is a plot of the strategies with respect to cost and project duration

TABLE 4.6

Project Compression Strategies

Project Duration	Crashing Strategy	Description of Crashing	Total Cost
$T = 11$	S_1	Activities at normal duration	$2,775
$T = 10$	S_2	Crash G by 1 unit	2,800
	S_3	Crash C by 1 unit	3,025
	S_4	Crash E by 1 unit	2,875
$T = 9$	S_5	Crash G and C by 1 unit	3,050
	S_6	Crash G and E by 1 unit	2,900
	S_7	Crash C and E by 1 unit	3,125
	S_8	Crash E by 2 units	2,975
$T = 8$	S_9	Crash G, C, and E by 1 unit	3,150
	S_{10}	Crash G by 1 unit, E by 2 units	3,000
	S_{11}	Crash C by 1 unit, E by 2 units	3,225
$T = 7$	S_{12}	Crash G and C by 1 unit, and E by 2 units	3,250

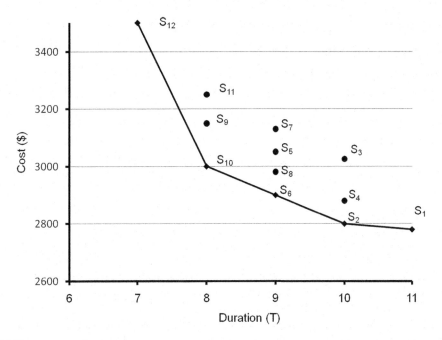

FIGURE 4.18

Time–cost plots for the strategies in Table 4.6.

values. The optimal strategy for each T value is the strategy with the minimum cost. Optimal strategies are connected in Figure 4.18. This piecewise linear and convex curve is referred to as the time–cost trade-off curve. Chapter 7 outlines an algorithm for a time–cost trade-off problem.

Several other approaches exist for determining which activities to crash in a project network. Two alternate approaches are presented in the following text for computing the crashing ratio, r. The first one directly uses the criticality of an activity to determine its crashing ratio, while the second one uses a computational expression shown in the following:

$$r = \text{Criticality index}$$

$$r = \frac{\text{Crash cost} - \text{Normal cost}}{(\text{Normal duration} - \text{Crash duration})(\text{Criticality index})}$$

The first approach gives crashing priority to the activity with the highest probability of being on the critical path. In deterministic networks, this refers to the critical activities. In stochastic networks, an activity is expected to fall on the critical path only a percentage of time. The second approach is a combination of the approach used for the illustrative example and the criticality index approach. It reflects the process of selecting the least-cost expected value. The denominator of the expression represents the expected number of days by which the critical path can be shortened. For different project networks, different crashing approaches should be considered, and the one that best fits the nature of the network should be selected.

PERT Network Analysis

PERT is an extension of CPM that incorporates variabilities in activity durations into project network analysis. PERT has been used extensively and successfully in practice.

In real life, activities are often prone to uncertainties that determine the actual durations of activities. In CPM, activity durations are assumed to be deterministic. In PERT, the potential uncertainties in activity durations are accounted for by using three time estimates for each activity. The three time estimates represent the spread of the estimated activity duration. The greater the uncertainty of an activity, the wider the range of estimates.

PERT Estimates and Formulas

PERT uses the three time estimates and simple equations to compute the expected duration and variance for each activity. The PERT formulas are based on a simplification of the expressions for the mean and variance of a beta distribution. The approximation formula for the mean is a simple weighted average of three time estimates, with the end points assumed to be equally likely and the mode four times as likely. The approximation formula for PERT is based on the recognition that most of the observations from a distribution will lie within plus or minus three standard deviations or a spread of six standard deviations. This leads to the simple method of setting the PERT formula for standard deviation equal to one-sixth of the estimated duration range. While there is no theoretical validation for these approximation approaches, the PERT formulas do facilitate ease of use. The formulas are presented as follows:

$$t_e = \frac{a + 4m + b}{6}$$

$$s^2 = \frac{(b - a)^2}{36}$$

where

a is the optimistic time estimate
m is the most likely time estimate
b is the pessimistic time estimate ($a < m < b$)
t_e is the expected time for the activity
s^2 is the variance of the duration of the activity

After obtaining the estimate of the duration for each activity, the network analysis is carried out in the same manner previously illustrated for the CPM approach. The major steps in PERT analysis are as follows:

1. Obtain three time estimates a, m, and b for each activity.
2. Compute the expected duration for each activity by using the formula for t_e.
3. Compute the variance of the duration of each activity from the formula for s^2. It should be noted that CPM analysis cannot calculate variance of activity duration, since it uses a single time estimate for each activity.
4. Compute the expected project duration, T_e. As in the case of CPM, the duration of a project in PERT analysis is the sum of durations of activities on the critical path.
5. Compute the variance of the project duration as the sum of variances of activities on the critical path. The variance of the project duration is denoted by S^2. It should be recalled that CPM cannot compute the variance of the project duration, since variances of activity durations are not computed.
6. If there are two or more critical paths in the network, choose the one with the largest variance to determine the project duration and the variance of the project duration. Thus, PERT is pessimistic with respect to the variance of project duration when there are multiple critical paths in the project network. For some networks, it may be necessary to perform a mean–variance analysis to determine the relative importance of the multiple paths by plotting the expected project duration versus the path duration variance.
7. If desired, compute the probability of completing the project within a specified time period. This is not possible under CPM.

In practice, a question often arises as to how to obtain good estimates of a, m, and b. Several approaches can be used in obtaining the required time estimates for PERT. Some of the approaches are as follows:

1. Estimates furnished by an experienced person
2. Estimates extracted from standard time data

3. Estimates obtained from historical data

4. Estimates obtained from simple regression and/or forecasting

5. Estimates generated by simulation

6. Estimates derived from heuristic assumptions

7. Estimates dictated by customer requirements

The pitfall of using estimates furnished by an individual is that they may be inconsistent, since they are limited by the experience and personal bias of the person providing them. Individuals responsible for furnishing time estimates are usually not experts in estimation, and they generally have difficulty in providing accurate PERT time estimates. There is often a tendency to select values of a, m, and b that are optimistically skewed. This is because a conservatively large value is typically assigned to b by inexperienced individuals.

The use of time standards, on the other hand, may not reflect the changes occurring in the current operating environment due to new technology, work simplification, new personnel, and so on. The use of historical data and forecasting is very popular because estimates can be verified and validated by actual records. In the case of regression and forecasting, there is the danger of extrapolation beyond the data range used for fitting the regression and forecasting models. If the sample size in a historical data set is sufficient and the data can be assumed to reasonably represent prevailing operating conditions, the three PERT estimates can be computed as follows:

$$\hat{a} = \bar{t} - kR$$

$$\hat{m} = \bar{t}$$

$$\hat{b} = \bar{t} + kR$$

where
  R is the range of the sample data
  $\bar{t}$ is the arithmetic average of the sample data
  $k = 3/d_2$
  d_2 is the an adjustment factor for estimating the standard deviation of a population

If $kR > \bar{t}$, then set $a = 0$ and $b = 2\bar{t}$. The factor d_2 is widely tabulated in the quality control literature as a function of the number of sample points, n. Selected values of d_2 are presented next.

n	5	10	15	20	25	30	40	50	75	100
d_2	2.326	3.078	3.472	3.735	3.931	4.086	4.322	4.498	4.806	5.015

In practice, probability distributions of activity times can be determined from historical data. The procedure involves three steps:

1. Appropriate organization of the historical data into histograms.

2. Determination of a distribution that reasonably fits the shape of the histogram.

3. Testing the goodness of fit of the hypothesized distribution by using an appropriate statistical model. The chi-square test and the Kolmogrov–Smirnov test are two popular methods for testing goodness of fit. Most statistical texts present the details of how to carry out goodness-of-fit tests.

Beta Distribution

PERT analysis assumes that the probabilistic properties of activity duration can be modeled by the beta probability density function. The beta distribution is defined by two end points and two shape parameters. The beta distribution was chosen by the original developers of PERT as a reasonable distribution to model activity times because it has finite end points and can assume a variety of shapes based on different shape parameters. While the true distribution of activity time will rarely ever be known, the beta distribution serves as an acceptable model. Figure 4.19 shows examples of alternate shapes of the standard beta distribution between 0 and 1. The uniform distribution between 0 and 1 is a special case of the beta distribution with both shape parameters equal to one.

The standard beta distribution is defined over the interval 0–1, while the general beta distribution is defined over any interval *a–b*. The general beta probability density function is given by

$$f(t) = \frac{\Gamma(\alpha+\beta)}{\Gamma(\alpha)\Gamma(\beta)} \frac{1}{(b-a)^{\alpha+\beta-1}} (t-a)^{\alpha-1}(b-t)^{\beta-1} \quad a \leq t \leq b; \quad \alpha > 0; \quad \beta > 0$$

where
  a is the lower end point of distribution
  b is the upper end point of distribution
  α, β are the shape parameters for distribution

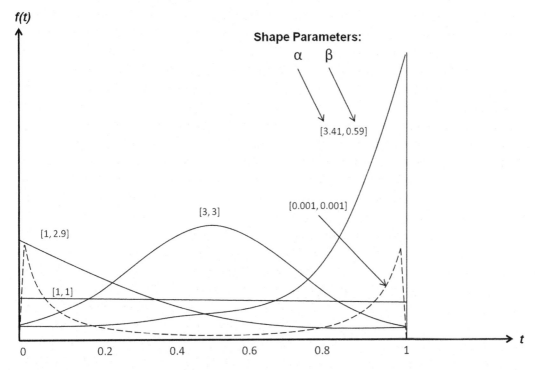

FIGURE 4.19
Alternate shapes of the beta distributions.

The mean, variance, and mode of the general beta distribution are defined as

$$\mu = a + (b-a)\frac{\alpha}{\alpha + \beta}$$

$$\sigma^2 = \frac{\alpha\beta}{(\alpha + \beta + 1)(\alpha + \beta)^2}$$

$$m = \frac{a(\beta - 1) + b(\alpha - 1)}{\alpha + \beta - 2}$$

The general beta distribution can be transformed into a standardized distribution by changing its domain from [a, b] to the unit interval [0, 1]. This is accomplished by using the relationship $t_g = a + (b-a)t_s$, where t_s is the standard beta random variable between 0 and 1. This yields the standardized beta distribution, given by

$$f(t) = \frac{\Gamma(\alpha + \beta)}{\Gamma(\alpha)\Gamma(\beta)} \quad 0 \le t \le 1; \quad \alpha > 0; \quad \beta > 0$$

$$= 0; \quad \text{elsewhere}$$

with mean, variance, and mode defined as

$$\mu = \frac{\alpha}{\alpha + \beta}$$

$$\sigma^2 = \frac{\alpha\beta}{(\alpha + \beta + 1)(\alpha + \beta)^2}$$

$$m = \frac{\alpha - 1}{\alpha + \beta - 2}$$

Triangular Distribution

The triangular probability density function has been used as an alternative to the beta distribution for modeling activity times. The triangular density has three essential parameters: minimum value (a), mode (m), and maximum (b). The triangular density function is defined mathematically as

$$f(t) = \frac{2(t - a)}{(m - a)(b - a)}; \quad a \le t \le m$$

$$= \frac{2(b - t)}{(b - m)(b - a)}; \quad m \le t \le b$$

with mean and variance defined, respectively, as

$$\mu = \frac{a + m + b}{3}$$

$$\sigma^2 = \frac{a(a - m) + b(b - a) + m(m - b)}{18}$$

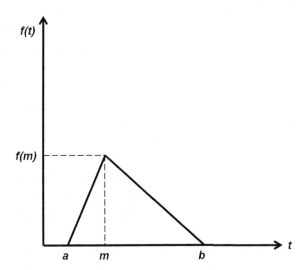

FIGURE 4.20
Triangular probability density function.

Figure 4.20 presents a graphical representation of the triangular density function. The three time estimates of PERT can be inserted into the expression for the mean of the triangular distribution to obtain an estimate of the expected activity duration. Note that in the conventional PERT formula, the mode (*m*) is assumed to carry four times as much weight as either *a* or *b* when calculating the expected activity duration. By contrast, under the triangular distribution, the three time estimates are assumed to carry equal weights.

Uniform Distribution

For cases where only two time estimates instead of three are to be used for network analysis, the uniform density function may be assumed for the activity times. This is acceptable for situations where the extreme limits of activity duration can be estimated, and it can be assumed that the intermediate values are equally likely to occur. The uniform distribution is defined mathematically as

$$f(t) = \frac{1}{b-a}; \quad a \leq t \leq b$$

$$= 0; \quad \text{elsewhere}$$

with mean and variance defined, respectively, as

$$\mu = \frac{a+b}{2}$$

$$\sigma^2 = \frac{(b-a)^2}{12}$$

Figure 4.21 presents a graphical representation of the uniform distribution. In the case of uniform distribution, the expected activity duration is computed as the average of the upper and lower limits of distribution. The appeal of using only two time estimates *a* and

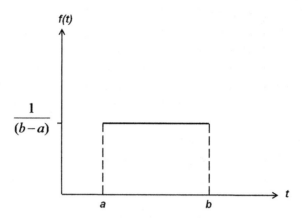

FIGURE 4.21
Uniform probability density function.

b is that the estimation error due to subjectivity can be reduced and the estimation task simplified. Even when a uniform distribution is not assumed, other statistical distributions can be modeled over the range of a and b.

Other distributions that have been explored for activity time modeling include the normal distribution, lognormal distribution, truncated exponential distribution, and Weibull distribution. Once the expected activity durations have been computed, the analysis of the activity network is carried out just as in the case of single-estimate CPM network analysis.

Distribution of Project Duration

Regardless of the distribution assumed for activity durations, the *central limit theorem* suggests that the distribution of the project duration will be approximately normally distributed. The theorem says that the distribution of averages obtained from any probability density function will be approximately normally distributed if the sample size is large and the averages are independent. In mathematical terms, the theorem is stated as follows:

Central Limit Theorem

Let X_1, X_2, ..., X_N be independent and identically distributed random variables. Then the sum of the random variables tends to be normally distributed for large values of N. The sum is defined as

$$T = X_1 + X_2 + \cdots + X_N$$

In activity network analysis, T represents the total project length as determined by the sum of the durations of the activities on the critical path. The mean and variance of T are expressed as

$$\mu = \sum_{i=1}^{N} E[X_i]$$

$$\sigma^2 = \sum_{i=1}^{N} V[X_i]$$

where
   $E[X_i]$ is the expected value of random variable X_i
   $V[X_i]$ is the variance of random variable X_i

When applying the central limit theorem to activity networks, it should be noted that the assumption of independent activity times may not always be satisfied. Because of precedence relationships and other interdependencies of activities, some activity durations may not be independent.

Probability Calculation

If the project duration T_e can be assumed to be approximately normally distributed based on the central limit theorem, then the probability of meeting a specified deadline T_d can be computed by finding the area under the standard normal curve to the left of T_d. Figure 4.22 shows an example of a normal distribution describing the project duration.

Using the following familiar transformation formula, a relationship between the standard normal random variable z and the project duration variable can be obtained:

$$z = \frac{T_d - T_e}{S}$$

where
   T_d is the specified deadline
   T_e is the expected project duration based on network analysis
   S is the standard deviation of project duration

The probability of completing a project by the deadline T_d is then computed as

$$(T \le T_d) = P\left(z \le \frac{T_d - T_e}{S}\right)$$

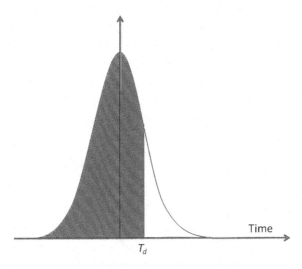

FIGURE 4.22
Area under the normal curve.

The probability is obtained from any standard normal table. The following example illustrates the procedure for probability calculations in PERT.

PERT Network Example

Suppose we have the project data presented in Table 4.7. The expected activity durations and variances, as calculated by the PERT formulas, are shown in the two right-hand columns of the table. Figure 4.23 shows the PERT network. Activities C, E, and G are shown to be critical, and the project completion time is 11 time units.

The probability of completing the project on or before a deadline of 10 time units (i.e., $T_d = 10$) is calculated as

$$T_e = 11$$

$$S^2 = V[C] + V[E] + V[G]$$

$$= 0.25 + 0.25 + 0.1111 = 0.6111$$

$$S = \sqrt{0.6111} = 0.7817$$

TABLE 4.7

PERT Project Data

Activity	Predecessors	A	m	b	t_e	s^2
A	—	1	2	4	2.17	0.2500
B	—	5	6	7	6.00	0.1111
C	—	2	4	5	3.83	0.2500
D	A	1	3	4	2.83	0.2500
E	C	4	5	7	5.17	0.2500
F	A	3	4	5	4.00	0.1111
G	B, D, E	1	2	3	2.00	0.1111

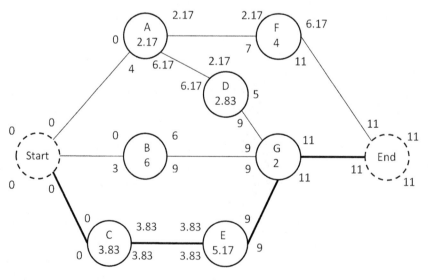

FIGURE 4.23
PERT network example.

$$P(T \leq T_d) = P(T \leq 10)$$

$$= P\left(z \leq \frac{10 - T_e}{S}\right)$$

$$= P\left(z \leq \frac{10 - 11}{0.7817}\right)$$

$$= P(z \leq -1.2793)$$

$$= 1 - P(z \leq 1.2793)$$

$$= 1 - 0.8997 = 0.1003$$

Thus, there is just over 10% probability of finishing the project within 10 days. By contrast, the probability of finishing the project in 13 days is calculated as

$$P(T \leq 13) = P\left(z \leq \frac{13 - 11}{0.7817}\right)$$

$$= P(z \leq 2.5585) = 0.9948$$

This implies that there is over 99% probability of finishing the project within 13 days. Note that the probability of finishing the project in exactly 13 days will be 0. An exercise at the end of this chapter requires the reader to show that $P(T = T_d) = 0$. If we desire the probability that the project can be completed within a certain lower limit (T_L) and a certain upper limit (T_U), the computation will proceed as follows: Let $T_L = 9$ and $T_U = 11.5$. Then,

$$P(T_L \leq T \leq T_u) = P(9 \leq T \leq 11.5)$$

$$= P(T \leq 11.5) - P(T \leq 9)$$

$$= P\left(z \leq \frac{11.5 - 11}{0.7817}\right) - P\left(z \leq \frac{9 - 11}{0.7817}\right)$$

$$= P(z \leq 0.6396) - P(z \leq -2.5585)$$

$$= P(z \leq 0.6396) - [1 - P(z \leq 2.5585)]$$

$$= 0.7389 - [1 - 0.9948] = 0.7337$$

Precedence Diagramming Method

The PDM was developed in the early 1960s as an extension of the basic PERT/CPM network analysis. PDM permits mutually dependent activities to be performed partially in parallel instead of serially. The usual finish-to-start dependencies between activities are relaxed to allow activities to be overlapped. This facilitates schedule compression. An example is the requirement that concrete should be allowed to dry for a number of days before drilling holes for handrails. That is, drilling cannot start until so many days after the completion of

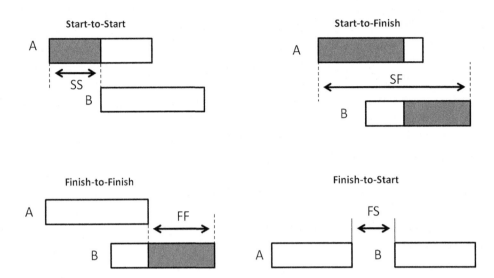

FIGURE 4.24
Lead–lag relationships in PDM.

concrete work. This is a finish-to-start constraint. The time between the finishing time of the first activity and the starting time of the second activity is called the *lead–lag* require-ment between the two activities. Figure 4.24 shows the graphical representation of the basic lead–lag relationships between activities A and B.

The terminology presented in Figure 4.24 is explained as follows:

1. SS_{AB} (*start-to-start*) *lead*: This specifies that activity B cannot start until activity A has been in progress for at least SS time units.

2. FF_{AB} (*finish-to-finish*) *lead*: This specifies that activity B cannot finish until at least FF time units after the completion of activity A.

3. FS_{AB} (*finish-to-start*) *lead*: This specifies that activity B cannot start until at least FS time units after the completion of activity A. Note that PERT/CPM approaches use $FS_{AB} = 0$ for network analysis.

4. SF_{AB} (*start-to-finish*) *lead*: This specifies that there must be at least SF time units between the start of activity A and the completion of activity B.

The leads or lags may, alternately, be expressed in percentages rather than time units. For example, we may specify that 25% of the work content of activity A must be completed before activity B can start. If the percentage of work completed is used for determining lead–lag constraints, then a reliable procedure must be used for estimating the percent completion. If the project work is broken up properly using work breakdown structure, it will be much easier to estimate percent completion by evaluating the work completed at the elementary task levels. The lead–lag relationships may also be specified in terms of *at most* relationships instead of *at least* relationships. For example, we may have at most an FF lag requirement between the finishing time of one activity and the finishing time of another activity. Splitting activities often simplifies the implementation of PDM, as will be shown later with some examples. Some of the factors that will determine whether or not an activity can be split are technical limitations affecting splitting of a task, morale of the person working on the split task, setup times required to restart split tasks, difficulty involved in managing resources for split tasks, loss of consistency of work, and management policy about splitting jobs.

Figure 4.25 presents a simple CPM network consisting of three activities. The activities are to be performed serially, and each has an expected duration of 10 days. The conventional CPM network analysis indicates that the duration of the network is 30 days. The earliest and latest times are shown in the figure.

The Gantt chart for the example is shown in Figure 4.26. For comparison, Figure 4.27 shows the same network but with some lead–lag constraints. For example, there is an SS constraint of 2 days and an FF constraint of 2 days between activities A and B. Thus, activity B can start as early as 2 days after activity A starts, but it cannot finish until 2 days after the completion of A. In other words, *at least* 2 days must be between the starting times of

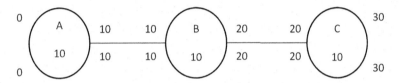

FIGURE 4.25
Serial activities in CPM network.

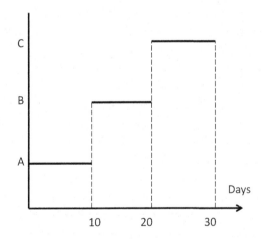

FIGURE 4.26
Gantt chart of serial activities in CPM example.

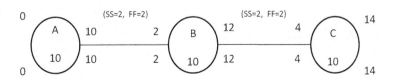

FIGURE 4.27
PDM network example.

A and B. Likewise, *at least* 2 days must separate the finishing time of A and the finishing time of B. A similar precedence relationship exists between activities B and C. The earliest and latest times obtained by considering the lag constraints are indicated in Figure 4.27.

The calculations show that if B is started just 2 days after A is started it can be completed as soon as 12 days as opposed to the 20 days obtained in the case of conventional CPM. Similarly, activity C is completed at time 14, which is considerably less than the 30 days calculated by conventional CPM. The lead–lag constraints allow us to compress or overlap activities. Depending on the nature of tasks involved, an activity does not have to wait until its predecessor finishes before it can start. Figure 4.28 shows the Gantt chart for the example incorporating the lead–lag constraints. It should be noted that a portion of a succeeding activity can be performed simultaneously with a portion of the preceding activity.

A portion of an activity that overlaps with a portion of another activity may be viewed as a distinct portion of the required work. Thus, partial completion of an activity may be elevated. Figure 4.29 shows how each of the three activities is partitioned into contiguous parts. Even though there is no physical break or termination of work in any activity, the distinct parts (beginning and ending) can still be identified. This means that there is no physical splitting of

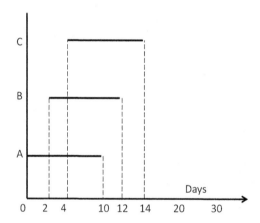

FIGURE 4.28
Gantt chart for PDM example.

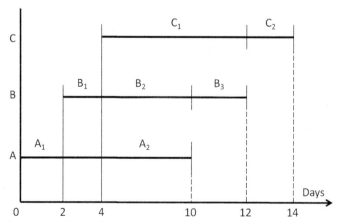

FIGURE 4.29   Partitioning of activities in PDM example.

the work content of any activity. The distinct parts are determined on the basis of the amount of work that must be completed before or after another activity, as dictated by the lead–lag relationships. In Figure 4.29, activity A is partitioned into parts A_1 and A_2. The duration of A_1 is 2 days because there is an SS = 2 relationship between activities A and B. Since the original duration of A is 10 days, the duration of A_2 is then calculated to be $10 - 2 = 8$ days.

Likewise, activity B is partitioned into parts B_1, B_2, and B_3. The duration of B_1 is 2 days because there is an SS = 2 relationship between activities B and C. The duration of B_3 is also 2 days because there is an FF = 2 relationship between activities A and B. Since the original duration of B is 10 days, the duration of B_2 is calculated to be $10 - (2 + 2) = 6$ days. In a similar fashion, activity C is partitioned into C_1 and C_2. The duration of C_2 is 2 days because there is an FF = 2 relationship between activities B and C. Since the original duration of C is 10 days, the duration of C_1 is then calculated to be $10 - 2 = 8$ days. Figure 4.30 shows a conventional CPM network drawn for the three activities after they are partitioned into distinct parts. The conventional forward and backward passes reveal that all the activity parts are on the critical path. This makes sense, since the original three activities are performed serially and no physical splitting of activities has been performed. Note that there are three critical paths in Figure 4.30, each with a length of 14 days. It should also be noted that the distinct parts of each activity are performed contiguously.

Figure 4.31 shows an alternate example of three serial activities. The conventional CPM analysis shows that the duration of the network is 30 days. When lead–lag constraints

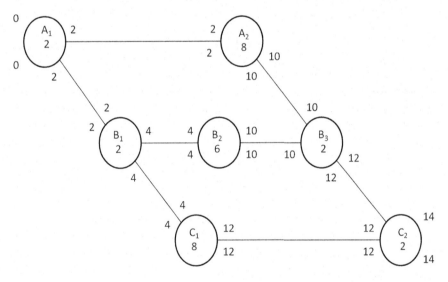

FIGURE 4.30
CPM network of partitioned activities.

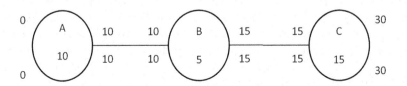

FIGURE 4.31
Another CPM example of serial activities.

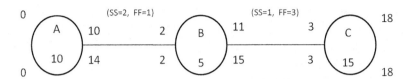

FIGURE 4.32
Compressed PDM network.

are introduced into the network as shown in Figure 4.32, the network duration is compressed to 18 days.

In the forward pass computations in Figure 4.32, note that the EC time of B is time 11, because there is an FF = 1 restriction between activities A and B. Since A finishes at time 10, B cannot finish until at least time 11. Even though the ES time of B is time 2 and its duration is 5 days, its EC time cannot be earlier than time 11. Also note that C can start as early as time 3 because there is an SS = 1 relationship between B and C. Thus, given a duration of 15 days for C, the EC time of the network is 3 + 15 = 18 days. The difference between the EC time of C and that of B is 18 − 11 = 7 days, which satisfies the FF = 3 relationship between B and C.

In the backward pass, the LC time of B is 15 (i.e., 18 − 3 = 15), since there is an FF = 3 relationship between activities B and C. The LS time for B is time 2 (i.e., 3 − 1 = 2), since there is an SS = 1 relationship between activities B and C. If we are not careful, we may erroneously set the LS time of B to 10 (i.e., 15 − 5 = 10). But that would violate the SS = 1 restriction between B and C. The LC time of A is found to be 14 (i.e., 15 − 1 = 14), since there is an FF = 1 relationship between A and B. All the earliest times and latest times at each node must be evaluated to ensure that they conform to all the lead–lag constraints. When computing ES or EC times, the smallest possible value that satisfies the lead–lag constraints should be used. By the same reasoning, when computing the LS or LC times, the largest possible value that satisfies the lead–lag constraints should be used.

Manual evaluations of the lead–lag precedence network analysis can become very tedious for large networks. A computer program may be used to simplify the implementation of PDM. If manual analysis must be done for PDM computations, it is suggested that the network be partitioned into more manageable segments. The segments may then be linked after the computations are completed. The expanded CPM network in Figure 4.33 was developed on the basis of the precedence network in Figure 4.32. It is seen that activity A is partitioned into two parts, activity B is partitioned into three parts, and activity C is partitioned into two parts. The forward and backward passes show that only the first parts of activities A and B are on the critical path. Both parts of activity C are on the critical path.

Figure 4.34 shows the corresponding earliest-start Gantt chart for the expanded network. Looking at the ES times, one can see that activity B is physically split at the boundary of B_2 and B_3 in such a way that B_3 is separated from B_2 by 4 days. This implies that work on activity B is temporarily stopped at time 6 after B_2 is finished and is not started again until time 10. Note that despite the 4-day delay in starting B_3, the entire project is not delayed. This is because B_3, the last part of activity B, is not on the critical path. In fact, B_3 has a TS of 4 days. In a situation like this, the duration of activity B can actually be increased from 5 to 9 days without any adverse effect on the project duration. It should be recognized, however, that increasing the duration of an activity may have negative implications for project cost and personnel productivity.

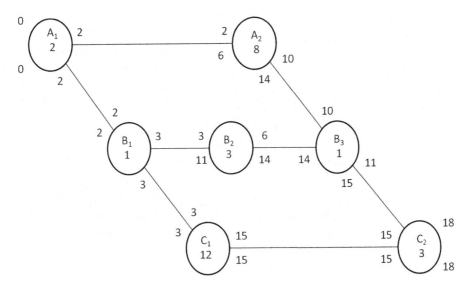

FIGURE 4.33
CPM expansion of second PDM example.

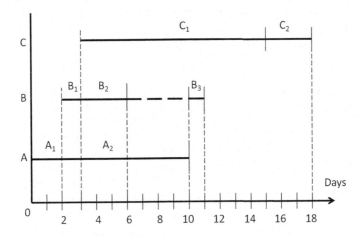

FIGURE 4.34
Compressed PDM schedule based on ES times.

If the physical splitting of activities is not permitted, then the best option available in Figure 4.34 is to stretch the duration of B_2 so as to fill up the gap from time 6 to 10. An alternative is to delay the starting time of B_1 until time 4 so as to use up the 4-day delay slack right at the beginning of activity B. Unfortunately, delaying the starting time of B_1 by 4 days will delay the overall project by 4 days, since B_1 is on the critical path as shown in Figure 4.33. The project analyst will need to evaluate the appropriate trade-offs among splitting activities, delaying activities, increasing activity durations, and incurring higher project costs. The prevailing project scenario should be considered when making such trade-off decisions. Figure 4.35 shows the Gantt chart for the compressed PDM schedule based on LS times. In this case, it will be necessary to split both activities A and B even though the total project duration remains the same at 18 days.

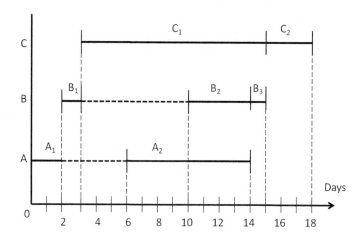

FIGURE 4.35
Compressed PDM schedule based on LS times.

If activity splitting is to be avoided, then we can increase the duration of activity A from 10 to 14 days and the duration of B from 5 to 13 days without adversely affecting the entire project duration. The important benefit of precedence diagramming is that the ability to overlap activities facilitates some flexibilities in manipulating individual activity times and compressing the project duration.

Anomalies in PDM Networks

Care must be exercised when working with PDM networks because of the potential for misuse or misinterpretation. Because of the lead and lag requirements, activities that do not have any slacks may appear to have generous slacks. Also, *reverse critical activities* may occur in PDM. Reverse critical activities are activities that can cause a decrease in project duration when their durations are increased. This may happen when the critical path enters the completion of an activity through a finish lead–lag constraint. Also, if a finish-to-finish dependency and a start-to-start dependency are connected to a reverse critical task, a reduction in the duration of the task may actually lead to an increase in the project duration. Figure 4.36 illustrates this anomalous situation. The finish-to-finish constraint between A and B requires that B should finish no earlier than 20 days. If the duration of task B is reduced from 10 to 5 days, the start-to-start constraint between B and C forces the starting time of C to be shifted forward by 5 days, thereby resulting in a 5-day increase in the project duration.

The preceding anomalies can occur without being noticed in large PDM networks. One safeguard against their adverse effects is to make only one activity change at a time and document the resulting effect on the network structure and length. The following categorizations are often used for the unusual characteristics of activities in PDM networks.

1. *Normal critical (NC)*: This refers to an activity for which the project duration shifts in the same direction as the shift in the duration of the activity.
2. *Reverse critical (RC)*: This refers to an activity for which the project duration shifts in the reverse direction to the shift in the duration of the activity.

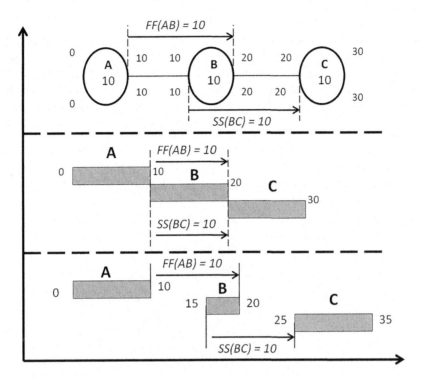

FIGURE 4.36
Reverse critical activity in a PDM network.

3. *Bicritical (BC)*: This refers to an activity for which the project duration increases as a result of any shift in the duration of the activity.

4. *Start critical (SC)*: This refers to an activity for which the project duration shifts in the direction of the shift in the start time of the activity but is unaffected (neutral) by a shift in the overall duration of the activity.

5. *Finish critical (FC)*: This refers to an activity for which the project duration shifts in the direction of the shift in the finish time of the activity but is unaffected (neutral) by a shift in the overall duration of the activity.

6. *Mid-Normal Critical (MNC)*: This refers to an activity whose midportion is normal critical.

7. *Mid-Reverse Critical (MRC)*: This refers to an activity whose midportion is reverse critical.

8. *Mid-BiCritical (MBC)*: This refers to an activity whose midportion is bicritical.

Complexity of Project Networks

The performance of a scheduling heuristic will be greatly influenced by the complexity of the project network. The more activities there are in the network and the more resource types are involved, the more complex the scheduling effort. Numerous analytical experiments have revealed the lack of consistency in heuristic performances. Some heuristics perform well for both small and large projects. Some perform well only for small projects.

Still, some heuristics that perform well for certain types of small projects may not perform well for other projects of comparable size. The implicit network structure based on precedence relationships and path interconnections influences network complexity and, hence, the performance of scheduling heuristics. The complexity of a project network may indicate the degree of effort that has been devoted to planning the project. The better the planning for a project, the lower the complexity of the project network can be expected to be. This is because many of the redundant interrelationships among activities can be identified and eliminated through better planning.

There have been some attempts to quantify the complexity of project networks. Since the structures of projects vary from simple to complex, it is desirable to have a measure of how difficult it will be to schedule a project. Some of the common measures of project network complexity are presented as follows:

For PERT networks,

$$C = \frac{(\text{Number of activities})^2}{(\text{Number of events})}$$

where an event is defined as an end point (or node) of an activity.

For precedence networks,

$$C = \frac{(\text{Preceding work items})^2}{(\text{Total number of work items})}$$

The aforementioned expressions represent simple measures of the degree of interrelationship of the project network. Another measure uses the following expression:

$$C = \frac{2(A - N + 1)}{(N - 1)(N - 2)}$$

where

A is the number of activities
N is the number of nodes in the project network

One complexity measure is defined as the total activity density, D, of a project network.

$$D = \sum_{i=1}^{N} \text{Max}\{0, (p_i - s_i)\}$$

where

N is the number of activities
p_i is the number of predecessor activities for activity i
s_i is the number of successor activities for activity i

Alternate measures of network complexity often found in the literature are presented in the following expressions:

w_j is a measure of total work content for resource type j

O is a measure of obstruction factor, which is a measure of the ratio of excess resource requirements to total work content

O_{est} is the adjusted obstruction per period based on earliest start time schedule

O_{lst} is the adjusted obstruction per period based on LS time schedule

U is a resource utilization factor

$$C = \frac{\text{Number of activities}}{\text{(Number of nodes)}}$$

$$D = \frac{\text{Sum of job durations}}{\text{(Sum of job durations + total free slack)}}$$

$$w_j = \sum_{i=1}^{N} d_i r_{ij}$$

$$= \sum_{t=1}^{CP} r_{jt}$$

where

d_i is the duration of job i

r_{ij} is the per-period requirement of resource type j by job i

t is the time period

N is the number of jobs

CP is the original critical path duration

r_{jt} is the total resource requirements of resource type j in time period t

$$O = \sum_{j=1}^{M} O_j$$

$$= \sum_{j=1}^{M} \left(\frac{\sum_{t=1}^{CP} \text{Max}\{0, r_{jt} - A_j\}}{w_j} \right)$$

where

O_j is the obstruction factor for resource type j

CP is the original critical path duration

A_j is the unit of resource type j available per period

M is the number of different resource types

w_j is the total work content for resource type j

r_{jt} is the total resource requirement of resource type j in time period t

$$O_{est} = \sum_{j=1}^{m} \left[\frac{\sum_{t=1}^{CP} \text{Max}\{0, r_{jt(est)} - A_j\}}{(M)(CP)} \right]$$

where $r_{jt(\text{est})}$ is the total resource requirement of resource type j in time period t based on ES times.

$$O_{\text{lst}} = \sum_{j=1}^{m} \left[\frac{\sum_{t=1}^{\text{CP}} \text{Max}\{0, r_{jt(\text{lst})} - A_j\}}{(M)(\text{CP})} \right]$$

where $r_{jt(\text{lst})}$ is the total resource requirements of resource type j in time period t based on LS times. The measures O_{est} and O_{lst} incorporate the calculation of excess resource requirements adjusted by the number of periods and the number of different resource types.

$$U = \text{Max}_j\{f_j\}$$

$$= \text{Max}_j\left\{ \frac{w_j}{(\text{CP})(A_j)} \right\}$$

where f_j is the resource utilization factor for resource type j. This measures the ratio of the total work content to the total work initially available. Another comprehensive measure of the complexity of a project network is based on the resource intensity (λ) of the network:

$$\lambda = \frac{p}{d}\left[\left(1 - \frac{1}{L}\right)\sum_{i=1}^{L} t_i + \sum_{j=1}^{R}\left(\frac{\sum_{i=1}^{L} t_i x_{ij}}{Z_j} \right) \right]$$

where
  λ is the project network complexity
  L is the number of activities in the network
  t_i is the expected duration for activity i
  R is the number of resource types
  x_{ij} is the unit of resource type j required by activity i
  Z_j is the maximum units of resource type j available
  p is the maximum number of immediate predecessors in the network
  d is the PERT duration of the project with no resource constraint

The terms in the expression for the complexity are explained as follows: The maximum number of immediate predecessors, p, is a multiplicative factor that increases the complexity and potential for bottlenecks in a project network. The $(1 - 1/L)$ term is a fractional measure (between 0.0 and 1.0) that indicates the time intensity or work content of the project. As L increases, the quantity $(1 - 1/L)$ increases, and a larger fraction of the total time requirement (sum of t_i) is charged to the network complexity. Conversely, as L decreases, the network complexity decreases proportionally with the total time requirement. The sum of $(t_i x_{ij})$ indicates the time-based consumption of a given resource type j relative to the maximum availability. The term is summed over all the different resource types. Having PERT duration in the denominator helps to express the complexity as a dimensionless quantity by canceling out the time units in the numerator. In addition, it gives the network complexity per unit of total project duration.

In addition to the approaches presented previously, organizations often use their own internal qualitative and quantitative assessment methods to judge the complexity of projects. Budgeting requirements are sometimes incorporated into the complexity measure. There is always a debate as to whether or not the complexity of a project can be accurately quantified. There are several quantitative and qualitative factors with unknown interactions that are present in any project network. As a result, any measure of project complexity should be used as a relative measure of comparison rather than as an absolute indication of the difficulty involved in scheduling a given project. Since the performance of a scheduling approach can deteriorate sometimes with the increase in project size, a further comparison of the rules may be done on the basis of a collection of large projects. A major deficiency in the existing measures of project network complexity is that there is a lack of well-designed experiments to compare and verify the effectiveness of the measures. Also, there is usually no guideline as to whether a complexity measure should be used as an ordinal or a cardinal measure, as is illustrated in the following example.

Example of Complexity Computation

Table 4.8 presents a sample project for illustrating the network complexity measure for the λ expression presented previously. For the data in the table, we obtain the following:

$$p = 1; \quad L = 6; \quad d = 6.33$$

$$\sum_{i=1}^{6} t_i = 13.5$$

$$\sum_{i=1}^{6} t_i x_{i1} = 22.5$$

$$\sum_{i=1}^{6} t_i x_{i2} = 6.3$$

$$\lambda = \frac{1}{6.33}\left[\left(\frac{6-1}{6}\right)(13.5)+\left(\frac{22.58}{5}+\frac{6.25}{2}\right)\right] = 2.99$$

TABLE 4.8

Data for Project Complexity Example

Activity Number	PERT Estimates (a, m, b)	Preceding Activities	Required Resources (x_{i_1}, x_{i_2})
1	1, 3, 5	—	1, 0
2	0.5, 1, 3	—	1, 1
3	1, 1, 2	—	1, 1
4	2, 3, 6	Activity 1	2, 0
5	1, 3, 4	Activity 2	1, 0
6	1.5, 2, 2	Activity 3	4, 2
			$Z_1 = 5, Z_2 = 2$

If the aforementioned complexity measure is to be used as an ordinal measure, then it must be used to compare and rank alternate project networks. For example, when planning a project, one may use the complexity measure to indicate the degree of simplification that has been achieved in each iteration of the project plan. Similarly, when evaluating project options, one may use the ordinal complexity measure to determine which network option will be easiest to manage. If the complexity measure is to be used as a cardinal measure, then a benchmark value must be developed. In other words, control limits will be needed to indicate when a project network can be classified as simple, medium, or complex. The following classification ranges are suggested:

1. Simple network: $0 \leq \lambda \leq 5.0$
2. Medium network: $5.0 < \lambda \leq 12.0$
3. Complex network: $\lambda > 12.0$

The aforementioned ranges are based on the result of an experimental investigation involving 30 alternate projects of various degrees of network complexity. Of course, the ranges cannot be said to be universally applicable, because consistency of measurement cannot be assured from one project network to another. Users can always determine what ranges will best suit their needs. The complexity measure can then be used accordingly. Perhaps the greatest utility of a complexity measure is obtained when evaluating computer implementation of network analysis. This is addressed further by the discussion in the next section.

Evaluation of Solution Time

By using solution time as a performance measure, another comparison of the scheduling heuristics may be conducted. Computer processing time should be recorded for each scheduling rule under each test problem. The following procedure may then be used to perform the solution time analysis. We let τ_{mn} denote computer processing time for scheduling heuristic m for test problem n. The sum of the processing times over the set of test problems is expressed as

$$\Psi_n = \sum_m \tau_{mn}, \quad n = 1, 2, \ldots, N; m = 1, 2, \ldots, M$$

where m, n, M, and N are as previously defined. Then,

$$\mu_n = \frac{\Psi_n}{M}$$

denotes the average time for scheduling project n, where M is the number of rules considered. The normalized solution time for heuristic m under test problem n can then be denoted as

$$\Omega_{mn} = \frac{\tau_{mn}}{\mu_n}$$

Each heuristic m is ranked on the basis of the sum of normalized solution times over all test problems. That is,

$$\theta_m = \sum_n \Omega_{mn}, \quad m = 1, 2, \ldots, M; n = 1, 2, \ldots, N$$

It is obvious that the solution time of each scheduling heuristic depends on its computational complexity. The computations for some heuristics (e.g., highest number of immediate successors) do not require a prior analysis of the PERT network. For heuristics that consider several factors in the activity sequencing process, their computations will be more complex than other scheduling heuristics. The solution time analysis results should thus be coupled with schedule effectiveness in order to judge the overall acceptability of any given heuristic. Prior analysis and selection of the most effective scheduling heuristic for a given project can help minimize schedule changes and delays often encountered in impromptu scheduling practices.

Performance of Scheduling Heuristics

In addition to comparing scheduling heuristics on the basis of project durations, the following aggregate measures may also be used. The first one is an evaluation of the ratio of the minimum project duration observed to the project duration obtained under each heuristic: For each heuristic m, the ratio under each test problem n is computed as

$$\rho_{mn} = \frac{q_n}{PL_{mn}}, \quad m = 1, 2, \ldots, M; n = 1, 2, \ldots, N$$

where
 ρ_{mn} is the efficiency ratio for heuristic m under test problem n
 PL_{mn} is the project duration for test problem n under heuristic m
 $q_n = Min_m\{PL_{mn}\}$; minimum duration observed for test problem n
 M is the number of scheduling heuristics considered
 N is the number of test problems

From the definitions given earlier, the maximum value for the ratio is 1.0. Thus, it is alternately referred to as the rule efficiency ratio. The value q_n is, of course, not necessarily the global minimum project duration for test problem n. Rather, it represents the local minimum based on the particular scheduling heuristics considered. If the global minimum duration for a project is known (probably from an optimization model), then it should be used in the expression for ρ_{mn}. Rules can be compared on the basis of the absolute values for ρ_{mn} or on the basis of the sums of ρ_{mn}. The sums of ρ_{mn} over the index n are defined as

$$\Phi_m = \sum_{n=1}^{N} \rho_{mn}, \quad m = 1, 2, \ldots, M$$

The use of the sums of ρ_{mn} is a practical approach to comparing scheduling heuristics. It is possible to have a scheduling rule that will consistently yield near-minimum project durations for all test problems. On the other hand, there may be another rule that performs very well for some test problems while performing poorly for other problems. A weighted sum helps to average out the overall performance over several test problems. The other comparison measure involves the calculation of percentage deviations from the observed minimum project duration. The deviations are computed as

$$S_{mn} = \left(\frac{\text{PL}_{mn} - q_n}{q_n} \right)(100), \quad m = 1, 2, \ldots, M; n = 1, 2, \ldots, N$$

which denotes the percentage deviation from the minimum project duration for rule m under test problem n. A project analyst will need to consider several factors discussed earlier when selecting and implementing scheduling heuristics. The potential variability in the work rates of resources is another complicating factor in the project scheduling problem.

Formulation of Project Graph

As discussed previously, a project network can be represented in two different modes: a precedence (AON) diagram and an arrow (activity-on-arc) diagram. While the first mode is frequently used in practical situations, the second mode is often used in optimization. In this section, we present the formulation of project graph using AOA representation. The concept of precedence is the key to the construction of project networks. The following rules apply:

If activity a precedes activity b, then it is indicated as $a \leftarrow b$.

The precedence relationship is transitive.

If $a \leftarrow b$ and $b \leftarrow c$, then $a \leftarrow c$

Regardless of the representation mode, a project network has the following structural properties:

1. It is acyclic. In other words, there does not exist a directed path $i-k_1-k_2-k_3 \ldots k_p-i$ meaning an activity precedes itself!

2. Each node should have at least one arc directed into the node and one arc directed out of the node with the exception of the start and finish nodes. The start node does not have any arc into it, and the finished node has no arc out of it.

3. All of the nodes and arcs of the network have to be visited (that is realized) in order to complete the project.

Since project networks are acyclic, the existence of a cycle in the network due to an error in the network construction or due to an error in entering the data into a computer will lead to cycling in the procedures. More specifically, critical path calculations will not terminate. Therefore, one may need to detect the cycles in the network. One way to accomplish that is to number the nodes such that for each arc$(i, j)i < j$. If such a numbering is possible, then the network is acyclic. However, this may not be an easy task when thousands of nodes are involved. A depth-first search (*dfs*) procedure can be used to detect cycles in a project network.

Depth-First Search Method

The procedure starts with all nodes unvisited and unmarked. The source node is visited first and its *dfs* number is set equal to 1 since it is the first visited node in the depth search. In general, the procedure locates an arc(i, j) emanating from a visited node i going to an unvisited node j. Existence of such an arc leads to node j being visited and initiates a new search at node j. The *dfs* number of node j will be one more than the *dfs* number of node i. If no arc connects node i to unvisited nodes, then the search from i is completed, and the node is marked. The procedure then backs up to a node with the *dfs* number, one less than the *dfs* number of node i, and initiates a new search process. The procedure terminates when the source node is marked. Depth-first traversal of the network **G** leads to four different arc definitions. A *tree arc* is an arc that connects a visited node to an unvisited node. A *forward arc* connects a low-numbered node to a high-numbered node. A *cross arc* connects a visited node to a visited and marked node. A *back arc* connects a visited node to a visited but unmarked node. Existence of back arcs indicates that cycles exist in the network. The procedure for detecting cycles in a network is summarized.

Procedure to Detect Cycles in a Network *G* = *(N, A)* Using the dfs Method

1. Initialize all nodes as unvisited and unmarked. $A_1 = \varnothing$. Set the *dfs* number of node 1, $dfs(1) = 1$. Node 1 is now visited. Set $i = 1$. Go to Step 2.
2. If all the arcs emanating from node i are explored, then go to Step 3. Otherwise, let (i, j) be an unexplored arc connecting node i to node j. If node j is unvisited, then set $dfs(j) = dfs(i) + 1$. Node j is now visited. Update arc set as $A_1 = A_1 + (i, j)$. Set $i = j$. Return to Step 2. If node j is visited and marked, then $A_1 = A_1 + (i, j)$. Return to Step 2. If node j is visited and unmarked, then stop. The network contains cycles. Arc(i, j) is an arc of the cycle. If the set of other arcs leading to a cycle needs to be detected, then return to Step 2 without adding (i, j) to A_1 instead of stopping.
3. Mark node i. Stop, if node i is node 1. The set of arcs in $A - A_1$ forms cycles in the network. Otherwise, choose node k with $dfs(k) = dfs(i) - 1$. Set $i = k$. Return to Step 2.

The dfs method applied to the network in Figure 4.37 produces arc(5, 2) as the back arc, indicating that it causes a cycle in the network.

Activity-on-Arc Representation

In this mode, the nodes represent the realizations of some milestones (events) of the project and the arcs represent activities. An arc connecting nodes i and j represented by an arrow from node i to node j is referred to as arc(i, j) or activity (i, j). Node i, the immediate predecessor node of arc(i, j), is the start node for the activity, and node j, the immediate successor node of arc(i, j), is the end node for the activity. This is shown in Figure 4.38.

Suppose we want to draw the activity network of the daily set of chores to be undertaken by a family of three. Table 4.9 outlines the tasks to be performed and the precedence relationship for the project. As shown in Figure 4.39, the project can be represented as a directed graph with 12 nodes and 15 arcs. Arcs(8, 10) and (9, 10) are dummy arcs

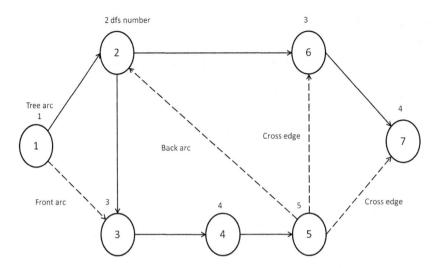

FIGURE 4.37
Illustration of dfs for cycles in network.

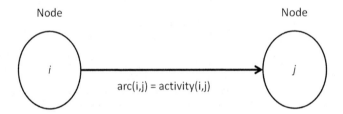

FIGURE 4.38
A generic AOA element.

TABLE 4.9

Sample Project for After-Work-Hours Chores

Activity	Activity Description	Immediate Predecessors
A	Dad, mom, and son arrive home in the same car.	—
B	Dad and mom change clothes	a
C	Son watches TV	a
d	Mom warms up the food	bb
e	Dad sets the table	b
f	Son does homework	c
g	Mom fixes salad	d
h	The family eats dinner	e, f, g
i	Dad loads the dishwasher	h
j	Mom checks son's homework	h
k	Son practices piano	h
l	All go to son's basketball game	i, j, k
m	All wash up and go to bed	l

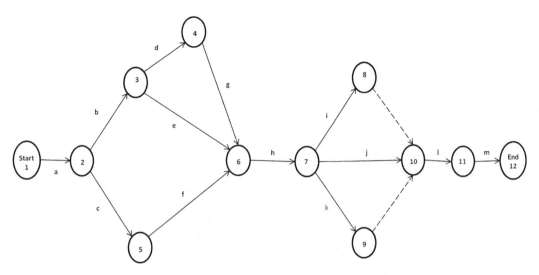

FIGURE 4.39
A project graph of the after-work-hours chores.

that do not consume any resources, but they are needed to represent the precedence relationship.

Since activities *i*, *j*, and *k* have the same immediate predecessor, activity h, and the same immediate successor, activity l, this necessitates three parallel arcs between nodes 7 and 10. An activity network allows only one arc between any two nodes, and hence nodes 8 and 9 are drawn and connected to node 10 via dummy arcs to overcome the problem. In general, dummy arcs are needed when

1. A set of jobs has the same set of immediate predecessors and the same set of immediate successors
2. Two or more jobs have the same set of immediate predecessors, some of which are also immediate predecessors to other jobs
3. Two or more jobs have the same set of immediate successors, some of which are immediate successors to other jobs

As indicated earlier, AOA representation is assumed by most of the project optimization models. It is, therefore, desired that redundant dummy arcs be removed from the network in order to reduce the problem size. Redundant dummy arcs exist when they are the only immediate predecessors of a node or the only immediate successors of a node. Also, if a dummy arc represents a precedence relationship which has already been represented by the project arc, then it is redundant. The inclusion of the redundant arcs in the network does not affect the optimization process but may affect the solution time and the memory requirements once the procedure is coded. One should be more careful not to omit any existing precedence relationship since failure to do so leads to incorrect decisions. Any cycles created signal an incorrect diagram and can be detected using the dfs procedure explained earlier. Figure 4.40 is the AON network corresponding to the project given in Table 4.9.

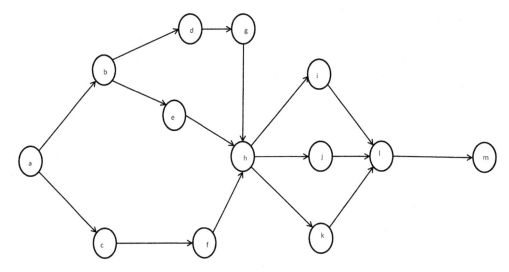

FIGURE 4.40
AON graph of the after-work-hours chores.

In the AON mode, the nodes represent the activities, and the arcs are drawn to indicate the precedence relationship. If node i is connected to node j through arc(i, j), then activity i is an immediate predecessor of activity j. The in degree (out degree) of a node is the number of arcs leading into (out of) the node. If the in degree of node j is 2 and the out degree is 3, then activity j has two immediate predecessors and three immediate successors. The dummy arcs are not needed in this mode. However, a dummy start node is needed when more than one activity does not have immediate predecessors. Similarly, a dummy end node is needed when two or more activities do not have immediate successors. The dummy start (end) node is connected to the set of initial (last) activities using dummy arcs. This representation mode is easier to construct and is similar to bar chart representations often used in industry.

Exercises

4.1 Show that, for any PERT project network, the probability of finishing the project in *exactly* x time units is zero.

4.2 Repeat the crashing exercise presented in Table 4.5 by crashing only the three critical activities C, E, and G. What is the new project duration? What is the new total project cost?

4.3 The following historical data was compiled for the time (days) it takes to perform a task: 11, 18, 25, 30, 8, 42, 25, 30, 13, 21, 26, 35, 15, 23, 29, 36, 17, 25, 30, 21. Based on the given data, determine reasonable values for the PERT estimates a, m, and b for the task.

4.4 Answer the following questions using the tabulated PERT project data.

Activity	Predecessor	Duration (Days) (a, m, b)
A	—	7, 9, 11
B	—	7, 8, 10
C	A	2, 4, 6
D	A	1, 3, 7
E	B	1, 3, 7
F	B	2, 4, 6

a. Draw the AON CPM network for the project, perform the forward and backward computations, and determine the project duration.

b. Draw the Gantt chart using the ES times.

c. Draw the Gantt chart using the LS times.

d. Compare and discuss the relative differences, advantages, and disadvantages between the two Gantt charts in parts (b) and (c).

4.5 The following CPM network is subject to the following two simultaneous restrictions:

a. The project cannot start until 5 days from time 0 (i.e., ES of the start node = 5).

b. There is a deadline of 35 days (i.e., LC of the finish node = 35). Perform the forward and backward CPM computations and show ES, LS, EC, LC, TS, and FS for each activity. What is the TS for each activity on the critical path?

Activity	Predecessor	Duration (Days)
A	—	2
B	A	4
C	B	6
D	A	3
E	B	2
F	E	1
G	C, D	2
H	G	3
I	H	4

4.6 Using the mathematical expressions for FS and TS, show that FS cannot be larger than TS.

4.7 Show that even if TS > 0, FS may or may not be greater than 0.

4.8 Suppose five activities are on a CPM critical path and each activity has TS = 2. Determine the overall effect on the critical path length if two of the critical activities are delayed by 2 time units each. Discuss.

4.9 Give practical examples of SS, SF, FF, and FS precedence constraints.

4.10 Suppose two serial activities A and B are the only activities in a precedence diagramming network. Activity A precedes activity B. There is an SS restriction of 4 time units and an FF restriction of 3 time units between A and B.

a. Determine the mathematical relationship between the duration of $A(t_A)$ and the duration of $B(t_B)$.

b. Verify your response to part (a) by considering the following alternate cases:

Case 1: $t_A = 0$

Case 2: $0 < t_A <= 1$

Case 3: $1 < t_A < 4$

Case 4: $t_A = 4$

Case 5: $t_A > 4$

4.11 Perform a PERT analysis of the following project and find the probability of finishing the project between 16 and 20 days inclusive.

Activity	Predecessors	Duration Estimates (Days) (a, m, b)
1	—	1, 4, 5
2	—	2, 3, 4
3	1	6, 10, 13
4	1	6, 6, 7
5	2	2, 2, 2
6	3	1, 2, 3
7	4, 5	5, 8, 9
8	2	12, 16, 19

4.12 Answer the following questions using the tabulated project data.

Project Phase	Preceding Phase	Normal Time	Normal Cost	Crash Time	Crash Cost	Variance of Duration
A	—	40	$9,000	30	$12,000	10
B	A	53	15,000	50	15,300	9
C	A	60	7,500	30	6,000[a]	600
D	A	35	20,000	30	22,000	25
E	C, D	28	12,000	20	15,000	40
F	B, E	30	6,000	27	7,000	30

[a] Crashing of this activity results in significant savings in direct labor cost.

a. There is a penalty of $3,000/day beyond the normal PERT duration. Perform a crashing analysis to compare total normal cost to total crash cost. First crash only the critical activities; then crash all activities.

b. Find the probability that the total lateness penalty will be less than $3,000.

c. Find the probability that the total lateness penalty will be $3,000 or less.

d. Find the probability that the total lateness penalty will be greater than $3,000.

e. Find the probability that the total lateness penalty will be less than $9,000.

4.13 Suppose it is known that the duration of a project has a triangular distribution with a lower limit of 24 months, an upper limit of 46 months, and a mode of 36 months. Find the probability that the project can be completed in 40 months or less.

4.14 Suppose it is known that the duration of a project is uniformly distributed between 24 and 46 months. Find the probability that the project can be completed in 40 months or less.

4.15 You are given the following project data:

Activity	Predecessor	Estimated CPM Duration, t (Days)
A	—	6
B	—	5
C	—	2
D	A, B	4
E	B	5
F	C	6
G	B, F	1
H	D, E	7

If the PERT time estimates for each activity are defined as

$$a = t - 0.3t$$
$$m = t$$
$$b = t + \sqrt{t}$$

find a deadline, T_d, such that there is 0.85 probability of finishing the project on or before the deadline. *Note*: Carry all computations to four decimal places.

4.16 Suppose we are given the following data for a project:

Activity	Predecessor	PERT Duration	Duration
A	—	4	0.5000
B	A	5	1.0000
C	A	7	0.2500
D	A	2	0.3000
E	C, D	8	0.3750
F	B, E	9	0.4375

There is no penalty if the project is completed on or before the normal PERT duration. However, a penalty of $3,000 is charged for each day that the project lasts beyond the normal PERT duration. Using standard probability approach, find the *expected dollar penalty* for this project. *Note*: Carry all intermediate computations to four decimal places.

4.17 Presented in the following are the data for a precedence diagram problem. The SS, SF, and FF relationships between pairs of activities are given.

Activity	Predecessor	Duration
A	—	5
B	A	10
C	A	4
D	B, C	15

SS(between A and B) = 1

F(between A and B) = 8

SS(between A and C) = 2

SF(between A and C) = 8

SS(between B and D) = 5

SS(between C and D) = 7

FF(between C and D) = 10

a. Perform forward and backward computations *without* considering the precedence constraints. What is the project duration?

b. Perform the forward and backward computations considering the precedence constraints. What is the project duration?

4.18

a. What potential effect(s) does the crashing of a noncritical activity have on the critical path and the project duration?

b. Use a hypothetical example of a CPM project network consisting of seven activities to support your answer given earlier. First show the network computations without crashing. Identify the critical path. Then show the computations with the crashing of one noncritical activity.

4.19 Assume that you are the manager responsible for the project whose data is presented as follows:

Activity	Description	Duration (h)	Preceding Activities
A	Develop required material list	8	—
B	Procure pipe	200	A
C	Erect scaffold	12	—
D	Remove scaffold	4	I, M
E	Deactivate line	8	—
F	Prefabricate sections	40	B
G	Install new pipes	32	F, L
H	(Deleted from work plan)	—	—
I	Fit-up pipes and valves	8	G, K
J	Procure valves	225	A
K	Install valves	8	J, L
L	Remove old pipes and valves	35	C, E
M	Insulate	24	G, K
N	Pressure test	6	I
O	Cleanup and start-up	4	D, N

a. Draw the CPM diagram for this project.

b. Perform the forward and backward pass calculations on the network and indicate the project duration.

c. List all the different paths in the network in *decreasing* order of criticality index in the tabulated form shown as follows. Define the criticality index as follows:

Let S_k = sum of the *TS*s on path number k (i.e., sum of TS values for all activities on path k).

S_{max} = maximum value of S (i.e., Max$\{S_k\}$).

$$C_k = \left[(S_{max} - S_k)/S_{max}\right] \times 100\%.$$

Define the most critical path as the one with $C_k = 100\%$.
Define the least critical path as the one with $C_k = 0\%$.

k	Activities in Path	Sum of TS on Path (S_k)	Path Criticality Index (C_k)
1			100%
2			
...	...	...	...
n			0%

4.20 Assume that we have the following precedence diagramming network data:

Activity	Predecessor	Duration
A	—	8
B	A	12
C	B	4
D	B	7
E	C, D	8
F	D	12
G	E, F	3

The following PDM restrictions are applicable to the network:

$SS_{AB} = 3$
$FF_{AB} = 12$
$FS_{BC} = 1$
$SS_{BD} = 6$
$SF_{DE} = 13$
$FS_{FG} = 3$

If the project start time is $t = 5$ and the project due date is $t = 52$, perform conventional CPM analysis of the network without considering the lead–lag PDM restrictions. Show both forward and backward passes. Identify the critical path. Tabulate the TS and FS for the activities in the network.

4.21 Repeat Exercise 4.20 and consider the lead–lag PDM restrictions. Draw the Gantt chart schedule that satisfies all the lead–lag constraints.

4.22 Draw an expanded CPM network for the Gantt chart in Exercise 4.21 considering the need to split activities.

4.23 Would you classify the CPM procedure as a management by objective approach or a management by exception approach? Discuss.

4.24 In activity scheduling, which of the following carries the higher priority?

    i. Activity precedence constraint

    ii. Resource constraint

4.25 For the project data in Exercise 4.20, assume that the given activity durations represent the most likely PERT estimates (m). Define the other PERT estimates a and b as follows:

$$a = m - 1$$
$$b = m + 1$$

Disregarding the PDM lead–lag constraints, find the probability of finishing the project within 51 time units.

4.26 The *median rule* of project control refers to the due date that has a 0.50 probability of being achieved. Suppose the duration, T, of a project follows a triangular distribution with end points a and b. If the mode of the distribution is closer to b than it is to a, find a general expression for the median (denoted by m_d) such that

$$P(T \leq m_d) = 0.50$$

4.27 Draw an AOA network for the project described in Exercise 4.19. Use as few dummy arcs as possible.

4.28 For the problem given in Exercise 4.12, assuming that the variance of duration is 0 for all the project phases, determine the time–cost trade-off curve by defining the strategies and calculating crashing cost for each strategy for $T_{max} \leq T \leq T_{min}$.

5

Resource Allocation Systems

Resource Allocation and Management

Whenever one person is found adequate to the discharge of a duty by close application thereto, it is worse executed by two persons, and scarcely done at all if three or more are employed therein.

George Washington

Resource availability and shortages permeate organizations. Even in times of extreme resource shortage, pockets of limited availability reside in nooks and corners of any organization. To find and redistribute such constrained resources require an enterprise-wide systems approach. Foraging for resource to reallocate is an essential part of project management. Where you think none exists, there may actually be plenty. You find whatever is available through systematic search and realignment. This chapter addresses systematic resource allocation and management strategies. The differences between unconstrained program evaluation and review technique (PERT)/critical path method (CPM) networks and resource-constrained networks are discussed. The resource loading and resource leveling strategies are presented. The resource idleness graph is introduced as a measure of the level of resource idleness in resource-constrained project schedules. Various resource allocation heuristics are presented. A procedure for calculating resource work rates to assess project productivity is presented. An example of a resource-constrained precedence diagramming method (PDM) network is presented. The chapter also illustrates the use of new graphical tools, referred to as the critical resource diagram (CRD) and the resource schedule (RS) Gantt chart. The CRD is used to represent the interrelationships among resource units as they perform their respective tasks. It is also used to identify bottlenecked resources in a project network. The RS Gantt chart indicates time intervals of allocation for specific resource types. The chapter concludes with examples of probabilistic evaluation of resource utilization levels.

Resource management is a crucial aspect of project management, particularly in a systems environment. Sawhney et al. (2004) suggest a model for integrating and managing resources considering training requirements. Basic CPM and PERT approaches assume unlimited resource availability in project network analysis. In this chapter, both time and resource requirements of activities are considered in developing network schedules. Projects are subject to three major constraints: time limitations, resource constraints, and performance requirements. Since these constraints are difficult to satisfy simultaneously, trade-offs must be made. The smaller the resource base, the longer the project schedule. The quality of work may also be adversely affected by poor resource allocation strategies.

TABLE 5.1

Sample Format for Resource Availability Database

Resource ID	Brief Description	Special Skills	When Available	Duration of Availability	How Many
Type 1	Manager	Planning	Jan 1, 2015	10 months	1
Type 2	Analyst	Scheduling	Dec 25, 2014	Indefinite	5
Type 3	Engineer	Design	Now	36 months	20
⋮	⋮	⋮	⋮	⋮	⋮
Type $n-1$	Operator	Machining	Immediate	Indefinite	10
Type n	Programmer	Software tools	Sept 2, 2014	12 months	2

Good planning, scheduling, and control strategies must be developed to determine what the next desired state of a project is and when the next state is expected to be reached, and the external factors will determine the nature of progress of a project from one state to another. Network diagrams, Gantt charts, progress charts, and resource loading graphs are visual aids for resource allocation strategies. One of the first requirements for resource management is to determine what resources are required versus what resources are available. Table 5.1 shows a model of a resource availability database. The database is essential when planning resource loading strategies for resource-constrained projects.

Resource-Constrained Scheduling

A resource-constrained scheduling problem arises when the available resources are not enough to satisfy the requirements of activities that can be performed concurrently. To satisfy this constraint, *sequencing rules* (also called priority rules, activity urgency factors, scheduling rules, or scheduling heuristics) are used to determine which of the competing activities will have priority for resource allocation. Several optimum-yielding techniques are available for generating resource-constrained schedules. Unfortunately, the optimal techniques are not generally used in practice because of the complexity involved in implementing them for large projects.

Even using a computer to generate an optimal schedule is sometimes cumbersome because of the modeling requirements, the drudgery of lengthy data entry, and the combinatorial nature of interactions among activities. However, whenever possible, effort should be made in using these methods, since they provide the best solution.

Most of the available mathematical techniques are based on integer programming that formulates the problem using 0 and 1 indicator variables. The variables indicate whether or not an activity is scheduled in specific time periods. Three of the common objectives in project network analysis are to minimize project duration, minimize total project cost, and maximize resource utilization. One or more of these objectives are attempted subject to one or more of the following constraints:

1. Limitation on resource availability
2. Precedence restrictions
3. Activity-splitting restrictions

4. Non-preemption of activities

5. Project deadlines

6. Resource substitutions

7. Partial resource assignments

8. Mutually exclusive activities

9. Variable resource availability

10. Variable activity durations

Instead of using mathematical formulations, a scheduling heuristic uses logical rules to prioritize and assign resources to competing activities. Many scheduling rules have been developed over the years. Some of the most popular ones are discussed in this chapter.

Resource Allocation Examples

Table 5.2 shows a project with resource constraints. There is one resource type (operators) in the project data, and there are only 10 units available. The PERT estimates for the activity durations are expressed in terms of days. It is assumed that the resource units are reusable. Each resource unit is reallocated to a new activity at the completion of its previous assignment. Resource units can be idle if there are no eligible activities for scheduling or if enough units are not available to start a new activity. For simplification, it is assumed that the total units of resource required by an activity must be available before the activity can be scheduled. If partial resource allocation is allowed, then the work rate of the partial resources must be determined. A methodology for determining resource work rates is presented in a later section of this chapter.

The unconstrained PERT duration was found earlier to be 11 days. The resource limitations are considered when creating the Gantt chart for the resource-constrained schedule. For this example, we will use the *longest-duration-first* heuristic to prioritize activities for resource allocation. Other possible heuristics are *shortest-duration-first*, *critical-activities-first*, *maximum-predecessors-first*, and so on. For very small project networks, many of the heuristics will yield identical schedules.

TABLE 5.2

PERT Project Data with Resource Requirements $Z_1 = 10$

Activity	Predecessor	PERT Estimates			Number of Operators Required
		a	m	B	
A	—	1	2	4	3
B	—	5	6	7	5
C	—	2	4	5	4
D	A	1	3	4	2
E	C	4	5	7	4
F	A	3	4	5	2
G	B, D, E	1	2	3	6

Longest-Duration-First

The initial step is to rank the activities in decreasing order of their PERT durations, t_e. This yields the following priority order:

$$B, E, F, C, D, A, G$$

At each scheduling instant, only the eligible activities are considered for resource allocation. Eligible activities are those whose preceding activities have been completed. Thus, even though activity B has the highest priority for resource allocation, it can compete for resources only if it has no pending predecessors. Referring to the PERT network shown in the preceding chapter, note that activities A, B, and C can start at time 0 since they all have no predecessors. These three activities require a total of 12 operators (3 + 5 + 4) altogether, but we have only 10 operators available. So, a resource allocation decision must be made. We check our priority order and find that B and C have priority over A. So, we schedule B with five operators and C with four operators. The remaining one operator is not enough to meet the need of any of the remaining activities. The two scheduled activities are drawn on the Gantt chart, as shown in Figure 5.1.

We have one operator idle from time 0 until time 3.83, when activity C finishes and releases four operators. At time 3.83, we have five operators available. Since activity E can start after activity C, it has to compete with activity A for resources. According to the established priority, E has priority over A, so four operators are assigned to E. The remaining one operator is not enough to perform activity A, so it has to wait and one operator remains idle. If E had required more operators than were available, activity A would have been able to get resources at time 3.83. No additional scheduling is done until time 6, when activity B finishes and releases five operators. So, we now have six operators available, and there are no activities to compete with A for resources. Thus, activity A is finally scheduled at time 6, and we are left with three idle operators. Even though three operators are enough to start either activity D or activity F, neither of these activities can start until activity A finishes, because of the precedence requirement.

When activity A finishes at time 8.17, F and D are scheduled in that order. Activity G is scheduled at time 11 and finishes at time 13 to complete the project. Figure 5.1 shows

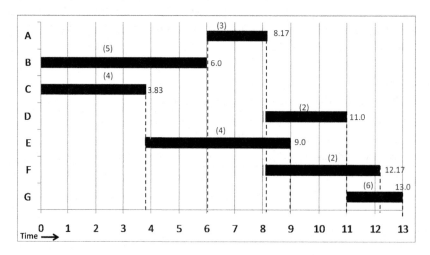

FIGURE 5.1
Resource-constrained PERT schedule.

the complete project schedule. The numbers in parentheses in the figure are the resource requirements. Note that our assumption is that activity splitting and partial resource assignments are not allowed. An activity cannot start until all the units of resources required are available. In real project situations, this assumption may be relaxed so that partial resource assignments are permissible. If splitting and partial assignments are allowed, the scheduling process will still be the same, except that more record keeping will be required to keep track of pending jobs.

Resource allocation may be affected by several factors including duration of availability, skill level required, cost, productivity level, and priority strategy. Ranking activities for resource allocation may be done under *parallel priority* or *serial priority*. In serial priority, the relative ranking of all activities is done at the beginning before starting the scheduling process. The activities maintain their relative priority ranking throughout the scheduling process. In the parallel priority approach, relative ranking is done at each scheduling instant, and it is done only for the activities that are eligible for scheduling at that instant. Thus, under the parallel priority approach, the relative ranking of an activity may change at any time, depending on which activities it is competing with for resources at that time. The illustrative example hereafter uses the serial priority approach. If desired, any other resource allocation heuristic could be used for the scheduling example. Some of the available resource allocation rules are presented in the following section.

Resource Allocation Heuristics

Resource allocation heuristics facilitate scheduling large projects subject to resource limitations. Some heuristics are very simple and intuitive, while others require computer implementations. Several scheduling heuristics have been developed over the years. While many of the early techniques are now defunct, many still remain in use in real projects, particularly by purists in the back alleys of project management.

Many project management software packages use proprietary resource allocation rules that are not transparent to the user. Microsoft Project® has emerged as a general-platform software tool for project management at the low end of the software spectrum. Large and complex software tools are also available on higher computing platforms. But no matter what the platform and software size, they all rely on basic activity scheduling algorithms and heuristics. A good scheduling heuristic should be simple, unambiguous, easily understood, and easily executable by those who must use it. The heuristic must be flexible and capable of resolving schedule conflicts quickly. When users trust and use a scheduling heuristic, the project scheduling process can become an effective communication tool in project management. Some of the classical and modern scheduling rules are presented in the next section. Even nowadays, the classical rules are still called upon in the search for quick and effective ways to schedule and allocate resources.

Activity Time

Activity time (ACTIM) is one of the classical activity sequencing rules. Although more robust scheduling algorithms have been developed since then, Brooks' algorithm provided the early foundational technique for project resource allocation following a systematic pattern. Even today, the technique can serve as the framework over which new scheduling heuristics can be developed. The original algorithm considered only the single-project, single-resource case, but it lends itself to extensions for multiresource cases. The ACTIM scheduling heuristic represents the maximum time that an activity controls in the project network on any one path. This is represented as the length of time between when the activity finishes and when the project finishes. It is computed for each project activity by subtracting the activity's latest start time from the critical path time, as shown in the following:

$$ACTIM = (Critical\ path\ time) - (Activity\ latest\ start\ time)$$

Activity Resource

Activity resource (ACTRES) is a scheduling heuristic that has been around since the early 1970s. This is a combination of ACTIM and resource requirements. It is computed as

$$ACTRES = (Activity\ time) * (Resource\ requirement)$$

For multiple resources, the computation of ACTRES can be modified to account for various resource types. For this purpose, the resource requirement can be replaced by a scaled sum of resource requirements over different resource types.

Time Resources

Time resources (TIMRES) is another priority rule that is composed of equally weighted portions of ACTIM and ACTRES. It is expressed as

$$TIMRES = 0.5(ACTIM) + 0.5(ACTRES)$$

GENRES

GENRES is a search technique developed as a modification of TIMRES with various weighted combinations of ACTIM and ACTRES. GENRES is implemented as a computer search technique whereby iterative weights (w) between 0 and 1 are used in the expression.

$$GENRES = (w)(ACTIM) + (1 - w)(ACTRES)$$

Resource over Time

Resource over time (ROT) is a scheduling criterion that is calculated as resource requirement divided by ACTIM:

$$ROT = \frac{Resource\ requirement}{Activity\ time}$$

The resource requirement can be replaced by the scaled sum of resource requirements in the case of multiple resource types with different units.

Composite Allocation Factor

Composite allocation factor (CAF) is a comprehensive rule that incorporates time and resource requirements. For each activity i, CAF is computed as a weighted and scaled sum of two components RAF (resource allocation factor) and SAF (stochastic activity duration factor) as

$$CAF_i = (w)RAF_i + (1 - w)SAF_i$$

where w is a weight between 0 and 1. RAF is defined for each activity i as

$$RAF_i = \frac{1}{t_i} \sum_{j=1}^{R} \frac{x_{ij}}{y_j}$$

where
  x_{ij} is the number of resource type j units required by activity i
  y_j is the $Max_j\{x_{ij}\}$, maximum units of resource type j required
  t_i is the expected duration of activity i
  R is the number of resource types

RAF is a measure of the expected resource consumption per unit time. In the case of multiple resource types, the different resource units are scaled by the y_j component in the formula for RAF. This yields dimensionless quantities that can be summed in the formula for RAF. The RAF formula yields real numbers that are expressed per unit time. To eliminate the time-based unit, the following scaling method is used:

$$Scaled\ RAF_i = \frac{RAF_i}{Max\{RAF_i\}}(100)$$

The aforementioned scaling approach yields unitless values of RAF between 0 and 100 for the activities in the project. Resource-intensive activities have larger magnitudes of RAF and, therefore, require a higher priority in the scheduling process. To incorporate the stochastic nature of ACTIMs in a project schedule, SAF is defined for each activity i as

$$SAF_i = t_i + \frac{s_i}{t_i}$$

where
  t_i is the expected duration for activity i
  s_i is the standard deviation of duration for activity i
  s_i/t_i is the coefficient of variation of the duration of activity i

It should be noted that the formula for SAF has one component (t_i) with units of time and one component s_i/t_i with no units. To facilitate the required arithmetic operation, each component is scaled as shown in the following:

$$Scaled\ t_i = \frac{t_i}{Max\{t_i\}}(50)$$

$$Scaled\left(\frac{s_i}{t_i}\right) = \frac{(s_i/t_i)}{Max\{s_i/t_i\}}(50)$$

When these scaled components are plugged into the formula for SAF, we automatically obtain unitless scaled SAF values that are on a scale of 0–100. However, the 100 weight will be observed only if the same activity has the highest scaled t_i value and the highest scaled s_i/t_i value at the same time. Similarly, the 0 weight will be observed only if the same activity has the lowest scaled t_i and the lowest scaled s_i/t_i value at the same time. The scaled values of SAF and RAF are now inserted in the formula for CAF as

$$CAF_i = (w)\{\text{scaled RAF}_i\} + (1 - w)\{\text{scaled SAF}_i\}$$

To ensure that the resulting CAF values range from 0 to 100, the following final scaling approach is applied:

$$\text{Scaled CAF}_i = \frac{CAF_i}{\text{Max}\{CAF_i\}}(100)$$

It is on the basis of the magnitudes of CAF that an activity is assigned a priority for resource allocation in the project schedule. Activities with larger values of CAF have higher priorities for resource allocation. An activity that lasts longer, consumes more resources, and varies more in duration will have a larger magnitude of CAF.

Resource Scheduling Method

Resource scheduling method (RSM) is a rule that gives priority to the activity with the minimum value of d_{ij}, which is calculated as

$$d_{ij} = \text{Increase in project duration when activity } j \text{ follows activity } i$$

$$= \text{Max}\{0, (EC_i - LS_j)\}$$

where
  EC_i is the earliest completion time of activity i
  LS_j is the latest start time of activity j

Competing activities are compared two at a time in the resource allocation process.

Greatest Resource Demand

This rule gives priority to the activity with the largest total resource-unit requirements. The greatest resource demand measure is calculated as

$$g_j = d_j \sum_{i=1}^{n} r_{ij}$$

where
  g_j is the priority measure for activity j
  d_j is the duration of activity j
  r_{ij} is the units of resource type i required by activity j per period
  n is the number of resource types (resource units are expressed in common units)

Greatest Resource Utilization

The greatest resource utilization rule assigns priority to activities that, if scheduled, will result in maximum utilization of resources or minimum idle time. For large problems, computer procedures are often required to evaluate the various possible combinations of activities and the associated utilization levels.

Most Possible Jobs

This approach assigns priority in such a way that the greatest number of activities is scheduled in any period.

Other Scheduling Rules

1. Most total successors
2. Most critical activity
3. Most immediate successors
4. Any activity that will finish first
5. Minimum activity latest start (Min LS)
6. Maximum activity latest start (Max LS)
7. Minimum activity earliest start (Min ES)
8. Maximum activity latest completion (Max LC)
9. Minimum activity earliest completion (Min EC)
10. Maximum activity earliest completion (Max EC)
11. Minimum activity latest completion (Min LC)
12. Maximum activity earliest start (Max ES)
13. Minimum activity total slack (Min TS)
14. Maximum activity total slack (Max TS)
15. Any activity that can start first
16. Minimum activity duration
17. Maximum activity duration

The project analyst must carefully analyze the prevailing project situation and decide which of the several rules will be most suitable for the resource allocation requirements. Since there are numerous rules, it is often difficult to know which rule to apply. Experience and intuition are often the best guides for selecting and implementing scheduling heuristics. Some of the shortcomings of heuristics include subjectivity of the technique, arbitrariness of the procedures, lack of responsiveness to dynamic changes in the project scenario, and oversimplified assumptions.

There are advantages and disadvantages to using specific heuristics. For example, the shortest duration heuristic is useful for quickly reducing the number of pending activities. This may be important for control purposes. The smaller the number of activities to be tracked, the lower the burden of project control. The longest duration heuristic, by contrast, has the advantage of scheduling the biggest tasks in a project first. This permits the lumping of the smaller activities into convenient work packages later on in the project. Decomposition of large projects into subprojects can enhance the application of heuristics that are only effective for small project networks.

Example of ACTIM

Brooks's algorithm uses ACTIM to determine which activities should receive limited resources first. The algorithm considers the single-project, single-resource case. The following example is used to illustrate the use of the ACTIM scheduling heuristic. Figure 5.2 presents a project network based on the activity-on-arrow (AOA) convention. The arrows represent activities, while the nodes represent activity end points. Each activity is defined by its end points as $i - j$. The two numbers within parentheses represent activity duration and resource requirement (t, r), respectively. The network consists of seven actual activities and one dummy activity. The dummy activity (3–4) is required to show that activities (1–3) and (2–3) are predecessors for activity (4–5). Table 5.3 presents the tabular implementation of the steps in the algorithm for 3 units of resource.

1. *Step 1*: Develop the project network as in CPM. Identify activities, their estimated durations, and resource requirements.
2. *Step 2*: Determine for each activity the maximum time it controls through the network on any one path. This is equivalent to the critical path time minus the latest start time of the starting node of the activity. These times are then scaled from 0 to 100. This scaled network control time is designated as ACTIM.
3. *Step 3*: Rank the activities in decreasing order of ACTIM as shown in Table 5.3. The duration and resources required for each activity are those determined in the first step.

The rows TEARL, TSCHED, TFIN, and TNOW are explained as follows: TEARL is the earliest time of an activity determined by traditional CPM calculations. TSCHED is the

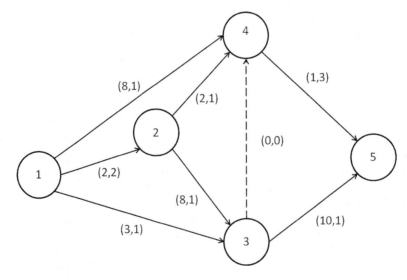

FIGURE 5.2
AOA network for ACTIM example.

TABLE 5.3

Scheduling with ACTIM Heuristic When $Z_1 = 3$

	Activity						
	1–2	2–3	1–3	3–5	1–4	2–4	4–5
Duration	2	8	3	10	8	2	1
ACTIM	20	18	13	10	9	3	1
Scaled ACTIM	100	90	65	50	45	15	5
Resources required	2	1	1	1	1	1	3
TEARL	0	2	0	10	0	2	10
TSCHED	0	2	0	10	2	3	20
TFIN	2	10	3	20	10	5	**21**
TNOW	0	2	3	5	10	20	
Resources available	3, 1, 0	2, 1, 0	1, 0	1	3, 2	3, 0	
ACT. ALLOW.	1–2	2–3	2–4	—	3–5	4–5	
	1–3	1–4			4–5		
	1–4	2–4					
Iteration number	1	2	3	4	5	6	

actual scheduled starting time of an activity as determined by Brooks' algorithm. TFIN is the completion time of each activity. TNOW is the time at which the resource allocation is being made.

1. *Step 4*: Set TNOW to 0. The allowable activities (ACT. ALLOW.) to be considered for scheduling at TNOW of 0 are those activities with TEARL of 0. These are 1–2, 1–3, and 1–4. These are placed in the ACT. ALLOW. row in decreasing order of ACTIM. Ties are broken by scheduling the activity with longest duration first. The number of resources initially available (i.e., 3) is placed in the resources available column.

2. *Step 5*: Determine whether the first activity in ACT. ALLOW. (i.e., 1–2) can be scheduled. Activity 1–2 requires two resource units and three are available. So, 1–2 is scheduled. A line is drawn through 1–2 to indicate that it has been scheduled and the number of resources available is decreased by two. TSCHED and TFIN are then set for activity 1–2. This process is repeated for the remaining activities in ACT. ALLOW. until the resources available are depleted.

3. *Step 6*: TNOW is raised to the next TFIN time of 2, which occurs at the completion of activity 1–2. The resources available are now 2. ACT. ALLOW. includes those activities not assigned at the previous TNOW (i.e., 1–4) and those new activities whose predecessors have been completed (i.e., 2–3 and 2–4).

4. *Step 7*: Repeat this assignment process until all activities have been scheduled. The latest TFIN gives the duration of the project. For this example, the duration is 21 days. Figure 5.3 presents the Gantt chart for the final schedule.

Comparison of ACTIM, ACTRES, and TIMRES

The classical Brooks' algorithm can be implemented with any other heuristic apart from ACTIM. The project network in Figure 5.4 is used to compare the schedules generated by

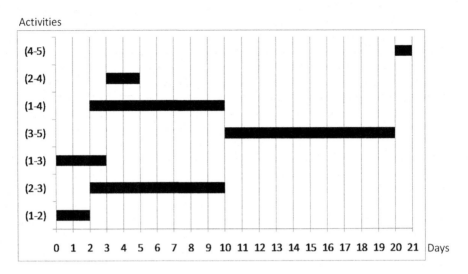

FIGURE 5.3
Gantt chart of ACTIM schedule.

ACTIM, ACTRES, and TIMRES. Tables 5.4–5.6 present the tabular stages of the implementation of Brooks' algorithm with scaled values of ACTIM, ACTRES, and TIMRES. Note that each heuristic yields a different project schedule. Even though the project durations obtained from ACTRES and TIMRES are the same (13 days), the specific scheduled times are different for the activities in each schedule. The larger the project network, the more the differences that can be expected from the schedules generated by different heuristics. Figure 5.5 presents the Gantt charts comparing the schedules generated by ACTIM, ACTRES, and TIMRES.

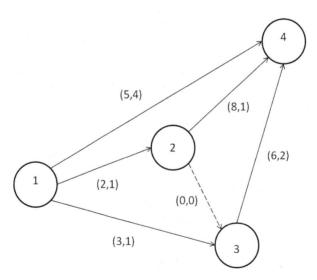

FIGURE 5.4
Network for ACTIM, ACTRES, and TIMRES comparison.

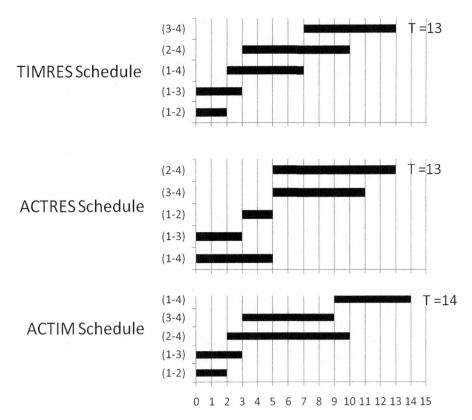

FIGURE 5.5
Gantt charts comparing ACTIM, ACTRES, and TIMRES schedules.

TABLE 5.4

ACTIM Schedule for Comparative Network When $Z_1 = 5$

	Activity				
	1–2	**1–3**	**2–4**	**3–4**	**1–4**
Duration	2	3	8	6	5
ACTIM	10	9	8	6	5
Scaled ACTIM	100	90	80	60	50
Resources required	1	1	1	2	4
TEARL	0	0	2	3	0
TSCHED	0	0	2	3	9
TFIN	2	3	10	9	**14**
TNOW	0	2	3	9	
Resources available	5, 4, 3	4, 3	4, 2	4	
ACT. ALLOW.	1–2	2–4	3–4	1–4	
	1–3	1–4	1–4		
	1–4				
Iteration number	1	2	3	4	

TABLE 5.5

ACTRES Schedule for Comparative Network $Z_1 = 5$

	Activity				
	1–4	1–3	1–2	3–4	2–4
Duration	5	3	2	6	8
ACTRES	20	3	2	12	8
Scaled ACTRES	100	15	10	60	40
Resources required	4	1	1	2	1
TEARL	0	0	0	5	5
TSCHED	0	0	3	5	5
TFIN	5	3	5	11	13
TNOW	0	3	5		
Resources available	5, 1, 0	1, 0	5, 3, 2		
ACT. ALLOW.	1–4	1–2	3–4		
	1–3		2–4		
	1–2				
Iteration number	1	2	3		

TABLE 5.6

TIMRES Schedule for Comparative Network ($Z_1 = 5$)

	Activity				
	1–2	1–3	1–4	2–4	3–4
Duration	2	3	5	8	6
TIMRES	85	82.5	75	60	60
Scaled TIMRES	100	97	88	71	59
Resources required	1	1	4	1	2
TEARL	0	0	0	2	3
TSCHED	0	0	2	3	7
TFIN	2	3	7	11	13
TNOW	0	2	3	7	
Resources available	5, 4, 3	4, 0	1, 0	4, 2	
ACT. ALLOW.	1–2	1–4	2–4	3–4	
	1–3	2–4	3–4		
	1–4				
Iteration number	1	2	3	4	

Quantitative Modeling of Worker Assignment

Operations research techniques are frequently used to enhance resource allocation decisions. One common resource allocation tool is the resource assignment algorithm, which can be applied to enhance human resource management. Suppose that there are n tasks that must be performed by n workers. The cost of worker i performing task j is c_{ij}. It is desired

to assign workers to the tasks in a fashion that minimizes the cost of completing the tasks. This problem scenario is referred to as the *assignment problem*. The technique for finding the optimal solution to the problem is called the *assignment method*. Like the transportation method, the assignment method is an iterative procedure that arrives at the optimal solution by improving on a trial solution at each stage of the procedure. Conventional CPM and PERT can be used in controlling projects to ensure that the project will be completed on time, but both techniques do not consider the assignment of resources to the tasks that make up a project. The *assignment method* can be used to achieve an optimal assignment of resources to specific tasks in a project. Although the assignment method is cost-based, task duration can be incorporated into the modeling in terms of time–cost relationships. Of course, task precedence requirements and other scheduling constraints of the tasks will be factored into the computational procedure. The objective is to minimize the total cost of assigning workers to tasks. The formulation of the assignment problem is as follows:
   Let

$x_{ij} = 1$ if worker i is assigned to task j, where $i, j = 1, 2,..., n$

$x_{ij} = 0$ if worker i is not assigned to task j

c_{ij} = cost of worker i performing task j

$$\text{Minimize } z = \sum_{i=1}^{n} \sum_{j=1}^{n} c_{ij} x_{ij}$$

$$\text{Subject to } \sum_{j=1}^{n} x_{ij} = 1, \quad i = 1, 2,..., n$$

$$\sum_{i=1}^{n} x_{ij} = 1, \quad j = 1, 2,..., n$$

$$x_{ij} \geq 0, \quad i, j = 1, 2,..., n$$

The previous formulation is a transportation problem with $m = n$ and all supplies and demands (sources to targets) are equal to 1. Note that we have used the nonnegativity constraint, $x_{ij} \geq 0$, instead of the integer constraint, $x_{ij} = 0$ or 1. However, the solution of the model will still be integer valued. Hence, the assignment problem is a special case of the transportation problem with $m = n$; $S_i = 1$ (supplies); and $D_j = 1$ (demands). Conversely, the transportation problem can also be viewed as a special case of the assignment problem. The basic requirements of an assignment problem are as follows:

1. There must be two or more tasks to be completed.
2. There must be two or more resources that can be assigned to the tasks.
3. The cost of using any of the resources to perform any of the tasks must be known.
4. Each resource is to be assigned to one and only one task.

If the number of tasks to be performed is greater than the number of workers available, we will need to add *dummy workers* to balance the problem formulation. Similarly, if the number of workers is greater than the number of tasks, we will need to add *dummy tasks* to balance the formulation. If there is no problem of overlapping, a worker's time may be split into segments so that the worker can be assigned more than one task. In this case, each segment of the worker's time will be modeled as a separate resource in the assignment problem. Thus, the assignment problem can be extended to consider partial allocation of resource units to multiple tasks.

Although the assignment problem can be formulated for and solved by the simplex method or the transportation method, a more efficient algorithm is available specifically for the assignment problem. The method, known as the *Hungarian method*, is a simple iterative technique. Details of the assignment problem and its solution techniques can be found in operations research textbooks. As an example, suppose that five workers are to be assigned to five tasks on the basis of the cost matrix presented in Table 5.7. Task 3 is a machine-controlled task with a fixed cost of $800.00, regardless of which worker it is assigned to. Using the data, we obtain the assignment solution presented in Table 5.8, which indicates the following:

$$x_{15} = 1$$

$$x_{23} = 1$$

$$x_{31} = 1$$

$$x_{44} = 1$$

$$x_{52} = 1$$

TABLE 5.7

Cost Matrix for Worker Assignment Problem

Worker	Task 1	Task 2	Task 3	Task 4	Task 5
1	300	200	800	500	400
2	500	700	800	1,250	700
3	300	900	800	1,000	600
4	400	300	800	400	400
5	700	350	800	700	900

TABLE 5.8

Solution to Worker Assignment Problem

Worker	Task 1	Task 2	Task 3	Task 4	Task 5
1	0	0	0	0	1
2	0	0	1	0	0
3	1	0	0	0	0
4	0	0	0	1	0
5	0	1	0	0	0

Thus, the minimum total cost is given by

$$TC = c_{15} + c_{23} + c_{31} + c_{44} + c_{52} = (400 + 800 + 300 + 400 + 350) = \$2,250.00$$

The technique of work rate analysis can be used to determine the cost elements that go into a resource assignment problem. The solution of the assignment problem can then be combined with the technique of critical resource diagramming. This combination of tools and techniques can help enhance project resource management decisions from a quantitative modeling perspective.

Takt Time for Activity Planning

Activity planning and scheduling are not totally unlike production planning functions. Thus, activity planning can benefit from techniques typically used in the production environment. One such technique is takt time computation. "Takt" is the German word referring to how an orchestra conductor regulates the speed, beat, or timing so that the orchestra plays in unison. So, the idea of takt time is to regulate the rate time or pace of producing a completed product. This refers to the production pace at which workstations must operate in order to meet a target production output rate. In other words, it is the pace of production needed to meet customer demand. The production output rate is set based on product demand. In a simple sense, if 2,000 units of a widget are to be produced within an 8-h shift to meet a market demand, then 250 units must be produced per hour. That means, a unit must be produced every $60/250 = 0.24$ min (14.8 s). Thus, the *takt time* is 14.4 s. Lean production planning then requires that workstations be balanced such that the production line generates a product every 14.4 s. This is distinguished from the *cycle time*, which refers to the actual time required to accomplish each workstation task. Cycle time may be less than, more than, or equal to takt time. Takt is not a number that can be measured with a stop watch. It must be calculated based on the prevailing production needs and scenario. Takt time equation is

$$T = \frac{T_a}{T_d}$$

$$= \frac{\text{Available work time} - \text{Breaks}}{\text{Customer demand}}$$

$$= \frac{\text{Net available time per day}}{\text{Customer demand per day}}$$

where
  T is the takt time (in minutes of work per unit produced)
  T_a is the net time available to work (in minutes of work per day)
  T_d is the time demand (i.e., customer demand in units required per day)

Takt time is often expressed as "seconds per piece," indicating that customers are buying a product once every so many seconds. Takt time is not expressed as "pieces per second."

The objective of lean production is to bring the cycle time as close to the takt time as possible that is choreographed. In a balanced line design, the takt time is the reciprocal of the production rate.

Improper recognition of the role of takt time can make an analyst to overestimate the production rate capability of a line. Many manufacturers have been known to overcommit to customer deliveries without accounting for the limitations imposed by takt time. Since takt time is set based on customer demand, its setting may lead to an unrealistic expectation of workstations. For example, if the constraints of the prevailing learning curve will not allow sufficient learning time for new operators, then takt times cannot be sustained. This may lead to the need for buffers to temporarily accumulate units at some workstations. But this defeats the pursuits of lean production or just-in-time. The need for buffers is a symptom of imbalances in takt time. Some manufacturers build *takt gap* into their production planning for the purpose of absorbing nonstandard occurrences in the production line. However, if there are more nonstandard or random events than have been planned for, then production rate disruption will occur.

It is important to recognize that the maximum production rate determines the minimum takt time for a production line. When demand increases, takt time should be decreased. When demand decreases, takt time should be increased. Production crew size plays an important role in setting and meeting takt time. The equation for calculating the crew size for an assembly line doing one piece flow that is paced to takt time is presented as follows:

$$\text{Crew size} = \frac{\text{Sum of manual cycle times}}{\text{Takt time}}$$

Even though takt time is normally used in production operations, the concept and computations are applicable to resource allocation and usage in project management. It can be adapted for coordination project task completion rates. In that case, "customer demand" can be defined in terms of workflow requirements to keep the project schedule on track in accordance with the desired project completion target dates or volume of output. For example, if a construction project calls for laying 500 blocks within a workday of seven and a half hours, then takt time would be calculated as follows:

$$T = \frac{7.5\,\text{h}}{500\,\text{blocks}}$$

$$= 0.015\,\text{h/blocks}$$

$$= 0.9\,\text{min/block (i.e., 54 s/block)}$$

Depending on the experience of the masons and the desired quality of the output (excessive pace degrades quality), this pace may be judged unachievable. Thus, necessitating an adjustment of the goal, modifying the work schedule, or allocating more resources.

Tips for Using Takt Time

Carefully evaluate the demand (i.e., what does the downstream workflow really need). Downstream refers to activity sequence that occurs later on in a project schedule.

Accurately assess the available work time. In addition to approved breaks, consider other "time robbers" that workers experience during the workday, either as an imposed reality or as self-initiated distractions. Not all such "illegal" times can be controlled in the human-infested workplace.

Identify opportunities to rebalance the workload.

Eliminate unlean practices wherever they are identified.

Reduce worker idle times. Workers should not be waiting for work to do.

Coordinate work input versus expected work output.

Review and recalculate takt time regularly, particularly when work situations change (e.g., new operators, new equipment, new process, new design, new materials, etc.)

Recognize that, sometimes, slowing down may lead to more effective utilization of human resources and a balanced efficiency of capital assets.

If used properly, takt time helps to achieve *rhythm* of work in a project.

Resource Work Rate

Work rate and work time are essential components of estimating the cost of specific tasks in project management. Given a certain amount of work that must be done at a given work rate, the required time can be computed. Once the required time is known, the cost of the task can be computed on the basis of a specified cost per unit time. Work rate analysis is important for resource substitution decisions. The analysis can help identify where and when the same amount of work can be done with the same level of quality and within a reasonable time span by a less expensive resource. The results of learning curve analysis can yield valuable information about the expected work rate. The general relationship among work, work rate, and time is given by

$$\text{Work done} = (\text{Work rate})(\text{Time})$$

This is expressed mathematically as

$$w = rt$$

where
   w is the amount of actual work done expressed in appropriate units. Example of work units are miles of road completed, lines of computer code typed, gallons of oil spill cleaned, units of widgets produced, and surface area painted
   r is the rate at which work is accomplished (i.e., work accomplished per unit time)
   t is the total time required to perform the work excluding any embedded idle times

It should be noted that work is defined as a physical measure of accomplishment with uniform density. That means, for example, that one line of computer code is as complex and desirable as any other line of computer code. Similarly, cleaning 1 gal of oil spill is as good as cleaning any other gallon of oil spill within the same work environment. The production of one unit of a product is identical to the production of any other unit of the product. If uniform work density cannot be assumed for the particular work being analyzed, then the relationship presented previously may lead to erroneous conclusions. Uniformity can be enhanced if the scope of analysis is limited to a manageable size. The larger the scope of the analysis, the more the variability from one work unit to another and the less uniform the overall work measurement will be. For example, in a project involving the construction of 50 miles of surface road, the work analysis may be done in increments of 10 miles at a time rather than the total 50 miles. If the total amount of work to be analyzed is defined as one whole unit, then the relationship can be developed for the case of a single resource performing the work, with the parameters as follows:

Resource	Machine A
Work rate	R
Time	T
Work done	100% (1.0)

The work rate, r, is the amount of work accomplished per unit time. For a single resource to perform the whole unit (100%) of the work, we must have the following:

$$rt = 1.0$$

For example, if machine A is to complete one work unit in 30 min, it must work at the rate of 1/30 of the work content per unit time. If the work rate is too low, then only a fraction of the required work will be performed. The information about the proportion of work completed may be useful for productivity measurement purposes. In the case of multiple resources performing the work simultaneously, the work relationship is as presented in Table 5.9.

Even though the multiple resources may work at different rates, the sum of the work they all performed must equal the required whole unit. In general, for multiple resources, we have the following relationship:

$$\sum_{i=1}^{n} r_i t_i = 1.0$$

where
   n is the number of different resource types
   r_i is the work rate of resource type i
   t_i is the work time of resource type i

TABLE 5.9

Work Rate Tabulation for Multiple Resources

Resource, i	Work Rate, r_i	Time, t_i	Work Done, w
RESource 1	r_1	t_1	$(r_1)(t_1)$
RESource 2	r_2	t_2	$(r_2)(t_2)$
$\vdots$	$\vdots$	$\vdots$	$\vdots$
RESource n	r_n	t_n	$(r_n)(t_n)$
		Total	1.0

For partial completion of work, the relationship is

$$\sum_{i=1}^{n} r_i t_i = p$$

where p is the proportion of the required work actually completed.

Work Rate Examples

Machine A, working alone, can complete a given job in 50 min. After machine A has been working on the job for 10 min, machine B was brought in to work with machine A in completing the job. Both machines working together finished the remaining work in 15 min. What is the work rate for machine B?

Solution
The amount of work to be done is 1.0 whole unit.

The work rate of machine A is 1/50.

The amount of work completed by machine A in the 10 min it worked alone is $(1/50)$ $(10) = 1/5$ of the required total work.

Therefore, the remaining amount of work to be done is 4/5 of the required total work.
Table 5.10 shows the two machines working together for 15 min.

The computation yields

$$\frac{15}{50} + 15(r_2) = \frac{4}{5}$$

which yields $r_2 = 1/30$. Thus, the work rate for machine B is 1/30. That means machine B, working alone, could perform the same job in 30 min.

In this example, it is assumed that both machines produce an identical quality of work. If quality levels are not identical, then the project analyst must consider the potentials for quality/time trade-offs in performing the required work. The relative costs of the different resource types needed to perform the required work may be incorporated into the analysis as shown in Table 5.11.

Using the aforementioned relationship for work rate and cost, the work crew can be analyzed to determine the best strategy for accomplishing the required work, within the required time, and within a specified budget.

TABLE 5.10

Work Rate Tabulation for Machines A and B

Resource, i	Work Rate, r_i	Time, t_i	Work Done, w
Machine A	1/50	15	15/50
Machine B	r_2	15	$15(r_2)$
		Total	4/5

TABLE 5.11

Incorporation of Resource Cost into Work Rate Analysis

Resource, i	Work Rate, r_i	Time, t_i	Work Done, w	Pay Rate, p_i	Pay, P_i
Machine A	r_1	t_1	$(r_1)(t_1)$	p_1	P_1
Machine B	r_2	t_2	$(r_2)(t_2)$	p_2	P_2
$\vdots$	$\vdots$	$\vdots$	$\vdots$	$\vdots$	$\vdots$
Machine n	r_n	t_n	$(r_n)(t_n)$	p_n	P_n
		Total	1.0		Budget

For another simple example of possible application scenarios, consider a case where an Information Technology (IT) technician can install new IT software in three computers every 4 h. At this rate, it is desired to compute how long it would take the technician to install the same software in five computers. We know, from the information given, that we can write the proportion three computers is to 4 h as the proportion that five computers is to x hours, where x represents the number of hours the technician would take to install software in five computers. This gives the following ratio relationship:

$$\frac{3 \text{ computers}}{4\,\text{h}} = \frac{5 \text{ computers}}{x\,\text{h}}$$

which simplifies to yield $x = 6\,\text{h}$, 40 min. Now consider a situation where the technician's competence with the software installation degrades over time for whatever reason. We will see that the time requirements for the IT software installation will vary depending on the current competency level of the technician. Half-life analysis can help to capture such situations so that an accurate work time estimate can be developed.

Resource-Constrained PDM Network

The conventional precedence diagramming network with no resource limitations was presented in Chapter 4. In this section, we extend the previous example to a resource-constrained problem with probabilistic activity durations. Table 5.12 presents the modified

TABLE 5.12

Resource-Constrained PDM Network Data

Activity Number	Name	Predecessors	Duration	Duration Variance	Resource 1 Required	Resource 2 Required
1	A_1	—	1.79	0.11	3	0
2	A_2	A_1	8.17	0.10	5	4
3	B_1	A_1	0.63	0.11	4	1
4	B_2	B_1	3.01	0.12	2	0
5	B_3	A_2, B_2	0.75	0.11	4	3
6	C_1	B_1	11.20	0.13	2	7
7	C_2	B_3, C_1	2.29	0.09	6	2
					$Z_1 = 8$	$Z_2 = 10$

project data after the activities are partitioned into segments (or subactivities) based on the lead–lag PDM restrictions. There are three main activities.

These are partitioned into seven subactivities. Two resource types are involved in this example. There are 8 units of resource type 1 available and 10 units of resource type 2 available. The resource requirements for the individual segments of each activity are shown in the two right-hand columns of the table. The seven activity segments are ranked by the CAF heuristic in the following priority order for resource allocation purposes:

Activity 6 (C_1)	CAF = 100
Activity 2 (A_2)	CAF = 71.7
Activity 5 (B_3)	CAF = 65.2
Activity 3 (B_1)	CAF = 49.6
Activity 7 (C_2)	CAF = 45.2
Activity 4 (B_2)	CAF = 29.6
Activity 1 (A_1)	CAF = 28.7

Figure 5.6 presents the resulting Gantt chart for the resource-constrained schedule. Note that the expected project duration is 24.08 time units. This may be compared to the duration of 18 time units obtained in Chapter 4 without resource limitations.

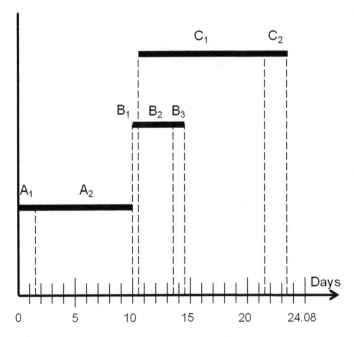

FIGURE 5.6
Resource-constrained PDM schedule.

Critical Resource Diagram

Resource management is a major function in any organization. In a project management environment, project goals are achieved through the strategic allocation of resources to tasks. Several analytical and graphical tools are available for activity planning, scheduling, and control. Examples are the CPM, the PERT, and the PDM. Unfortunately, similar tools are not available for resource management. There is a need for simple tools for resource allocation planning, scheduling, tracking, and control. In this section, a simple extension of the CPM diagram is developed for resource management purposes. The extension, called the CRD, is a graphical tool that brings the well-known advantages of the CPM diagram to resource scheduling. The advantages of CRD include simplified resource tracking and control, better job distribution, better information to avoid resource conflicts, and better tools for resource leveling.

Resource Management Constraints

Resource management is a complex task that is subject to several limiting factors including

1. Resource interdependencies
2. Conflicting resource priorities
3. Mutual exclusivity of resources
4. Limitations on resource availability
5. Limitations on resource substitutions
6. Variable levels of resource availability
7. Limitations on partial resource allocation

The aforementioned factors invariably affect the criticality of certain resource types. It is logical to expect different resource types to exhibit different levels of criticality in a resource allocation problem. For example, some resources are very expensive, some resources possess special skills, and some are in very limited supply. The relative importance of different resource types should be considered when carrying out resource allocation in activity scheduling. The CRD helps in representing resource criticality.

CRD Network Development

Figure 5.7 shows an example of a CRD for a small project requiring six different resource types. Each node identification, RES_j, refers to a task responsibility for resource type j. In a CRD, a node is used to represent each resource unit. Thus, there are eight nodes in Figure 5.7 because there are 2 units of resource type 1 and 2 units of resource type 4. The interrelationships between resource units are indicated by arrows. The arrows are referred to as resource–relationship (R–R) arrows. For example, if the job of resource 1 must precede the job of resource 2, then an arrow is drawn from the node for resource 1 to the node for resource 2. Task durations are included in a CRD to provide further details about resource relationships. Unlike activity diagrams, a resource unit may appear at more than one location in a CRD, provided there are no time or task conflicts.

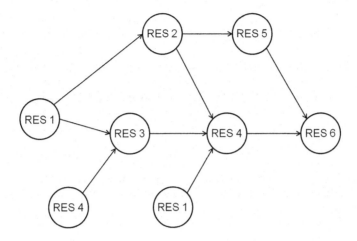

FIGURE 5.7
CRD network example.

Such multiple locations indicate the number of different jobs for which the resource is responsible. This information may be useful for task distribution and resource leveling purposes. In Figure 5.7, resource 1 (RES 1) and resource 4 (RES 4) appear at two different nodes, indicating that each is responsible for two different jobs within the same work scenario.

CRD Computations

The same forward and backward computations used in CPM are applicable to a CRD diagram. However, the interpretation of the critical path may be different, since a single resource may appear at multiple nodes. Figure 5.8 presents an illustrative computational

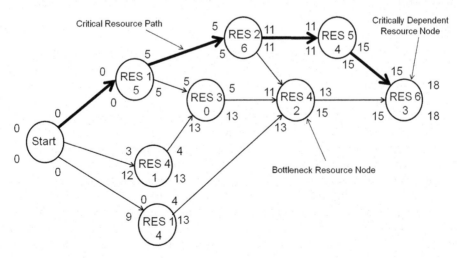

FIGURE 5.8
CRD network analysis.

analysis of the CRD network in Figure 5.7. Task durations (days) are given below the resource identifications. Earliest and latest times are computed and appended to each resource node in the same manner as in CPM analysis. RES 1, RES 2, RES 5, and RES 6 form the critical resource path. These resources have no slack times with respect to the completion of the given project. Note that only one of the two tasks of RES 1 is on the critical resource path. Thus, RES 1 has slack time for performing one job, while it has no slack time for performing the other. Neither of the two tasks of RES 4 is on the critical resource path. For RES 3, the task duration is specified as 0. Despite the favorable task duration, RES 3 may turn out to be a bottleneck resource. RES 3 may be a senior manager whose task is signing a work order. But if he or she is not available to sign at the appropriate time, then the tasks of several other resources may be adversely affected. A major benefit of CRD is that both senior-level and lower-level resources can be included in the resource planning network.

CRD Node Classifications

A *bottleneck* resource node is defined as a node at which two or more arrows merge. In Figure 5.8, RES 3, RES 4, and RES 6 have bottleneck resource nodes. The tasks to which bottleneck resources are assigned should be expedited to avoid delaying dependent resources. A *dependent* resource node is a node whose job depends on the job of immediately preceding nodes. A *critically dependent* resource node is defined as a node on the critical resource path at which several arrows merge. In Figure 5.8, RES 6 is both a critically dependent resource node and a bottleneck resource node. As a scheduling heuristic, it is recommended that activities that require bottleneck resources be scheduled as early as possible. A *burst* resource node is defined as a resource node from which two or more arrows emanate. Like bottleneck resource nodes, burst resource nodes should be expedited, since their delay will affect several following resource nodes.

RS Chart

The CRD has the advantage that it can be used to model partial assignment of resource units across multiple tasks in single or multiple projects. A companion chart for this purpose is the RS chart. Figure 5.9 shows an example of an RS chart based on the earliest times computed in Figure 5.8. A horizontal bar is drawn for each resource unit or resource type. The starting point and the length of each resource bar indicate the interval of work for the resource. Note that the two jobs of RES 1 overlap over a 4-day time period. By comparison, the two jobs of RES 4 are separated by a period of 6 days. If RES 4 is not to be idle over those 6 days, "fill-in" tasks must be assigned to it. For resource jobs that overlap, care must be taken to ensure that the resources do not need the same tools (e.g., equipment, computers, and lathe) at the same time. If a resource unit is found to have several jobs overlapping over an extensive period of time, then a task reassignment may be necessary to offer some relief for the resource. The RS chart is useful for a graphical representation of the utilization of resources. Although similar information can be obtained from a conventional resource loading graph, the RS chart gives a clearer picture of where and when resource commitments overlap. It also shows areas where multiple resources are working concurrently.

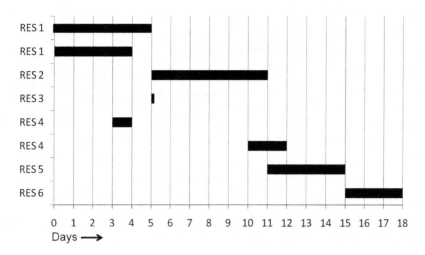

FIGURE 5.9
RS chart.

CRD and Work Rate Analysis

When resources work concurrently at different work rates, the amount of work accomplished by each may be computed by the procedure for work rate analysis. The CRD and the RS chart provide information to identify when, where, and which resources work concurrently.

Example 5.1

Suppose the work rate of RES 1 is such that it can perform a certain task in 30 days. It is desired to add RES 2 to the task so that the completion time of the task can be reduced. The work rate of RES 2 is such that it can perform the same task alone in 22 days. If RES 1 has already worked 12 days on the task before RES 2 comes in, find the completion time of the task. Assume that RES 1 starts the task at time 0.

Solution

The amount of work to be done is 1.0 whole unit (i.e., the full task).
The work rate of RES 1 is 1/30 of the task per unit time.
The work rate of RES 2 is 1/22 of the task per unit time.
The amount of work completed by RES 1 in the 12 days it worked alone is (1/30)
(12) = 2/5 (or 40%) of the required work.
Therefore, the remaining work to be done is 3/5 (or 60%) of the full task.
Let T be the time for which both resources work together.
The two resources work together to complete the task yield (Table 5.13).

TABLE 5.13

Tabulation of Resource Work Rates for RES 1 and RES 2

Resource Type i	Work Rate, r_i	Time, t_i	Work Done, w_i
RES 1	1/30	T	$T/30$
RES 2	1/22	T	$T/22$
		Total	3/5

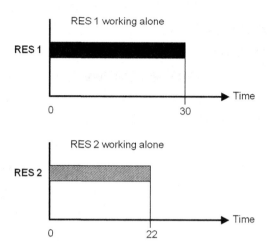

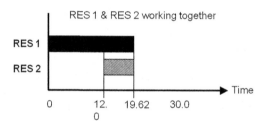

FIGURE 5.10
RS charts for RES 1 and RES 2.

Thus, we have

$$\frac{T}{30} + \frac{T}{22} = \frac{3}{5}$$

which yields $T = 7.62$ days. Thus, the completion time of the task is $(12 + T) = 19.62$ days from time 0. The results of this example are summarized graphically in Figure 5.10. It is assumed that both resources produce identical quality of work and that the respective work rates remain consistent.

The respective costs of the different resource types may be incorporated into the work rate analysis. The CRD and RS charts are simple extensions of familiar tools. They are simple to use and they convey resource information quickly. They can be used to complement existing resource management tools. Users can find innovative ways to modify or implement them for specific resource planning, scheduling, and control purposes. For example, resource-dependent task durations and resource cost can be incorporated into the CRS and RS procedures to enhance their utility for resource management decisions.

Resource Loading and Leveling

Resource loading refers to the allocation of resources to work elements in a project network. A resource loading graph is a graphical representation of resource allocation over time.

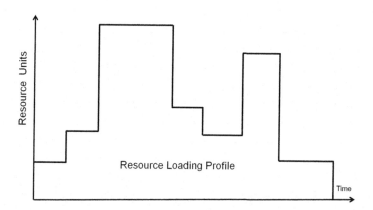

Resource Loading Profile

FIGURE 5.11
Example of resource loading graph.

Figure 5.11 shows an example of a resource loading graph. A resource loading graph may be drawn for the different resource types involved in a project.

The graph provides information useful for resource planning and budgeting purposes. A resource loading graph gives an indication of the demand a project will place on an organization's resources. In addition to resource units committed to activities, the graph may also be drawn for other tangible and intangible resources of an organization. For example, a variation of the graph may be used to present information about the depletion rate of the budget available for a project. If drawn for multiple resources, it can help identify potential areas of resource conflicts. For situations where a single resource unit is assigned to multiple tasks, a variation of the resource loading graph can be developed to show the level of load (responsibilities) assigned to the ROT.

Resource Leveling

Resource leveling refers to the process of reducing the period-to-period fluctuation in a resource loading graph. If resource fluctuations are beyond acceptable limits, actions are taken to move activities or resources around to level out the resource loading graph. For example, it is bad for employee morale and public relations when a company has to hire and lay people off indiscriminately. Proper resource planning will facilitate a reasonably stable level of workforce. Other advantages of resource leveling include simplified resource tracking and control, lower cost of resource management, and improved opportunity for learning. Acceptable resource leveling is typically achieved at the expense of longer project duration or higher project cost. Figure 5.12 shows a leveled resource loading.

When attempting to level resources, note that

1. Not all of the resource fluctuations can be eliminated
2. Resource leveling often leads to an increase in project duration

Resource leveling attempts to minimize fluctuations in resource loading by shifting activities within their available slacks. For small networks, resource leveling can be attempted manually through trial-and-error procedures. For large networks, resource leveling is best handled by computer software techniques. Most of the available commercial project management software packages have internal resource leveling routines. One heuristic

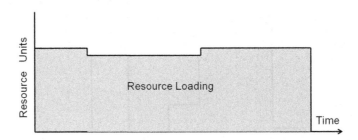

FIGURE 5.12
Resource leveling graph.

procedure for leveling resources, known as *Burgess's Method,* is based on the technique of minimizing the sum of the squares for the resource requirements in each period.

Resource Idleness Graph

A resource idleness graph is similar to a resource loading graph, except that it is drawn for the number of unallocated resource units over time. The area covered by the resource idleness graph may be used as a measure of the effectiveness of the scheduling strategy employed for a project. Suppose two scheduling strategies yield the same project duration, a measure of the resource utilization under each strategy is desired as a means to compare strategies. Figure 5.13 shows two hypothetical resource idleness graphs for the alternate strategies.
   The areas are computed as follows:

$$\text{Area } A = 6(5) + 10(5) + 7(8) + 15(6) + 5(16) = 306 \text{ resource} - \text{units} - \text{time}$$

$$\text{Area } B = 5(6) + 10(9) + 3(5) + 6(5) + 3(3) + 12(12) = 318 \text{ resource} - \text{units} - \text{time}$$

Since area A is less than area B, it is concluded that strategy A is more effective for resource utilization than strategy B. Similar measures can be developed for multiple resources. However, for multiple resources, the different resource units must all be scaled to dimensionless quantities before computing the areas bounded by the resource idleness graphs.

Probabilistic Resource Utilization

In a nondeterministic project environment, probability information can be used to analyze resource utilization characteristics. Suppose the level of availability of a resource is probabilistic in nature. For simplicity, we will assume that the level of availability, X, is a continuous variable whose probability density function is defined by $f(x)$. This is true for many resource types such as funds, nature resources, and raw materials. If we are interested in the probability that resource availability will be within a certain range of x_1 and x_2, then the required probability can be computed as

$$P(x_1 \leq X \leq x_2) = \int_{x_1}^{x_2} f(x)dx$$

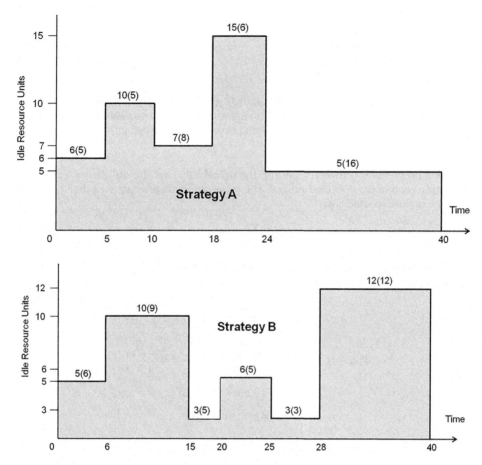

FIGURE 5.13
Resource idleness graphs for resource allocation.

Similarly, a probability density function can be defined for the utilization level of a particular resource. If we denote the utilization level by U and its probability density function by $f(u)$, then we can calculate the probability that the utilization will exceed a certain level, u_0, by the following expression:

$$P(U \geq u_0) = \int_{u_0}^{\infty} f(u)\,du$$

Example 5.2

Suppose a critical resource is leased for a large project. There is a graduated cost associated with using the resource at a certain percentage level, U. The cost is specified as $10,000 per 10% increment in the utilization level above 40%. A flat cost of $5,000 is charged for utilization levels below 40%. The utilization intervals and the associated costs are presented as follows:

1. $U < 40\%$, $5,000
2. $40\% \leq U < 50\%$, $10,000

3. $50\% \le U < 60\%$, \$20,000
4. $60\% \le U < 70\%$, \$30,000
5. $70\% \le U < 80\%$, \$40,000
6. $80\% \le U < 90\%$, \$50,000
7. $90\% \le U < 100\%$, \$60,000

Thus, a utilization level of 50% will cost \$20,000, while a level of 49.5% will cost \$10,000. Suppose the utilization level is a normally distributed random variable with a mean of 60% and a standard deviation of 4%. Find the expected cost of using this resource.

Solution

The solution procedure involves finding the probability that the utilization level will fall within each of the specified ranges. The expected value formula will then be used to compute the expected cost:

$$E[C] = \sum_k x_k P(x_k)$$

where x_k represents the kth interval of utilization. The standard deviation of utilization is 4%.

$$P(U < 40) = P\left(z \le \frac{40-60}{4}\right) = P(z \le -5) = 0.0$$

$$P(40 \le U < 50) = P\left(z < \frac{50-60}{4}\right) - P\left(z \le \frac{40-60}{4}\right)$$

$$= P(z \le -2.5) - P(z \le -5)$$

$$= 0.0062 - 0.0 = 0.0062$$

$$P(50 \le U < 60) = P\left(z < \frac{60-60}{4}\right) - P\left(z \le \frac{50-60}{4}\right)$$

$$= P(z \le 0) - P(z \le -2.5)$$

$$= 0.5000 - 0.0062 = 0.4938$$

$$P(60 \le U < 70) = P\left(z < \frac{70-60}{4}\right) - P\left(z \le \frac{60-60}{4}\right)$$

$$= P(z \le 2.5) - P(z \le 0)$$

$$= 0.9938 - 0.5000 = 0.4938$$

$$P(70 \le U < 80) = P\left(z < \frac{80-60}{4}\right) - P\left(z \le \frac{70-60}{4}\right)$$

$$= P(z \le 5) - P(z \le 2.5)$$

$$= 1.0 - 0.9938 = 0.0062$$

$$P(80 \le U < 90) = P\left(z < \frac{90-60}{4}\right) - P\left(z \le \frac{80-60}{4}\right)$$

$$= P(z \le 7.5) - P(z \le 5)$$

$$= 1.0 - 1.0 = 0.0$$

$$E(C) = \$5,000(0.0) + \$10,000(0.0062) + \$20,000(0.4938)$$

$$+ \$30,000(0.4938) + \$40,000(0.0062) + \$50,000(0.0)$$

$$= \$25,000$$

Thus, it can be expected that leasing this critical resource will cost $25,000 in the long run. A decision can be made whether to lease or buy the resource. Resource substitution may also be considered on the basis of the expected cost of leasing.

Learning Curve Analysis

Learning curves are important for resource allocation decisions (Badiru, 2010). Learning curves present the relationship between cost (or time) and level of activity on the basis of the effect of learning. An early study disclosed the "80% learning" effect, which indicates that a given operation is subject to a 20% productivity improvement each time the activity level or production volume doubles. A learning curve can serve as a predictive tool for obtaining time estimates for tasks in a project environment. Typical learning rates that have been encountered in practice range from 70% to 95%. A learning curve is also referred to as a *progress function*, a *cost–quantity relationship*, a *cost curve*, a *product acceleration curve*, an *improvement curve*, a *performance curve*, an *experience curve*, and an *efficiency curve*.

Several alternate models of learning curves have been presented in the literature. Some of the most notable models are the *log-linear model*, the *S-curve*, the *Stanford-B model*, *DeJong's learning formula*, *Levy's adaptation function*, *Glover's learning formula*, *Pegels' exponential function*, *Knecht's upturn model*, and *Yelle's product model*. The univariate learning curve expresses a dependent variable (e.g., production cost) in terms of some independent variable (e.g., cumulative production). The log-linear model is by far the most popular and most used of all the learning curve models.

The Log-Linear Model

The log-linear model states that the improvement in productivity is constant (i.e., it has a constant slope) as the output increases. There are two basic forms of the log-linear model: the average cost model and the unit cost model.

Average Cost Model

The average cost model is used more than the unit cost model. It specifies the relationship between the cumulative average cost per unit and cumulative production. The relationship indicates that cumulative cost per unit will decrease by a constant percentage as the cumulative production volume doubles. The model is expressed as

$$A_x = C_1 x^b$$

where
  A_x is the cumulative average cost of producing x units
  C_1 is the cost of the first unit

x is the cumulative production count

b is the learning curve exponent (i.e., constant slope of on log–log paper)

The relationship between the learning curve exponent, b, and the learning rate percentage, p, is given by

$$b = \frac{\log p}{\log 2}$$

$$p = 2^b$$

The derivation of the previous relationship can be seen by considering two production levels where one level is double the other, as shown next.

Let level I = x_1 and level II = x_2 = $2x_1$. Then

$$A_{x_1} = C_1 (x_1)^b$$

$$A_{x_2} = C_1 (2x_1)^b$$

The percent productivity gain is then computed as

$$p = \frac{C_1 (2x_1)^b}{C_1 (x_1)^b} = 2^b$$

When linear graph paper is used, the log-linear learning curve is a hyperbola of the form shown in Figure 5.14.

On log–log paper, the model is represented by the following straight line equation:

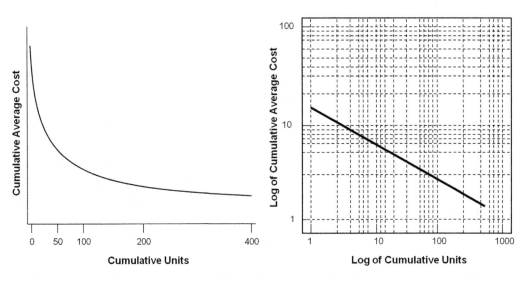

FIGURE 5.14
The log-linear learning curve.

$$\log A_x = \log C_1 + b \log x$$

where b is the constant slope of the line. It is from this straight line that the name *log-linear* was derived.

Example 5.3

Assume that 50 units of an item are produced at a cumulative average cost of $20 per unit. Suppose we want to compute the learning percentage when 100 units are produced at a cumulative average cost of $15 per unit. The learning curve analysis would proceed as follows:

Initial production level = 50 units; average cost = $20

Double production level = 100 units; cumulative average cost = $15

Using the log relationship, we obtain the following equations:

$$\log 20 = \log C_1 + b \log 50$$

$$\log 15 = \log C_1 + b \log 100$$

Solving the equations simultaneously yields

$$b = \frac{\log 20 - \log 15}{\log 50 - \log 100} = -0.415$$

Thus,

$$p = (2)^{-0.415} = 0.75$$

That is a 75% learning rate. In general, the learning curve exponent, b, may be calculated directly from actual data or computed analytically. That is,

$$b = \frac{\log A_{x_1} - \log A_{x_2}}{\log x_1 - \log x_2} \quad \text{(for *the case where two learning curve data points are available*)}$$

$$b = \frac{\ln(p)}{\ln(2)} \quad \text{(for *the case where p, the percentage learning, is known*)}$$

where
x_1 is the first production level
x_2 is the second production level
A_{x_1} is the cumulative average cost per unit at the first production level
A_{x_2} is the cumulative average cost per unit at the second production level
p is the learning rate percentage

Using the basic cumulative average cost function, the total cost of producing x units is computed as

$$\text{TC}_x = (x)A_x = (x)C_1x^b = C_1x^{(b+1)}$$

The unit cost of producing the xth unit is given by

$$U_x = C_1x^{(b+1)} - C_1(x-1)^{(b+1)}$$

$$= C_1x\left[x^{(b+1)} - (x-1)^{(b+1)}\right]$$

The marginal cost of producing the xth unit is given by

$$\text{MC}_x = \frac{d[\text{TC}_x]}{dx} = (b+1)C_1x^b$$

Example 5.4

Suppose, in a production run of a certain product, it is observed that the cumulative hours required to produce 100 units is 100,000 h with a learning curve effect of 85%. For project planning purposes, an analyst needs to calculate the number of hours spent in producing the 50th unit. Following the notation used previously, we have the following information:

$$p = 0.85$$

$$X = 100 \text{ units}$$

$$A_x = 100,000 \text{ h}/100 \text{ units} = 1,000 \text{ h}/\text{unit}$$

Now,

$$0.85 = 2^b$$

Therefore, $b = -0.2345$
    Also,

$$100,000 = C_1(100)^b$$

Therefore, $C_1 = 2{,}944.42$ h. Thus,

$$C_{50} = C_1(50)^b = 1{,}176.50 \text{ h}$$

That is, the cumulative average hours for 50 units is 1,176.50 h. Therefore, cumulative total hours for 50 units = 58,824.91 h. Similarly

$$C_{49} = C_1(49)^b = 1{,}182.09 \text{ h}$$

That is, the cumulative average hours for 49 units is 1,182.09 h. Therefore, cumulative total hours for 49 units = 57,922.17 h. Consequently, the number of hours for the 50th unit is given by

$$58,824.91\,h - 57,922.17\,h = 902.74\,h$$

Unit Cost Model

The unit cost model is expressed in terms of the specific cost of producing the xth unit. The unit cost formula specifies that the individual cost per unit will decrease by a constant percentage as cumulative production doubles. The formulation of the unit cost model is presented as follows. Define the average cost as A_x.

$$A_x = C_1 x^b$$

The total cost is defined as

$$TC_x = (x)A_x = (x)C_1 x^b = C_1 x^{(b+1)}$$

and the marginal cost is given by

$$MC_x = \frac{d[TC_x]}{dx} = (b+1)C_1 x^b$$

This is the cost of one specific unit. Therefore, define the marginal unit cost model as

$$U_x = (1+b)C_1 x^b$$

U_x is the cost of producing the xth unit. To derive the relationship between A_x and U_x

$$U_x = (1+b)C_1 x^b$$

$$\frac{U_x}{(1+b)} = C_1 x^b = A_x$$

$$A_x = \frac{U_x}{(1+b)}$$

$$U_x = (1+b)A_x$$

To derive an expression for finding the cost of the first unit, C_1, we will proceed as follows. Since $A_x = C_1 x^b$, we have the following expressions:

$$C_1 x^b = \frac{U_x}{(1+b)}$$

$$C_1 = \frac{U_x x^{-b}}{(1+b)}$$

Figure 5.15 presents a plot comparing the unit cost model to the average cost model for the previous example, where $C_1 = \$2,944$ and $b = -0.2345$.

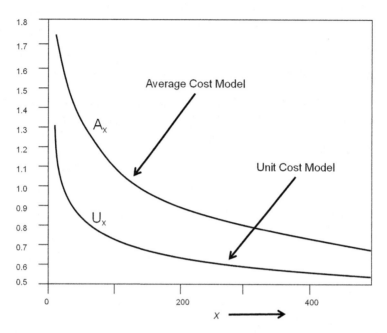

FIGURE 5.15
Comparison of the unit cost and average cost models.

For the case of continuous product volume (e.g., chemical processes), we have the following corresponding expressions:

$$TC_x = \int_0^x U(z)dz C_1 \int_0^x z^b \, dz = \frac{C_1 x^{(b+1)}}{b+1}$$

$$Y_x = \left(\frac{1}{x}\right)\frac{C_1 x^{(b+1)}}{b+1}$$

$$MC_x = \frac{d[TC_x]}{dx} = \frac{d\left[C_1 x^{(b+1)}/b+1\right]}{dx} = C_1 x^b$$

Graphical Analysis

Suppose an observation of the cumulative hours required to produce a unit of an item was recorded at irregular intervals during a production cycle. The recorded observations are presented in Table 5.14.

The project analyst would like to perform the following computational analyses:

1. Calculate the learning curve percentage when cumulative production doubles from 10 to 20 units.

2. Calculate the learning curve percentage when cumulative production doubles from 20 to 40 units.

3. Calculate the learning curve percentage when cumulative production doubles from 40 to 80 units.

TABLE 5.14

Learning Curve Observations

Cumulative Units Produced (X)	Cumulative Average Hours (A_x)
10	92.5
15	71.2
25	50.0
40	35.0
50	26.2
60	20.0
85	11.3
115	10.0
165	7.5
190	6.3

4. Calculate the learning curve percentage when cumulative production doubles from 80 to 160 units.

5. Compute the average learning curve percentage for a given operation.

6. Estimate a standard time for performing the given operation if the steady production level per cycle is 200 units.

A plot of the recorded data is shown in Figure 5.16. A regression model fitted to the data is also shown in the figure. The fitted model is expressed mathematically as

$$A_x = 634.22x^{-0.8206}$$

with an R^2 value of 98.6%. Thus, we have a highly significant model fit. The fitted model can be used for estimation and planning purposes.

Time requirements for the operation at different production levels can be estimated from the model. From the model, we have an estimated cost of the first unit as

$$C_1 = \$634.22$$

$$b = -0.8206$$

$$p = 2^{(-0.8206)} = 56.62\% \text{ learning rate}$$

By using linear interpolation for the recorded data, we can estimate the percentage improvement from one production level to another. For example, when production doubles from 10 to 20 units, we obtain an estimated cumulative average hours of 60.6 h by interpolating between 71.2 and 50 h. Similarly, cumulative average hours of 13.04 h were obtained for the production level of 80 units, and cumulative average hours of 7.75 h were obtained for the production level of 160 units. Now, these average hours are used to compute the percent improvement over the various production levels. For example, the percentage improvement when production doubles from 10 to 20 units is obtained as $p = 60.6/92.5 = 65.5\%$. The calculated percent improvement levels are presented in Table 5.15. The average percent is found to be 53.76%. This compares favorably with the 56.62% suggested by the fitted regression model.

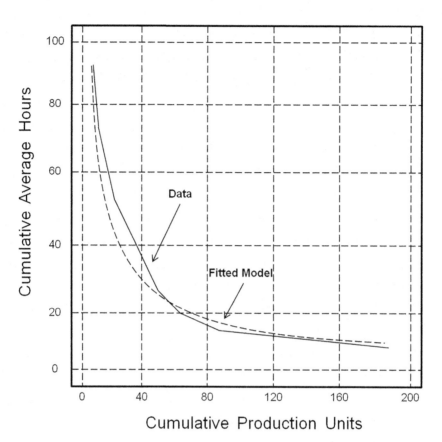

FIGURE 5.16
Graphical analysis of empirical learning curve.

TABLE 5.15

Learning Curve Percentage Analysis

Initial Production Level	Doubled Production Level	Learning Percentage
10	20	65.50
20	40	57.76
40	80	37.26
80	160	59.43
	Average	54.99

Using the fitted model, the estimated cumulative average hours per unit when 200 units are produced is estimated as

$$A_x = 634.22(200)^{-0.8206}$$

$$= 8.20\text{h}$$

Caution should be exercised in using the fitted learning curve for extrapolation beyond the range of the data used to fit the model.

Multivariate Learning Curves

Extensions of the single factor learning curve are important for realistic analysis of productivity gain. In project operations, several factors can intermingle to affect performance. Heuristic decision making, in particular, requires careful consideration of qualitative factors. There are numerous factors that can influence how fast, how far, and how well a worker learns within a given time span. Multivariate models are useful for performance analysis in project planning and control. One form of the multivariate learning curve is defined as

$$A_x = K \prod_{i=1}^{n} c_i x_i^{b_i}$$

where
A_x is the cumulative average cost per unit based on production level of x units
K is the cost of first unit of the product
x_i is the specific value of the ith factor
n is the number of factors in the model
c_i is the coefficient for the ith factor
b_i is the learning exponent for the ith factor

A bivariate form of the model is presented as follows:

$$C = \beta_0 x_1^{\beta_1} x_2^{\beta_2}$$

where
C is a measure of cost
x_1 and x_2 are independent variables

Using World War II data, researchers have experimented with alternate mathematical functions to estimate direct labor per pound of airframe needed to manufacture the Nth airframe in a cumulative production of N planes. The functions presented hereafter describe the relationships between direct labor per pound of airframe (m), cumulative production (N), time (T), and rate of production per month (DN):

1. $\log m = a_2 + b_2 T$
2. $\log m = a_3 + b_3 T + b_4 DN$
3. $\log m = a_4 + b_5(\log T) + b_6(\log DN)$
4. $\log m = a_5 + b_7 T + b_8(\log DN)$
5. $\log m = a_6 + b_9 T + b_{10}(\log N)$
6. $\log m = a_7 + b_{11}(\log N) + b_{12}(\log DN)$

The multiplicative power function, often referred to as the Cobb–Douglas function, has also been investigated as a model for learning curves. The model is of the general form

$$C = b_0 x_1^{b_1} x_2^{b_2} \ldots x_n^{b_n} \varepsilon$$

where

C is the estimated cost

b_0 is the model coefficient

x_i is the ith independent variable ($i = 1, 2, ..., n$)

b_n is the exponent of the ith variable ($i = 1, 2, ..., n$)

ε is the error term

For parametric cost analysis, we can use an additive model of the form

$$C = c_1 x_1^{b_1} + c_2 x_2^{b_2} + \cdots + c_n x_n^{b_n} + \varepsilon$$

where c_i ($i = 1, 2, ..., n$) is the coefficient of the ith independent variable. The model was reported to have been fitted successfully for missile tooling and equipment cost. A variation of the power model was used to study weapon system production. The bivariate model has also been used for the assessment of costs and benefits of a single-source versus multiple-source production decision with variations in quantity and production rate in major Department of Defense programs. The multiplicative power model seems to be effective for expressing program costs in terms of cumulative quantity and production rate to evaluate contractor behavior.

There is also a nonlinear cost–volume–profit model for learning curve analysis. The nonlinearity in the model is affected by incorporating a nonlinear cost function that expresses the effects of employee learning. The profit equation for the initial period of production for a product subject to the usual learning function is expressed as

$$P = px - c\left(ax^{b+1}\right) - f$$

where

P is the profit

p is the price per unit

x is the cumulative production

c is the labor cost per unit time

f is the fixed cost per period

b is the index of learning

The profit function for the initial period of production with n production processes operating simultaneously is given as follows:

$$P = px - nca\left(\frac{x}{n}\right)^{b+1} - f$$

where x is the number of units produced by n labor teams consisting of one or more employees each. Each team is assumed to produce x/n units. This model indicates that, when additional production teams are included, more units are produced over a given time period. However, the average time for a given number of units increases because more employees are producing while they are still learning. That is, more employees with low (but improving) productivity are engaged in production at the same time. The preceding model is extended to the case where employees with different skill levels produce different learning parameters between production runs. This is modeled as follows:

$$P = p\sum_{i=1}^{n} x_i - c\sum_{i=1}^{n} a_i x_i^{b_i+1} - f$$

where

a_i and b_i denote the parameters applicable to the average skill level of the *i*th production run

x_i represents the output of the *i*th run in a given time period

This model could be useful for multiple projects that require concurrent execution. Multivariate learning curve models are available for incorporating cumulative production, production rate, and program cost. The approach involves a production function that relates output rate to a set of inputs with variable utilization rates.

Bivariate Example

A bivariate model is used here to illustrate the nature and modeling approach for general multivariate models. An experiment conducted by the author models a learning curve containing two independent variables: *cumulative production* (x_1) and *cumulative training time* (x_2). The following model was chosen for illustration purposes:

$$A_{x_1 x_2} = K c_1 x_1^{b_1} c_2 x_2^{b_2}$$

where

A_x is the cumulative average cost per unit for a given set X, of factor values

K is the intrinsic constant

x_1 is the specific value of first factor

x_2 is the specific value of second factor

c_i is the coefficient for the *i*th factor

b_i is the learning exponent for the *i*th factor

The set of test data used for the modeling is shown in Table 5.16.

Two data replicates are used for each of the ten combinations of cost and time values. Observations are recorded for the number of units representing double production volumes. The model is transformed to the natural logarithmic form

$$\ln A_x = \left[\ln K + \ln(c_1 c_2)\right] + b_1 \ln x_1 + b_2 \ln x_2$$

$$= \ln a + b_1 \ln x_1 + b_2 \ln x_2$$

where a represents the combined constant in the model such that $a = (K)(c_1)(c_2)$. A regression approach yielded the fitted model

$$\ln A_x = 5.70 - 0.21(\ln x_1) - 0.13(\ln x_2)$$

$$A_x = 298.88 x_1^{-0.21} x_2^{-0.13}$$

with an R^2 value of 96.7%. The variables in the model are explained as follows:

$$\ln(a) = 5.70 \text{ (i.e., } a = 298.88)$$

TABLE 5.16

Data for Modeling Bivariate Learning Curve

Treatment Number	Observation Number	Cumulative Average Cost ($)	Cumulative Production (Units)	Cumulative Training Time (Hours)
1	1	120	10	11
	2	140	10	8
2	3	95	20	54
	4	125	20	25
3	5	80	40	100
	6	75	40	80
4	7	65	80	220
	8	50	80	150
5	9	55	160	410
	10	40	160	500
6	11	40	320	660
	12	38	320	600
7	13	32	640	810
	14	36	640	750
8	15	25	1,280	890
	16	25	1,280	800
9	17	20	2,560	990
	18	24	2,560	900
10	19	19	5,120	1,155
	20	25	5,120	1,000

x_1 is the cumulative production units

x_2 is the cumulative training time in hours

As in the univariate case, the bivariate model indicates that the cumulative average cost decreases as cumulative production and training time increase. For a production level of 1,750 units and a cumulative training time of 600 h, the fitted model indicates an estimated cumulative average cost per unit as

$$A_{(1750,600)} = (298.88)\left(1,750^{-0.21}\right)\left(600^{-0.13}\right) = 27.12$$

Similarly, a production level of 3,500 units and training time of 950 h yield the following cumulative average cost per unit:

$$A_{(3500,950)} = (298.88)\left(3,500^{-0.21}\right)\left(950^{-0.13}\right) = 22.08$$

To use the fitted model, consider the following problem. The standards department of a manufacturing plant has set a target average cost per unit of $12.75 to be achieved after 1,000 h of training. We want to find the cumulative units that must be produced to achieve the target cost. From the fitted model, the following expression is obtained:

$$\$12.75 = (298.88)\left(X^{-0.21}\right)\left(1,000^{-0.13}\right)$$

$$X = 46,409.25$$

On the basis of the large number of cumulative production units required to achieve the expected standard cost, the standards department may want to review the cost standard. The standard of $12.75 may not be achievable if there is a limited market demand (i.e., demand is much less than 46,409 units) for the particular product being considered. The relatively flat surface of the learning curve model as units and training time increase implies that more units will need to be produced to achieve any additional significant cost improvements. Thus, even though an average cost of $22.08 can be obtained at a cumulative production level of 3,500 units, it takes several thousand additional units to bring the average cost down to $12.75 per unit.

Interruption of Learning

Interruption of the learning process can adversely affect expected performance. An example of how to address this is the *manufacturing interruption ratio*, which considers the learning decay that occurs when a learning process is interrupted. One possible expression for the ratio is

$$Z = (C_1 - A_x)\frac{(t-1)}{11}$$

where
Z is the per product loss of learning costs due to manufacturing interruption
$t = 1, 2, \ldots, 11$ (months of interruption from 1 to 12 months)
C_1 is the cost of the first unit of the product
A_x is the cost of the last unit produced before a production interruption

The unit cost of the first unit produced after production begins again is given by

$$A_{(x+1)} = A_x + Z$$

$$= A_x + (C_1 - A_x)\frac{(t-1)}{11}$$

Interruption to the learning process can be modeled by incorporating forgetting functions into regular learning curves. In any practical situation, an allowance must be made for the potential impacts that forgetting may have on performance. Potential applications of the combined learning and forgetting models include design of training programs, manufacturing economic analysis, manpower scheduling, production planning, labor estimation, budgeting, and resource allocation.

Learning Curves in Health-Care Projects

As in the production environment, earning curves are applicable to the routine and repetitive aspects of surgical procedures. However, the difficulty is that each surgery is unique and may encounter things that had not been seen before. Even a doctor who has performed a particular surgery many, many times over several years may still have to deal with a new and unique procedure. The best way to apply learning curves for performance improvement in surgical operations is to first determine those functions that are routine or standardized. Those are the ones that we can expect to make learning curve improvements with. There still needs to be an allowance for procedures that are new or unique and may require more care and time on the part of the surgeon. In other words, there will be some incompressible functions in the surgical process. As in the case of unplanned production interruption, which is one of the things that may be experienced during surgery in a hospital, learning and forgetting do occur in health-care delivery projects (operations).

The theory of learning curve states that direct labor improves as it works longer with more repetitions. The more often a worker repeats a given task, the more efficient he or she will become. This phenomenon of learning also occurs for a doctor who performs a surgery through repetition in the healthcare industry. During the process of an operation scheduling, a hospital may need to assign the duration for every surgery in an operating room. It is known that different kinds of surgeries require different durations. Even for the same kind of surgery, if it is carried out by different doctors, the durations will be different. Besides, even for the same doctor, who performs the same kind of surgery, the durations can still be significantly different depending on the prevailing circumstances and needs. Hospital administrators often find out that there exists a lot of wasted time during surgical operations. If they hope to improve the efficiency of the surgical staff (surgeons, nurses, etc.), learning curves could be one useful approach. The goal is to find some analytical approach to determine a "standard duration" for a doctor for performing a certain kind of surgery. But questions to address include the following:

1. The theory and some models of learning curve have been widely used in low-skill and high-volume repetitive manufacture. Can theory, the conventional production-based theory, be applicable to high-skill and one-of-kind operation in the operating room?
2. There are incompressible task durations in the operating room regardless of the number of repetitions. Which learning curve models are relevant for predicting the duration of surgery?

As reiterated by Jaber and Guiffrida (2008), learning curves are applied to surgery, whether it is directly performed by a surgeon (hand–body contact) or through a median (telesurgery). The learning curves for both types of surgery are different. The time to complete a surgery reduces with time, but not always. A surgery involves a set of sequential steps where a surgeon allows some time for a feedback (reaction time) from the patient. If there are complications, that is, the patient's body is reacting negatively to the procedure performed by the surgeon, then a corrective action must be performed by the surgeon to alleviate any risk to the patient. Such complications can extend the time of a surgery; however, it does not necessarily mean that there is no learning on the part of the surgeon. These complications may bring learning opportunities for a surgeon.

Each type of surgery may follow a different learning curve. Also, if there is a knowledge sharing mechanism among the surgeons (i.e., supporting complementary mutual "surgeons' experiences" or an experience repository bank in a hospital, then the surgeons will always progress on their respective learning curves.

Exercises

5.1 For the project data in Table 5.2, redefine the activity durations in terms of the PERT estimates presented as follows. If the other activity data are the same as presented in Table 5.2, compute the scaled CAF for each activity in the project and use the CAF criterion to schedule the project. Compare the CAF schedule to the ACTIM schedule.

Activity	a, m, b
1–2	1, 2, 3
2–3	7, 8, 9
1–3	2, 3, 4
3–5	9, 10, 11
1–4	7, 8, 9
2–4	1, 2, 3
4–5	0, 1, 2

5.2 Presented in the table is the activity data for a project that is subject to variable resource availability. It is assumed that the scaled CAF weights for the activities are already known, as given in the last column of the tabulated data. There are two resource types. Resource availability varies from day to day on the basis of the RS tabulated after the project data.

Activity	Predecessor	Duration (Days)	Resource Requirement (Type 1, Type 2)	Scaled CAF
A	—	6	2, 5	50.1
B	—	5	1, 3	57.1
C	—	2	3, 1	56.8
D	A, B	4	2, 0	100
E	B	5	1, 4	53.2
F	C	6	3, 1	51.8
G	B, F	1	4, 3	44.9
H	D, E	7	2, 3	54.9

Variable resource availability for the project is as follows:

Time Period (Days)	Units Available (Type 1, Type 2)
0 to 5	3, 5
5+ to 10	2, 3
10+ to 14	3, 4
14+ to 20	4, 4
20+ to 22	5, 3
22+ to 30	4, 5
30+ to 99	6, 6

Use the CAF weights to develop the Gantt chart for the project considering the daily resource availability. A task that is already in progress will be temporarily suspended whenever there are not enough units of resources to continue it. The task will resume (without increasing its duration) whenever enough resources become available.

5.3 Use the minimum total slack (MinTS) heuristic to schedule the resource-constrained project. If a tie occurs, use minimum activity duration to break the tie.

Units of resource type 1 available = 10

Units of resource type 2 available = 15

					Resource Units	
Activity	Predecessor	*a*	*M*	*b*	Type 1	Type 2
A	—	1	2	4	3	0
B	—	5	6	7	5	4
C	—	2	4	5	4	1
D	A	1	3	4	2	0
E	C	4	5	7	4	3
F	A	3	4	5	2	7
G	B, D, E	1	2	3	6	2

5.4 For Exercise 5.3, draw the resource loading diagram for each resource type on the same graph. Which resource type exhibits more fluctuations in resource loading?

5.5 Using the computational approach presented in this chapter, compute the scaled CAF weight for each activity in Exercise 5.3.

5.6 Schedule the project in Exercise 5.3 using each of the following heuristics: CAF, LS, ES, LC, EC, TS, CRD, and ACTIM. What is the shortest project duration obtained? Which heuristics yield that shortest project duration?

5.7 Draw the resource loading graph for each schedule in Exercise 5.6.

5.8 Draw the resource idleness graph for each schedule in Exercise 5.6.

5.9 Use the computational measure presented in this chapter to compute the resource utilization level (area) for each schedule in Exercise 5.6.

5.10 Three project crews are awarded a contract to construct an automated plant. The crews can work simultaneously on the project if necessary. Crew 1, working alone, can complete the project in 82 days. Crew 2, working alone, can complete the project in 50 days. Crew 3, working alone, can complete the project in 64 days. Crew 2 started the project. Ten days after the project started, crew 1 joined the project. Fifteen days after crew 1 joined the project, crew 3 also joined the project. Crew 2 left the project 5 days before it was completed. Crew 1 is paid $500,000/day on the project. Crew 2 is paid $750,000/day. Crew 3 is paid $425,975/day.

a. Find how many days it took to complete the project (assume that fractions of a day are permissible).

b. Based on the project participation described earlier, find the minimum budget needed to complete the project.

5.11 A project involves laying 100 miles of oil distribution pipe in the Alaskan wilderness. It is desired to determine the lowest-cost crew size and composition that can get the job done within 90 days. Three different types of crew are available.

Each crew can work independently or work simultaneously with other crews. Crew 1 can lay pipes at the rate of 1 mile per day. If Crew 2 is to do the whole job alone, it can be finished in 120 days. Crew 3 can lay pipes at the rate of 1.7 miles per day, but this crew already has other imminent project commitments that will limit its participation in the piping project to only 35 contiguous days. All three crews are ready, willing, and available to accept their contracts at the beginning of the project. Crew 3 charges \$50,000/mile of pipe laid, Crew 2 charges \$35,000/mile, and Crew 1 charges \$40,000/mile. Contracts can be awarded only in increments of 10 miles. That is, a crew can get a contract only for 0, 10, 20, 30 miles, and so on. *Required*: Determine the best composition of the project crews and the minimum budget needed to finish the project in the shortest possible time.

5.12 Suppose the utilization level of a critical resource is defined by a normal distribution with a mean of 70% and a variance of 24% squared. Compute the probability that the utilization of the resource will exceed 85%.

5.13 Suppose rainfall is a critical resource for a farming project. The availability of rainfall in terms of inches during the project is known to be a random variable defined by a triangular distribution with a lower end point of 5.25 in., a mode of 6 in., and an upper end point of 7.5 in. Compute the probability that there will be between 5.5 and 7 in. of rainfall during the project.

5.14 A scarce resource is to be leased for an engineering project. There is a graduated cost associated with using the resource at a certain percentage level, U. A step function has been defined for the cost rates for different levels of utilization. The step cost function is presented as follows:

If $U < 50\%$, cost = \$0

If $50\% \leq U < 60\%$, \$5,000

If $60\% \leq U < 67\%$, \$20,000

If $67\% \leq U < 85\%$, \$45,000

If $U \geq 85\%$, \$55,000

Suppose the utilization level is a random variable following a triangular distribution with a lower limit of 40%, a mode of 70%, and an upper limit of 95%.

a. Plot the step function for the cost of leasing the resource.

b. Plot the triangular probability density function for the utilization level.

c. Find the expected cost of using this resource.

5.15 Suppose three resource types (RES 1, RES 2, and RES 3) are to be assigned to a certain task. RES 1 working alone can complete the task in 35 days, RES 2 working alone can complete the task in 40 days, and RES 3 working alone can complete the task in 60 days. Suppose 1 unit of RES 1 and 1 unit of RES 2 start working on the task together at time 0. Two units of RES 3 joined the task after 45% of the task has been completed. RES 2 quits the task when there is 15% of the task remaining. Determine the completion time of the task and the total number of days that each resource type worked on the task. Assume that work rates are constant and units of the same resource type have equal work rates.

5.16 Three resource types (RES 1, RES 2, and RES 3) are to be assigned to a certain task. RES 1 working alone can complete the task in 35 days, RES 2 working alone can complete the task in 40 days, and RES 3 working alone can complete the task in 60 days. Suppose 1 unit of RES 1 and 1 unit of RES 2 start working on the task together at time 0. Two units of RES 3 are brought in to join the task some-time after time 0. RES 2 quits the task when there is 20% of the task remaining. If the full task is desired to be completed 13 days from time 0, determine how many days from time 0 the 2 units of RES 3 should be brought in to join the task. Assume that work rates are constant and units of the same resource type have equal work rates.

5.17 Develop a quantitative methodology for incorporating different levels of quality of work by different resource types into the procedure for work rate analysis presented in this chapter.

5.18 Use the approach presented in Chapter 4 to compute the project network complexity for the data in Exercise 5.3.

5.19 For the project data in the table, use the comprehensive network complexity measure to compute the network complexity for the following alternate levels of resource availability: 2, 3, 4, 5, 6, 7, 8, 9, 10, 11, and 12. Plot the complexity measures against the respective resource availability levels. Discuss your findings.

Activity Number	PERT Estimates (a, m, b)	Preceding Activities	Required Resources
1	1, 3, 5	—	1
2	0.5, 1, 3	—	2
3	1, 1, 2	—	1
4	2, 3, 6	Activity 1	2
5	1, 3, 4	Activity 2	1
6	1.5, 2, 2	Activity 3	2

5.20 For the project data in the table, use the comprehensive network complexity approach to compute the network complexity for all possible combinations of resource availability (Z_1, Z_2), where the possible values of Z_1 are 4, 6, 7, 8, and 9 and the possible values of Z_2 are 2, 3, 4, 5, and 6. Note that this will generate 25 network complexity values. Fit an appropriate multiple regression function to the data you generated, where network complexity is the dependent variable and Z_1 and Z_2 are independent variables.

Activity Number	PERT Estimates (a, m, b)	Preceding Activities	Required Resources (x_{i_1}, x_{i_2})
1	1, 3, 5	—	1, 0
2	0.5, 1, 3	—	1, 1
3	1, 1, 2	—	1, 1
4	2, 3, 6	Activity 1	2, 0
5	1, 3, 4	Activity 2	1, 0
6	1.5, 2, 2	Activity 3	4, 2

5.21 There are three resource types available for a certain project. One unit of each resource type is available. The project manager wants to evaluate the project cost on the basis of how resource teams are made up. She has the option of using a resource team, where the team can consist of only one resource type or a combi-nation of resource types. If the team consists of more than one resource type, the resources will start and stop working at the same time. Fractions of a workday are permissive with prorated cost.

Resource 1 can complete the project alone in 50 days at a cost of $2,500/day.

Resource 2 can complete the project alone in 35 days at a cost of $5,750/day.

Resource 3 can complete the project alone in 75 days at a cost of $1,500/day. The resources produce identical quality of work.

a. Determine the team composition that will yield the minimum project cost.

b. What is the project duration corresponding to that minimum cost? Plot all the project costs versus the respective project durations.

5.22 Three resource types (RES 1, RES 2, and RES 3) are to be assigned to a certain task. RES 1 working alone can complete the task in 30 days, RES 2 working alone can complete the task in 45 days, and RES 3 working alone can complete the task in 50 days. Suppose 2 units of RES 1 and 1 unit of RES 2 start working on the task together at time 0. One unit of RES 3 is brought in to join the task sometime after time 0. RES 2 quits the task when there is 30% of the task remaining. If the full task is desired to be completed 20 days from time 0, determine when, if at all, RES 3 should be brought in to join the task. Assume that work rates are constant and units of the same resource type have equal work rates. Show all work completely and clearly.

5.23 Suppose snowfall is a critical resource for a skiing business. The availability of snowfall in inches during a season is known to be a random variable defined by a triangular distribution with a lower end point of 3 in., a mode of 6 in., and an upper end point of 10.5 in. Compute the probability that there will be between 5 and 9 in. of snowfall during a season.

5.24 Compute scaled CAF priority weight for each of the activities in the project described hereafter. Use the scaled values to schedule the activities. Compare the schedule obtained to the schedule in Exercise 5.3.

Units of resource type 1 available = 10

Units of resource type 2 available = 15

Activity	Predecessor	A	m	B	Resource Units Type 1	Type 2
A	—	1	2	4	3	0
B	—	5	6	7	5	4
C	—	2	4	5	4	1
D	A	1	3	4	2	0
E	C	4	5	7	4	3
F	A	3	4	5	2	7
G	B, D, E	1	2	3	6	2

5.25 In a certain project operation, it was noted that a total assembly cost of $2,000 was incurred for the production of 50 units of a product. Suppose an additional 100 units were produced at a cost of $1,000. Determine the learning curve percentage for the operation.

5.26 Suppose an operation is known to exhibit a learning curve rate of 85%. Production was interrupted for 3 months after 100 units had been produced. If the cost of the very first unit is $50, find

   a. Unit cost of the first unit produced after production begins again.

   b. Cost of unit 150 of the product, assuming no further production interruptions occur.

5.27 Suppose the first 20 unit batch of a product has an average cost of $72 per unit, the next 30 unit batch has an average cost of $60 per unit, and the next 50 unit batch has an average cost of $50 per unit. Based on this cost history, determine the appropriate learning curve percent to recommend for this operation.

5.28 The first performance of a task requires 8 h. The twentieth performance of the task requires only 2 h. If this task is subject to a conventional learning process, determine

   a. Learning rate associated with the task.

   b. Number of hours it will take to perform the task the 12th time.

5.29 Suppose unit 190 of a product requires 45 h to produce under a learning curve rate of 80%. Determine the number of hours required by the first unit of the product and the number of hours required by unit 250.

5.30 The first 50 unit order of a job shop costs $1,500. It is believed that the shop experiences a 75% learning curve rate. Determine a reasonable price quote for the next 80 unit order of the same job.

5.31 Suppose the 60th unit of a product produced under a learning rate of 80% is $30. If the production standard is $20 per unit, determine how many more units must be produced before the standard can be reached.

5.32 Pay and work rate distribution exercise: Suppose two construction workers did a job together over several days and agreed to split the profits as follows:

Contractor A: 40%

Contractor B: 60%

   Contractor B had to leave the job early on some days. So, contractor A ended up working 9 h more than contractor B. The job took 39 h to complete and the two contractors worked together side by side for 30 h. If the net profit was $6,000.00, how much should each receive based on the percentage split that they agreed to?

5.33 Calculate the takt time for a production project that calls for the production of 125 units of a product within an 8-h workday that includes two 15-min rest breaks and a 45-min lunch break.

References

Badiru, A.B. (2010). Half-life of learning curves for information technology project management, *International Journal of IT Project Management*, 1(3): 28–45.

Jaber, M.Y. and Guiffrida, A.L. (2008). Learning curves for imperfect production processes with reworks and process restoration interruptions, *European Journal of Operational Research*, 189(1): 93–104.

Sawhney, R., Badiru, A.B., and Niranjan, A. (2004). A model for integrating and managing resources for technical training programs, in *Internet Economy: Opportunities and Challenges for Developed and Developing Regions of the World*, Y.A. Hosni and T.M. Khalil (eds.), Elsevier, Boston, MA, pp. 337–351.

6

Project Control System

Elements of Project Control

If everything seems under control, you're just not going fast enough.

Mario Andretti

Everything needs control. Every project needs some aspect of control, whether relating to human resources, operating procedure, tools of operation, or work process. If the project moves along, as it should, it needs control either to keep performance on track or to correct unacceptable developments. Project control can be achieved through a combination of quantitative and qualitative tools. Computational and graphical techniques that are developed for other uses can be adopted and adapted for applications in project control. A project control system represents a collection of factors and their interrelationships affecting the performance of a project. This chapter presents some approaches to project monitoring and control. The steps required to carry out project control are discussed. Schedule control through progress review is presented. Guidelines for performance control are developed. Project information systems needed for control are discussed. An approach to terminating projects as a managerial control is also discussed in this chapter.

The three factors (time, budget, and performance) that form the basis for the operating characteristics of a project also help determine the basis for project control. Project control is the process of reducing the deviation between actual performance and planned performance. To control a project, we must be able to measure performance. The ability to measure accurately is a critical aspect of control. Measurements are taken on each of the three components of project constraints: time, performance, and cost. These constraints are often encountered in conflicting terms, and they cannot be fully satisfied simultaneously. The conflicting nature of the three constraints on a project is represented in Figure 6.1.

The higher the desired performance, the more time and resources (cost) will be required. It will be necessary to compromise one constraint in favor of another. There are some projects where performance is the sole focus. Time and cost may be of secondary importance in such projects. In other projects, cost reduction may be the main goal. Performance and time may be sacrificed to some extent in such projects. And, there are projects where schedule compression (time) is the ultimate goal. The specific nature of a project will determine which constraints must be satisfied and which can be sacrificed.

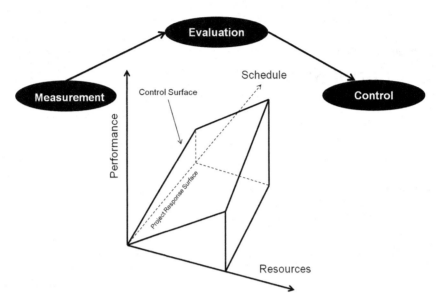

FIGURE 6.1
Axes of project control.

Some of the factors that can require control of a project are listed as follows:

Factors Affecting Time
1. Supply delays
2. Missed milestones
3. Delay of key tasks
4. Change in customer specifications
5. Change of due dates
6. Unreliable time estimates
7. Increased use of expediting
8. Time-consuming technical problems
9. Impractical precedence relationships
10. New regulations that need time to implement

Factors Affecting Performance
1. Poor design
2. Poor quality
3. Low reliability
4. Fragile components
5. Poor functionality
6. Maintenance problems
7. Complicated technology
8. Change in statement of work

9. Conflicting objectives
10. Restricted access to resources
11. Employee morale
12. Poor retention of experienced workforce
13. Sickness and injury

Factors Affecting Cost

1. Inflation
2. New vendors
3. Incorrect bids
4. High labor cost
5. Budget revisions
6. High overhead costs
7. Inadequate budget
8. Increased scope of work
9. Poor timing of cash flows

Project control may be handled in a hierarchical manner, starting with the global view of a project and ending with the elementary level of unit performance, as shown in Figure 6.2.

A product is project dependent. A process is product dependent. The performance of a unit depends on the process from which the unit is made. Such a control hierarchy makes the control process more adaptive, dynamic, and effective. The basic elements of adaptive project control are

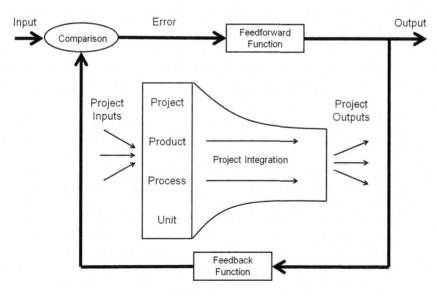

FIGURE 6.2
Control hierarchy from inputs to outputs.

1. Continual tracking and reporting
2. Modifying project implementation as objectives change
3. Replanning to deal with new developments
4. Evaluating achievement versus objectives
5. Documenting project success and failures as guides for the future

Control Steps

Parkinson's law states that a schedule will expand to fill available time, and cost will increase to consume the available budget. Project control prevents a schedule from expanding out of control. Project control also assures that a project can be completed within budget. A recommended project control process is presented as follows:

1. *Step 1*: Determine the criterion for control. This means that the specific aspects that will be measured should be specified.
2. *Step 2*: Set performance standards. Standards may be based on industry practice, prevailing project agreements, work analysis, forecasting, and so on.
3. *Step 3*: Measure actual performance. The measurement scale should be predetermined. The measurement approach must be calibrated and verified for accuracy. Quantitative and nonquantitative factors may require different measurement approaches. This step also requires reliable project tracking and reporting tools. Project status, no matter how unfavorable, must be reported.
4. *Step 4*: Compare actual performance with the specified performance standard. The comparison should be done objectively and consistently based on the specified control criteria. Meet periodically to determine
   a. What has been achieved?
   b. What remains to be done?
5. *Step 5*: Identify unacceptable variance from expectation.
6. *Step 6*: Determine the expected impact of the variance on overall project performance.
7. *Step 7*: Investigate the source of poor performance.
8. *Step 8*: Determine the appropriate control actions needed to overcome (nullify) the variance observed.
9. *Step 9*: Implement the control actions with total dedication.
10. *Step 10*: Ensure that poor performance does not occur elsewhere in the project.

The control steps can be carried out within the framework of the flowchart presented in Figure 6.3. The flow of capital, materials, and labor must be controlled throughout the project management process.

Formal and Informal Control

Informal control refers to the process of using unscheduled or unplanned approaches to assess project performance and using informal control actions. Informal control requires unscheduled visits and impromptu queries to track progress. The advantages of the informal control process are as follows:

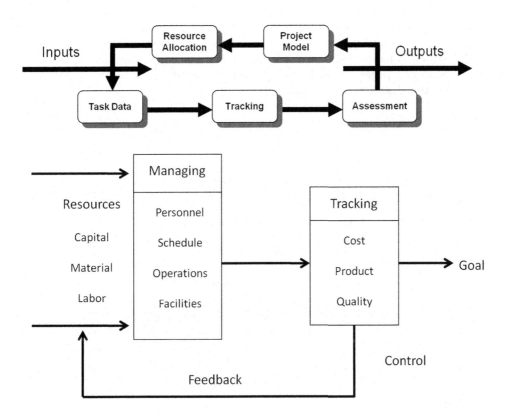

FIGURE 6.3
Flowchart for control process.

1. It allows the project manager to learn more about project progress.
2. It creates a surprise element that keeps workers on their toes.
3. It precludes the temptation for "doctored" progress reports.
4. It allows peers and subordinates to assume control roles.
5. It facilitates prompt appraisal of latest results.
6. It gives the project manager more visibility.

A formal control process deals with the process of achieving project control through formal and scheduled reports, consultations, or meetings. Formal control is typically more time consuming.

1. It can be used by only a limited (designated) group of people.
2. It reduces the direct visibility of the project manager.
3. It can impede the implementation of control actions.
4. It encourages bureaucracy and "paper pushing."
5. It requires a rigid structure.

Despite its disadvantages, formal control can be effective for all types of projects. With a formal control process, project responsibilities and accountability can be pursued in a structured manner. For example, standard audit questions may be posed to determine the current status of a project in order to establish the strategy for future performance. Examples of suitable questions are as follows:

1. Where are we today?
2. Where were we expected to be today?
3. What are the prevailing problems?
4. What problems are expected in the future?
5. Where shall we be at the next reporting time?
6. What are the major results since the last project review?
7. What is the ratio of percent completion to budget depletion?
8. Is the project plan still the same?
9. What resources are needed for the next stage of the project?

A formal structured documentation of what questions to ask can guide the project auditor in carrying out project audits in a consistent manner. The availability of standard questions makes it unnecessary for the auditor to guess or ignore certain factors that may be crucial for project control.

Schedule Control

The Gantt charts developed in the scheduling phase of a project can serve as the benchmark for measuring project progress. Project status should be monitored frequently. Symbols may be used to mark actual activity positions on the Gantt chart. A record should be maintained of the difference between the actual status of an activity and its expected status. This information should be conveyed to the appropriate personnel with a clear indication for the required control actions. The more milestones or control points there are in a project, the easier the control function. The larger number allows for more frequent and distinct avenues for monitoring the schedule. Problems can be identified and controlled before they accumulate into more serious problems. However, more control points mean higher cost of control.

Schedule variance magnitudes may be plotted on a timescale (e.g., on a daily basis). If the variance continues to get worse, drastic actions may be necessitated. Temporary deviations without a lasting effect on the project may not be a cause for concern. Some control actions that may be needed for project schedule delays are

1. Job redesign
2. Productivity improvement
3. Revision of project scope
4. Revision of project master plan
5. Expediting or activity crashing
6. Elimination of unnecessary activities
7. Reevaluation of milestones or due dates
8. Revision of time estimates for pending activities

Project Tracking and Reporting

Tracking and reporting provide the avenues for monitoring and evaluating project progress. This is an area where computerized tools are useful. Customized computer-based project tracking programs can be developed to monitor various aspects of project performance. An example is the tracking scheme presented in Table 6.1.

The model can be expanded to keep track of several parameters for each activity, including the following:

1. Days of work planned for an activity
2. Planned cost
3. Number of days worked
4. Number of days credited
5. Percent completion expected
6. Actual percent completion
7. Actual cost
8. Amount of budget shortfall or surplus

Workdays planned indicate the amount of work required for a task expressed in terms of the number of days required to perform the work. This is entered as an input by the user. In a more sophisticated model, the number of workdays required may be computed directly by the program based on a standard database or some internal estimation formulas.

Planned cost indicates the estimated cost of performing the full amount of work required by the activity. This can be entered as a user input or can be calculated internally by the program.

Task weight represents the relative importance of the activity based on some stated criteria. In the model in Table 6.1, the task weight is computed on the basis of the work content of the activity relative to the overall project. The work content is expressed in terms of days. This is computed as follows:

$$\text{Task weight} = \frac{(\text{Workdays for activity})}{(\text{Total workdays for project})}$$

TABLE 6.1

Data for Project Performance Monitoring

	Project Start Date: Jan 15, 2019 Current Date: July 15, 2019 Project Name: Installation of New Products Line				
Task Name	Workdays Planned	Planned Cost ($)	Task Weight	Workdays Completed	Workdays Credited
1. Define problem	16	100,000	0.1524	16	16
2. Interviews	38	100,000	0.3619	38	38
3. Network analysis	7	100,000	0.0667	7	7
4. Improve flow	7	150,000	0.0667	7	7
5. New process	7	200,000	0.0667	14	7
6. Expert system	30	42,000	0.2857	0	0
	105		1.000		75

For example, in Table 6.1, the total work content for the project is 105 days. The work content for task 2 is 38 days. So, the weight for task 2 is 38/105 = 0.3619.

Workdays completed indicate the amount of work actually completed out of the total planned workdays as of the current date. This value is based on actual measurement of work accomplishment. This may be done by monitoring the total number of cumulative hours worked and dividing that total by the number of hours per day to obtain the number of workdays.

Workdays credited indicate the number of productive workdays recognized out of the total workdays completed. For example, if no credit is given for setup times, then the workdays credited will be obtained by subtracting all setup times from workdays completed. For example, in Table 6.1, a total of 14 days were worked on task 5, but only 7 of the 14 days were credited to the task.

Expected % completion indicates the percentage of the required work that is expected to be completed by the current date. This is computed as shown in the following equation:

$$\text{Expected \% completion} = \frac{(\text{Workdays completed})}{(\text{Workdays planned})}$$

Expected relative % completion indicates the percentage of work expected to be completed on the given activity relative to the total work on the project. This is computed as follows:

$$\text{Expected relative \% completion} = (\text{Expected \% completion}) * (\text{Task weight})$$

For example, task 1 in Table 6.1 is expected to have completed 16 of the 105 days for the project by the current date. That translates to 16/105 = 0.1524 or 15.24%.

Actual % completion indicates the actual percentage of work completed by the current date. This value will normally be obtained by a direct observation of the goal of the task. Note that even though the number of planned workdays has been completed, the actual amount of work completed may not be worth the number of days worked. This may be due to low employee productivity. For example, if a certain amount of work is expected to take 10 days, an employee may work diligently on the task for 10 days, but he or she may accomplish only one-half of what is required to be done.

In Table 6.1, this occurs in the case of tasks 4 and 5. This is reflected in the actual % completion column. Even though the required 7 days for task 4 have been completed, only 65% of the required work has been done. The amount of work yet to be done on task 4 is equivalent to 35% of 7 days (i.e., 2.45 days' worth of work behind schedule). Similarly, the amount of work remaining to be done on task 5 is 90% of 7 days (i.e., 6.3 days' worth of work behind schedule). Consequently, the total project is behind schedule by 2.45 + 6.3 = 8.75 days. It should be noted that a project may be behind schedule either in terms of the physical schedule (i.e., number of days elapsed) or in terms of the amount of work accomplished. The 8.75 days is in terms of amount of work to be accomplished. That is, it will take 8.75 days to accomplish what is yet to be accomplished on tasks 4 and 5.

Actual relative % completion indicates the relative percentage of work actually completed on the given activity when compared with other activities in the project. This is computed as follows:

$$\text{Actual relative \% completion} = (\text{Actual \% completion}) * (\text{Task weight})$$

Actual cost indicates the actual total amount spent on the activity as of the current date.

Cost deviation indicates the difference between the actual and planned costs. Cost overrun occurs when actual cost exceeds planned cost. Cost saving occurs when actual cost falls below planned cost. In Table 6.1, brackets are used to indicate cost savings. It is noted that tasks 2, 3, and 5 have cost overruns.

In addition to the activity parameters explained earlier, the following parameters are associated with the overall project:

Planned project % completion indicates the percent of the project expected to be completed by the current date. This is computed as shown next:

$$\text{Planned project \% completion} = \frac{\text{(Workdays completed on project)}}{\text{(Total workdays planned)}}$$

This can also be obtained as the sum of the expected relative % completion for all activities. In Table 6.1, the planned % completion of the project is 71.43%.

Project tracking index is a relative measure of the performance of the project based on the amount of work actually completed rather than the number of days elapsed. It is computed as follows:

$$\text{Index} = \frac{\text{(Actual relative \% completion)}}{\text{(Expected relative \% completion)}} - 1$$

If the project is performing better than expected (e.g., ahead of schedule), then the index will be positive. The index will be negative if the project performance is below expectation. The index, expressed as a percentage, may be viewed as a measure of the criticality of the control action needed on the project. For example in Table 6.1, the project tracking index is computed as

$$\frac{(0.6310)}{(0.7143)} - 1 = -0.1167 \text{ or } -11.67\%$$

indicating that the project is behind schedule in work accomplishment. This implies that 11.67% of the work expected to be accomplished by the current date has not been accomplished.

Days ahead/behind schedule refer to the performance of the project schedule. It indicates the amount of work by which the project deviates from expectation. The deviation is expressed in terms of days of work. This measure is computed as follows:

$$\text{Schedule performance} = \text{(Project tracking index)} * \text{(Planned days credited)}$$

If this number is negative, then the project is behind schedule in terms of work accomplishment. If the number is positive, then the project is ahead of schedule. It should be recalled that a project may be physically on schedule but still be behind schedule in terms of actual work accomplishment. The measure used in the Baker model provides a better basis for evaluating project performance than the traditional approach of merely looking at the physical schedule. In Table 6.1, the project is behind schedule by the following amount:

$$(-11.67\%) * (75 \text{ days}) = -8.75 \text{ days' worth of work}$$

It should be recalled that this same number, −8.75 days, was obtained earlier by adding up the work deficiencies of individual activities. Even though this project has been credited with 75 workdays, the actual amount of work accomplished is only 75−8.75 days = 66.25 days.

Performance Control

Many project performance problems may not surface until after a project has been completed. This makes performance control very difficult. Effort should be made to measure all the interim factors that may influence final project performance. After-the-fact performance measurements are typically not effective for project control. Some of the performance problems may be indicated by time and cost deviations. So, when project time and cost have problems, an analysis of how the problems may affect performance should be made. Since project performance requirements usually relate to the performance of end products, controlling performance problems may necessitate altering product specifications. Performance analysis will involve checking key elements of the product, such as those discussed next.

1. *Scope*

   Is the scope reasonable based on the project environment?

   Can the required output be achieved with the available resources?

   The current wave of *downsizing* and *rightsizing* in industry may be an attempt to define the proper scope of operations.

2. *Documentation*

   Is the requirement specification accurate?

   Are statements clearly understood?

3. *Requirements*

   Is the technical basis for the specification sound?

   Are the requirements properly organized?

   What are the interactions among specific requirements?

   How does the prototype perform?

   Is the raw material appropriate?

   What is a reasonable level of reliability?

   What is the durability of the product?

   What are the maintainability characteristics?

   Is the product compatible with other products?

   Are the physical characteristics satisfactory?

4. *Quality assurance*

   Who is responsible for inspection?

   What are the inspection policies and methods?

   What actions are needed for nonconforming units?

5. *Function*

   Is the product usable as designed?

   Can the expected use be achieved by other means?

   Is there any potential for misusing the product?

Careful evaluation of performance on the basis of the previous questions throughout the life cycle of a project should help identify problems early so that control actions may be initiated to forestall greater problems later.

Continuous Performance Improvement

Continuous performance improvement (CPI) is an approach to obtaining a steady flow of improvement in a project. The approach is based on the concept of continuous process improvement, which is used in quality management functions. The iterative decision processes in project management can benefit quite well from the concept of CPI. CPI is a practical method of improving performance in business, management, or technical processes. The approach is based on the following key points:

1. Early detection of problems
2. Incremental improvement steps
3. Projectwide adoption of the CPI concept
4. Comprehensive evaluation of procedures
5. Prompt review of methods of improvement
6. Prioritization of improvement opportunities
7. Establishment of long-term improvement goals
8. Continuous implementation of improvement actions

A steering committee is typically set up to guide the improvement efforts. The typical functions of the steering committee with respect to performance improvement include the following:

1. Determination of organizational goals and objectives
2. Communication with the personnel
3. Team organization
4. Administration of CPI procedures
5. Education and guidance for companywide involvement
6. Allocation or recommendation of resource requirements

Figure 6.4 represents the conventional fluctuating approach to performance improvement. In the figure, the process starts with a certain level of performance. A certain performance level is specified as the target to be achieved by time T. Without proper control, the performance will gradually degrade until it falls below the lower control limit at time t_1. At that time, a drastic effort will be needed to raise the performance level. If neglected once again, the performance will go through another gradual decline until it again falls below the lower control limit at time t_2. Again, a costly drastic effort will be needed to improve the performance. This cycle of *degradation–innovation* may be repeated several times before time T is reached. At time T, a final attempt will be needed to suddenly raise the performance to the target level. But unfortunately, it may be too late to achieve the target performance.

There are many disadvantages of the conventional fluctuating approach to improvement. They are

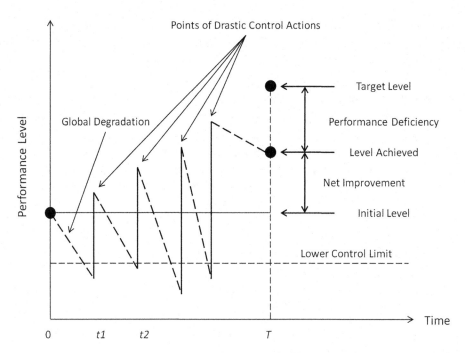

FIGURE 6.4
Conventional approach to control.

1. High cost of implementation
2. Need for drastic control actions
3. Potential loss of project support
4. Adverse effect on personnel morale
5. Frequent disruption of the project
6. Too much focus on short-term benefits
7. Need for frequent and strict monitoring
8. Opportunity cost during the degradation phase

Figure 6.5 represents the approach of continuous improvement. In the figure, the process starts with the same initial quality level, and it is continuously improved in a gradual pursuit of the target performance level. As opportunities to improve occur, they are immediately implemented. The rate of improvement is not necessarily constant over the planning horizon. Hence, the path of improvement is shown as a curve rather than a straight line.

The important aspect of the CPI is that each subsequent performance level is at least as good as the one preceding it.

The major advantages of continuous process of control include

1. Better client satisfaction
2. Clear expression of project expectations
3. Consistent pace with available technology
4. Lower cost of achieving project objectives

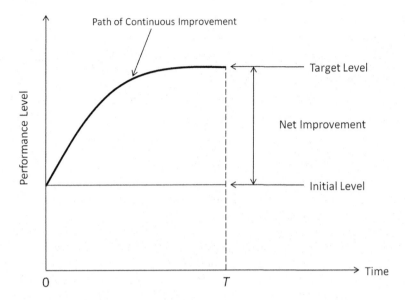

FIGURE 6.5
Continuous process of control.

5. Conducive environment for personnel involvement
6. Dedication to higher quality products and services

Cost Control

Several aspects of a project can contribute to the overall cost of the project. These aspects must be carefully tracked during the project to determine when control actions may be needed. Some of the important cost aspects of a project are

1. Cost estimation approach
2. Cost accounting practices
3. Project cash flow management
4. Company cash flow status
5. Direct labor costing
6. Overhead rate costing
7. Incentives, penalties, and bonuses
8. Overtime payments

The process of controlling project cost covers several key issues that management must address. These include

1. Proper planning of the project to justify the basis for cost elements
2. Reliable estimation of time, resources, and cost
3. Clear communication of project requirements, constraints, and available resources
4. Sustained cooperation of project personnel

5. Good coordination of project functions

6. Consistent policy for project expenditures

7. Timely tracking and reporting of time, materials, and labor transactions

8. Periodic review of project progress

9. Revision of project schedule to adapt to prevailing project scenarios

10. Evaluation of budget depletion versus project progress

These items must be evaluated as an integrated control effort rather than as individual functions. The interactions among the various actions needed may be so unpredictable that the success achieved at one level may be nullified by failure at another level. Such uncoordinated analysis makes cost control very difficult. Project managers must be alert and persistent in the cost monitoring function.

Some government agencies have developed cost control techniques aimed at managing large projects that are typical in government contracts. The cost and schedule control system is based on work breakdown structure (WBS), and it can quantitatively measure project performance at a particular point in a project. Another useful cost control technique is the accomplishment cost procedure (ACP). This is a simple approach for relating resources allocated to actual work accomplished. It presents costs based on scheduled accomplishments rather than as a function of time. In order to determine the progress of an individual effort with respect to cost, the cost/progress relationship in the project plan is compared with the cost/progress relationship actually achieved. The major aspect of the ACP technique is that it is not biased against high costs. It gives proper credit to high costs as long as comparable project progress is maintained. The cost analysis aspects of project management are covered in detail in a later chapter.

Information for Project Control

Complex projects require a well-coordinated communication system that can quickly reveal the status of each activity. Reports on individual project elements must be tied together in a logical manner to facilitate managerial control. The project manager must have prompt access to individual activity status as well as the status of the overall project. A critical aspect of this function is the prevailing level of communication, cooperation, and coordination in the project. The *project management information system* (*PMIS*) has evolved as a solution to the problem of monitoring, organizing, storing, and dissemination of project information. Many commercial computer programs have been developed for the implementation of PMIS. The basic reporting elements in a PMIS may include the following:

1. Financial reports

2. Project deliverables

3. Current project plan

4. Project progress reports

5. Material supply schedule

6. Client delivery schedule

7. Subcontract work and schedule

8 Project conference schedule and records

9. Graphical project schedule (Gantt chart)

10. Performance requirements evaluation plots

11. Time performance plots (plan versus actual)

12. Cost performance plots (expected versus actual)

Many standard forms have been developed to facilitate the reporting process. With the availability of computerized systems, manual project information systems are no longer used much in practice.

Measurement Scales

Project control requires data collection, measurement, and analysis. In project management, the manager will encounter different types of measurement scales depending on the particular items to be controlled. Data may need to be collected on project schedules, costs, performance levels, problems, and so on. The different types of data measurement scales that are applicable are discussed next.

Nominal scale of measurement. A *nominal scale* is the lowest level of measurement scales. It classifies items into categories. The categories are mutually exclusive and collectively exhaustive. That is, the categories do not overlap, and they cover all possible categories of the characteristics being observed. For example, in the analysis of the critical path in a project network, each job is classified as either critical or not critical. Gender, type of industry, job classification, and color are some examples of measurements on a nominal scale.

Ordinal scale of measurement. An *ordinal scale* is distinguished from a nominal scale by the property of order among the categories. An example is the process of prioritizing project tasks for resource allocation. We know that first is above second, but we do not know how far above. Similarly, we know that better is preferred to good, but we do not know by how much. In quality control, the ABC classification of items based on the Pareto distribution is an example of a measurement on an ordinal scale.

Interval scale of measurement. An *interval scale* is distinguished from an ordinal scale by having equal intervals between the units of measure. The assignment of priority ratings to project objectives on a scale of 0–10 is an example of a measurement on an interval scale. Even though an objective may have a priority rating of 0, it does not mean that the objective has absolutely no significance to the project team. Similarly, the scoring of 0 on an examination does not imply that a student knows absolutely nothing about the materials covered by the examination. Temperature is a good example of an item that is measured on an interval scale. Even though there is a zero point on the temperature scale, it is an arbitrary relative measure. Other examples of interval scales are IQ measurements and aptitude ratings.

Ratio scale of measurement. A *ratio scale* has the same properties of an interval scale but with a true zero point. For example, an estimate of a zero time unit for the duration of a task is a ratio scale measurement. Other examples of items measured on a ratio scale are cost, time, volume, length, height, weight, and inventory level. Many of the items measured in a project management environment will be on a ratio scale.

Another important aspect of data analysis for project control involves the classification scheme used. Most projects will have both *quantitative* and *qualitative* data. Quantitative data require that we describe the characteristics of the items being studied numerically.

Qualitative data, on the other hand, are associated with object attributes that are not measured numerically. Most items measured on the nominal and ordinal scales will normally be classified into the qualitative data category, while those measured on the interval and ratio scales will normally be classified into the quantitative data category.

The implication for project control is that qualitative data can lead to bias in the control mechanism, because qualitative data are subject to the personal views and interpretations of the person using the data. Whenever possible, data for project control should be based on quantitative measurements.

There is a class of project data referred to as *transient data*. This is defined as a volatile set of data that is used for onetime decision making and is not needed again. An example may be the number of operators that show up at a job site on a given day. Unless there is some correlation between the day-to-day attendance records of operators, this piece of information will have relevance only for that given day. The project manager can make his decision for that day on the basis of that day's attendance record. Transient data need not be stored in a permanent database unless it may be needed for future analysis or uses (e.g., forecasting, incentive programs, and performance review).

Recurring data refer to data that are encountered frequently enough to necessitate storage on a permanent basis. An example is a file containing contract due dates. This file will need to be kept at least through the project life cycle. Recurring data may be further categorized into *static data* and *dynamic data*. Recurring data that are static will retain their original parameters and values each time they are retrieved and used. Recurring data that are dynamic have the potential for taking on different parameters and values each time they are retrieved and used. Storage and retrieval considerations for project control should address the following questions:

1. What is the origin of the data?
2. How long will the data be maintained?
3. Who needs access to the data?
4. What will the data be used for?
5. How often will the data be needed?
6. Are the data for look-up purposes only (i.e., no printouts)?
7. Are the data for reporting purposes (i.e., generate reports)?
8. In what format are the data needed?
9. How fast will the data need to be retrieved?
10. What security measures are needed for the data?

Data Determination and Collection

It is essential to determine what data to collect for project control purposes. Data collection and analysis are the basic components of generating information for project control. The requirements for data collection are discussed next.

Choosing the data. This involves selecting data on the basis of their relevance and the level of likelihood that they will be needed for future decisions and whether or not they contribute to making the decision better. The intended users of the data should also be identified.

Collecting the data. This identifies a suitable method of collecting the data as well as the source from which the data will be collected. The collection method will depend on the particular operation being addressed. The common methods include manual tabulation, direct keyboard entry, optical character reader, magnetic coding, electronic scanner, and more recently, voice command. An input control may be used to confirm the accuracy of collected data. Examples of items to control when collecting data are the following:

Relevance check. This checks if the data are relevant to the prevailing problem. For example, data collected on personnel productivity may not be relevant for a decision involving marketing strategies.

Limit check. This checks to ensure that the data are within known or acceptable limits. For example, an employee overtime claim amounting to over 80 h/week for several weeks in a row is an indication of a record well beyond ordinary limits.

Critical value. This identifies a boundary point for data values. Values below or above a critical value fall in different data categories. For example, the lower specification limit for a given characteristic of a product is a critical value that determines whether or not the product meets quality requirements.

Coding the data. This refers to the technique used in representing data in a form useful for generating information. This should be done in a compact and yet meaningful format. The performance of information systems can be greatly improved if effective data formats and coding are designed into the system right from the beginning.

Processing the data. Data processing is the manipulation of data to generate useful information. Different types of information may be generated from a given data set, depending on how it is processed. The processing method should consider how the information will be used, who will be using it, and what caliber of system response time is desired. If possible, processing controls should be used. This may involve the following steps:

Control total. Check for the completeness of the processing by comparing accumulated results to a known total. An example of this is the comparison of machine throughput to a standard production level or the comparison of cumulative project budget depletion to a cost accounting standard.

Consistency check. Check if the processing is producing the same results for similar data. For example, an electronic inspection device that suddenly shows a measurement that is ten times higher than the norm warrants an investigation of both the input and the processing mechanisms.

Scales of measurement. For numeric scales, specify units of measurement, increments, the zero point on the measurement scale, and the range of values.

Using the information. Using information involves people. Computers can collect data, manipulate data, and generate information, but the ultimate decision rests with people, and decision making starts when information becomes available. Intuition, experience, training, interest, and ethics are just a few of the factors that determine how people use information. The same piece of information that is positively used to further the progress of a project in one instance may also be used negatively in another instance. To assure that data and information are used appropriately, computer-based security measures can be built into the information system.

Project data may be obtained from several sources. Some potential sources are

1. Formal reports
2. Interviews and surveys
3. Regular project meetings
4. Personnel time cards or work schedules

The timing of data is also very important for project control purposes. The contents, level of detail, and frequency of data can affect the control process. An important aspect of project management is the determination of the data required to generate the information needed for project control. The function of keeping track of the vast quantity of rapidly changing and interrelated data about project attributes can be very complicated. The major steps involved in data analysis for project control are

1. Data collection
2. Data analysis and presentation
3. Decision making
4. Implementation of action

Data are processed to generate information. Information is analyzed by the decision maker to make the required decisions. Good decisions are based on timely and relevant information, which in turn is based on reliable data. Data analysis for project control may involve the following functions:

1. Organizing and printing computer-generated information in a form usable by managers
2. Integrating different hardware and software systems to communicate in the same project environment
3. Incorporating new technologies such as expert systems into data analysis
4. Using graphics and other presentation techniques to convey project information

Proper data management will prevent misuse, misinterpretation, or mishandling. Data are needed at every stage in the life cycle of a project from the problem identification stage through the project phaseout stage. The various items for which data may be needed are project specifications, feasibility study, resource availability, staff size, schedule, project status, performance data, and phaseout plan. The documentation of data requirements should cover the following:

1. *Data summary.* A data summary is a general summary of the information and decision for which the data are required as well as the form in which the data should be prepared. The summary indicates the impact of data requirements on the organizational goals.
2. *Data processing environment.* The processing environment identifies the project for which the data are required, the user personnel, and the computer system to be used in processing the data. It refers to the project request or authorization and relationship to other projects and specifies the expected data communication needs and mode of transmission.
3. *Data policies and procedures.* Data handling policies and procedures describe policies governing data handling, storage, and modification and the specific procedures for implementing changes to the data. Additionally, they provide instructions for data collection and organization.
4. *Static data.* A static data description describes that portion of the data that are used mainly for reference purposes, and it is rarely updated.

5. *Dynamic data.* A dynamic data description describes that portion of the data that are frequently updated based on the prevailing circumstances in the organization.

6. *Data frequency.* The frequency of data update specifies the expected frequency of data change for the dynamic portion of the data, for example, quarterly. This data change frequency should be described in relation to the frequency of processing.

7. *Data constraints.* Data constraints refer to the limitations on the data requirements. Constraints may be procedural (e.g., based on corporate policy), technical (e.g., based on computer limitations), or imposed (e.g., based on project goals).

8. *Data compatibility.* Data compatibility analysis involves ensuring that data collected for project control needs will be compatible with future needs.

9. *Data contingency.* A data contingency plan concerns data security measures in case of accidental or deliberate damage or sabotage affecting hardware, software, or personnel.

Data Analysis and Presentation

Data analysis refers to the various mathematical and graphical operations that can be performed on data to elicit the inherent information contained in the data. The manner in which project data are analyzed and presented can affect how the information is perceived by the decision maker. The examples presented in this section illustrate how basic data analysis techniques can be used to convey important information for project control.

In many cases, data are represented as an answer to direct questions such as: When is the project deadline? Who are the people assigned to the first task? How many resource units are available? Are enough funds available for the project? What are the quarterly expenditures on the project for the past 2 years? Is personnel productivity low, average, or high? Who is the person in charge of the project? Answers to these types of questions constitute data of different forms or expressed on different scales. The resulting data may be qualitative or quantitative. Different techniques are available for analyzing the different types of data. This section discusses some of the basic techniques for data analysis. The data presented in Table 6.2 is used to illustrate the data analysis techniques.

Raw Data

Raw data consist of ordinary observations recorded for a decision variable or factor. Examples of factors for which data may be collected for decision making are revenue, cost, personnel productivity, task duration, project completion time, product quality, and resource availability. Raw data should be organized into a format suitable for visual review and computational analysis. The data in Table 6.2 represent the quarterly revenues from

TABLE 6.2

Quarterly Revenue from Four Projects (in $1,000s)

Project	Quarter 1	Quarter 2	Quarter 3	Quarter 4	Row Total
A	3,000	3,200	3,400	2,800	12,400
B	1,200	1,900	2,500	2,400	8,000
C	4,500	3,400	4,600	4,200	16,700
D	2,000	2,500	3,200	2,600	10,300
Column total	10,700	11,000	13,700	12,000	47,400

projects A, B, C, and D. For example, the data for quarter 1 indicate that project C yielded the highest revenue of $4,500,000, while project B yielded the lowest revenue of $1,200,000. Figure 6.6 presents the raw data of project revenue as a line graph. The same information is presented as a multiple bar chart in Figure 6.7.

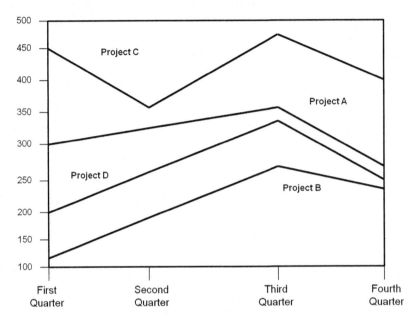

FIGURE 6.6
Line graph of quarterly project revenues.

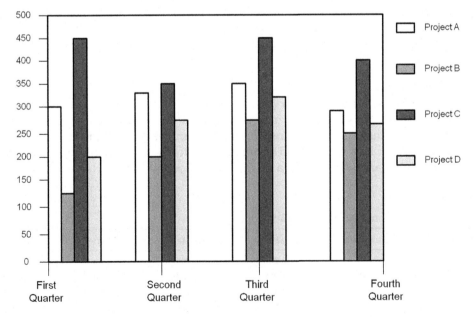

FIGURE 6.7
Multiple bar chart of quarterly project revenues.

Total Revenue

A total or sum is a measure that indicates the overall effect of a particular variable. If X_1, X_2, X_3, ..., X_n represent a set of n observations (e.g., revenues), then the total is computed as

$$T = \sum_{i=1}^{n} X_i$$

For the data in Table 6.2, the total revenue for each project is shown in the last column. The totals indicate that project C brought in the largest total revenue over the four quarters under consideration, while project B produced the lowest total revenue. The last row of the table shows the total revenue for each quarter. The totals reveal that the largest revenue occurred in the third quarter. The first quarter brought in the lowest total revenue. The grand total revenue for the four projects over the four quarters is shown as $47,400,000 in the last cell in the table. Figure 6.8 presents the quarterly total revenues as stacked bar charts. Each segment in a stack of bars represents the revenue contribution from a particular project. The total revenues for the four projects over the four quarters are shown in a pie chart in Figure 6.9. The percentage of the overall revenue contributed by each project is also shown on the pie chart.

Average Revenue

Average is one of the most used measures in data analysis. Given n observations (e.g., revenues), X_1, X_2, X_3, ..., X_n, the average of the observations is computed as

$$\bar{X} = \frac{\sum_{i=1}^{n} X_i}{n}$$

$$= \frac{T_x}{n}$$

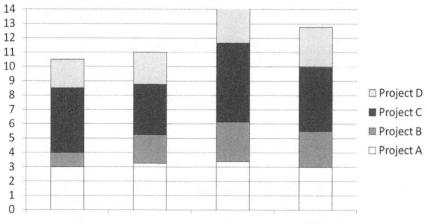

FIGURE 6.8
Stacked bar graph of quarterly total revenues.

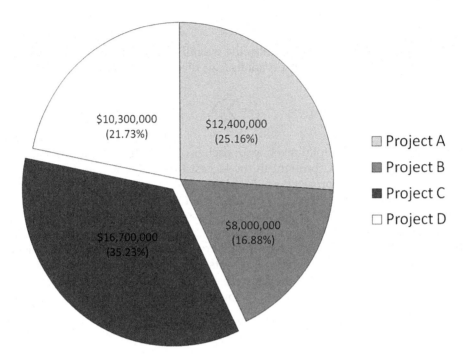

FIGURE 6.9
Pie chart of total revenue per project.

where T_x is the sum of n revenues. For our sample data, the average quarterly revenues for the four projects are computed as follows:

$$\bar{X}_A = \frac{(3,000+3,200+3,400+2,800)(\$1,000)}{4} = \$3,100,000.00$$

$$\bar{X}_B = \frac{(1,200+1,900+2,500+2,400)(\$1,000)}{4} = \$2,000,000.00$$

$$\bar{X}_C = \frac{(4,500+3,400+4,600+4,200)(\$1,000)}{4} = \$4,175,000.000$$

$$\bar{X}_D = \frac{(2,000+2,500+3,200+2,600)(\$1,000)}{4} = \$2,575,000.00$$

Similarly, the expected average revenues per project for the four quarters are computed as follows:

$$\bar{X}_1 = \frac{(3,000+1,200+4,500+2,000)(\$1,000)}{4} = \$2,675,000.00$$

$$\bar{X}_2 = \frac{(3,200+1,900+3,400+2,500)(\$1,000)}{4} = \$2,750,000.00$$

$$\bar{X}_3 = \frac{(3{,}400 + 2{,}500 + 4{,}600 + 3{,}200)(\$1{,}000)}{4} = \$3{,}425{,}000.00$$

$$\bar{X}_4 = \frac{(2{,}800 + 2{,}400 + 4{,}200 + 2{,}600)(\$1{,}000)}{4} = \$3{,}000{,}000.00$$

The aforementioned values are shown in a bar chart in Figure 6.10.

The average revenue from any of the four projects in any given quarter is calculated as the sum of all the observations divided by the number of observations. That is,

$$\bar{X} = \frac{\displaystyle\sum_{i=1}^{N}\sum_{j=1}^{M} X_{ij}}{K}$$

where
N is the number of projects
M is the number of quarters
K is the total number of observations $(K = NM)$

The overall average per project per quarter is computed as follows:

$$\bar{\bar{X}} = \frac{\$47{,}400{,}000}{16} = \$47{,}400.00$$

As a cross-check, the sum of the quarterly averages should be equal to the sum of the project revenue averages, which is equal to the grand total divided by 4.

$$(2{,}675 + 2{,}750 + 3{,}425 + 3{,}000)(\$1{,}000) = (3{,}100 + 2{,}000 + 4.175 + 2{,}575)(\$1{,}000)$$

$$= \$11{,}800.000 = \frac{\$47{,}400{,}000}{4}$$

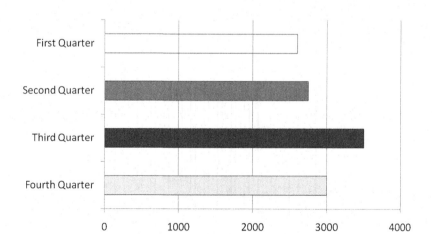

FIGURE 6.10
Average revenue per project for each quarter.

The cross-check procedure described earlier works because we have a balanced table of observations. That is, we have four projects and four quarters. If there were only three projects, for example, the sum of the quarterly averages would not be equal to the sum of the project averages.

Median Revenue

The median is the value that falls in the middle of a group of observations arranged in order of magnitude. One-half of the observations are above the median and the other half are below the median. The method of determining the median depends on whether or not the observations are organized into a frequency distribution. For unorganized data, it is necessary to arrange the data in an increasing or decreasing order before finding the median. Given K observations (e.g., revenues), X_1, X_2, X_3, ..., X_K, arranged in increasing or decreasing order, the median is identified as the value in position $(K+1)/2$ in the data arrangement, if K is an odd number. If K is an even number, then the average of the two middle values is considered to be the median. If the sample data are arranged in increasing order, we would get the following:

$$1200, 1900, 2000, 2400, 2500, 2500, 2600, 2800, 3000, 3200, 3200, 3400, 3400, 4200,$$
$$4500, 4600$$

The median is then calculated as (2,800+3,000)/2 = 2,900. Half of the recorded revenues are expected to be above \$2,900,000 while half are expected to be below that amount. Figure 6.11 presents a bar chart of the revenue data arranged in increasing order. The median is anywhere between the eighth and ninth values in the ordered data.

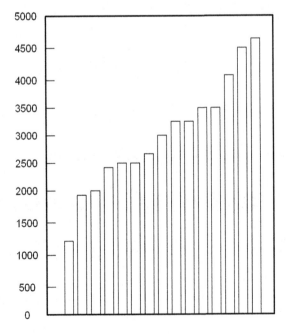

FIGURE 6.11
Bar chart of ordered data.

Quartiles and Percentiles

The median is a position measure, because its value is based on its position in a set of observations. Other measures of position are *quartiles* and *percentiles*. There are three quartiles that divide a set of data into four equal categories. The first quartile, denoted Q_1, is the value below which one-fourth of all the observations in the data set fall. The second quartile, denoted, Q_2, is the value below which two-fourths or one-half of all the observations in the data set fall. The third quartile, denoted Q_3, is the value below which three-fourths of the observations fall. The second quartile is identical to the median. It is technically incorrect to talk of the fourth quartile, because that will imply that there is a point within the data set below which all the data points fall: a contradiction! A data point cannot lie within the range of observations and at the same time exceed all the observations, including itself.

The concept of percentiles is similar to the concept of quartiles, except that reference is made to percentage points. There are 99 percentiles that divide a set of observations into 100 equal parts. The X percentile is the value below which X percent of the data fall. The 99 percentile refers to the point below which 99% of the observations fall. The three quartiles discussed previously are regarded as the 25th, 50th, and 75th percentiles. It would be technically incorrect to talk of the 100 percentile. In performance ratings, such as on an examination or product quality rating, the higher the percentile of an individual or product, the better the performance. In many cases, recorded data are classified into categories that are not indexed to numerical measures. In such cases, other measures of central tendency or position will be needed. An example of such a measure is the mode.

Mode

The mode is defined as the value that has the highest frequency in a set of observations. When the recorded observations can be classified into categories, the mode can be particularly helpful in describing the data. Given a set of K observations (e.g., revenues), X_1, X_2, X_3, ..., X_K, the mode is identified as the value that occurs more than any other value in the set. Sometimes, the mode is not unique in a set of observations. For example, in Table 6.2, $2,500, $3,200, and $3,400 all have the same number of occurrences. Each of them is a mode of the set of revenue observations. If there is a unique mode in a set of observations, then the data are said to be unimodal. The mode is very useful in expressing the central tendency for observations with qualitative characteristics such as color, marital status, or state of origin. The three modes in the raw data can be identified in Figure 6.11.

Range of Revenue

The range is determined by the two extreme values in a set of observations. Given K observations (e.g., revenues), X_1, X_2, X_3, ..., X_K, the range of observations is simply the difference between the lowest and highest observations. This measure is useful when the analyst wants to know the extent of extreme variations in a parameter. The range of revenues in our sample data is ($4,600,000 – $1,200,000) = $3,400,000. Because of its dependence on only two values, the range tends to increase as the sample size increases. Furthermore, it does not provide a measurement of the variability of the observations relative to the center of the distribution. This is why the standard deviation is normally used as a more reliable measure of dispersion than the range.

The variability of a distribution is generally expressed in terms of the deviation of each observed value from the sample average. If the deviations are small, the set of data are said to have low variability. The deviations provide information about the degree of dispersion in a set of observations. A general formula to evaluate the variability of data cannot be based on deviations. This is because some of the deviations are negative while some are positive and the sum of all the deviations is equal to zero. One possible solution to this is to compute the average deviation.

Average Deviation

The average deviation is the average of the absolute values of the deviations from the sample average. Given K observations (e.g., revenues), X_1, X_2, X_3, ..., X_K, the average deviation of the data is computed as follows:

$$\bar{D} = \frac{\sum\limits_{i=1}^{K} |X_i - \bar{X}|}{K}$$

Table 6.3 shows how the average deviation is computed for our sample data. One aspect of the average deviation measure is that the procedure ignores the sign associated with each deviation. Despite this disadvantage, its simplicity and ease of computation make it useful. In addition, knowledge of the average deviation helps in understanding the standard deviation, which is the most important measure of dispersion available.

TABLE 6.3

Computation of Average Deviation, Standard Deviation, and Variance

Observation Number (i)	Recorded Observation X_i	Deviation from Average $X_i - \bar{X}$	Absolute Value $|X_i - \bar{X}|$	Square of Deviation $(X_i - \bar{X})^2$
1	3,000	37.5	37.5	1,406.25
2	1,200	−1,762.5	1,762.5	3,106,406.30
3	4,500	1,537.5	1,537.5	2,363,906.30
4	2,000	−962.5	962.5	926,406.25
5	3,200	237.5	237.5	56,406.25
6	1,900	−1,062.5	1,062.5	1,128,906.30
7	3,400	437.5	437.5	191,406.25
8	2,500	−462.5	462.5	213,906.25
9	3,400	437.5	437.5	191,406.25
10	2,500	−462.5	462.5	213,906.25
11	4,600	1,637.5	1,637.5	2,681,406.30
12	3,200	237.5	237.5	56,406.25
13	2,800	−162.5	162.5	26,406.25
14	2,400	−562.5	562.5	316,406.25
15	4,200	1,237.5	1,237.5	1,531,406.30
16	2,600	−362.5	362.5	131,406.25
Total	47,400.0	0.0	11,600.0	13,137,500.25
Average	2,962.5	0.0	725.0	821,093.77
Square root				906.14

Sample Variance

Sample variance is the average of the squared deviations computed from a set of observations. If the variance of a set of observations is large, the data are said to have a large variability. For example, a large variability in the levels of productivity of a project team may indicate a lack of consistency or improper methods in the project functions. Given K observations (e.g., revenues), X_1, X_2, X_3, ..., X_K, the sample variance of the data is computed as

$$s^2 = \frac{\sum_{i=1}^{K}(X_i - \bar{X})^2}{K-1}$$

The variance can also be computed by the following alternate formulas:

$$s^2 = \frac{\sum_{i=1}^{K}\left(X_i^2 - (1/K)\right)\left[\sum_{i=1}^{K}X_i\right]^2}{K-1}$$

$$s^2 = \frac{\sum_{i=1}^{K}X_i^2 - K\left(\bar{X}^2\right)}{K-1}$$

Using the first formula, the sample variance of the data in Table 6.3 is calculated as

$$s^2 = \frac{13{,}137{,}500.25}{16-1} = 875{,}833.33$$

The average calculated in the last column of Table 6.3 is obtained by dividing the total for that column by 16 instead of $16 - 1 = 15$. That average is not the correct value of the sample variance. However, as the number of observations gets very large, the average as computed in the table will become a close estimate for the correct sample variance. Analysts make a distinction between the two values by referring to the average calculated in the table as the population variance when K is very large and referring to the average calculated by the previous formulas as the sample variance particularly when K is small. For our example, the population variance is given by

$$\sigma^2 = \frac{\sum_{i=1}^{K}(X_i - \bar{X})^2}{K}$$

$$= \frac{13{,}137{,}500.25}{16} = 821{,}093.77$$

while the sample variance, as shown previously for the same data set, is given by

$$\sigma^2 = \frac{\sum_{i=1}^{K}\left(X_i - \bar{X}\right)^2}{K-1}$$

$$= \frac{13{,}137{,}500.25}{(16-1)} = 875{,}833.33$$

Standard Deviation

The sample standard deviation of a set of observations is the positive square root of the sample variance. The use of variance as a measure of variability has some drawbacks. For example, the knowledge of the variance is helpful only when two or more sets of observations are compared. Because of the squaring operation, the variance is expressed in square units rather than the original units of the raw data. To get a reliable feel for the variability in the data, it is necessary to restore the original units by performing the square root operation on the variance. This is why standard deviation is a widely recognized measure of variability. Given K observations (e.g., revenues), $X_1, X_2, X_3, \ldots, X_K$, the sample standard deviation of the data is computed as

$$s = \sqrt{\frac{\sum_{i=1}^{K}\left(X_i - \bar{X}\right)^2}{K-1}}$$

As in the case of the sample variance, the sample standard deviation can also be computed by the following alternate formulas:

$$s = \sqrt{\frac{\sum_{i=1}^{K}X_i^2 - (1/K)\left[\sum_{i=1}^{K}X_i\right]^2}{K-1}}$$

$$s = \sqrt{\frac{\sum_{i=1}^{K}X_i^2 - K\left(\bar{X}\right)^2}{K-1}}$$

Using the first formula, the sample standard deviation of the data in Table 6.3 is calculated as follows:

$$s = \sqrt{\frac{13{,}137{,}500.25}{(16-1)}} = \sqrt{875{,}833.33} = 935.8597$$

We can say that the variability in the expected revenue per project per quarter is $935,859.70. The population sample standard deviation is given by

$$\sigma = \sqrt{\frac{\sum_{i=1}^{K}\left(X_i - \bar{X}\right)^2}{K}}$$

$$= \sqrt{\frac{13{,}137{,}500.25}{16}} = 906.1423$$

while the sample standard deviation is given by

$$s = \sqrt{\frac{\sum_{i=1}^{K} \left(X_i - \bar{X} \right)^2}{K-1}}$$

$$= \sqrt{\frac{13,137,500.25}{(16-1)}} = 935.8597$$

The results of data analysis can be reviewed directly to determine where and when project control actions may be needed. The results can also be used to generate control charts as discussed in the next section.

Control Charts

Control charts may be used to track project performance before deciding what control actions are needed. Control limits are incorporated into the charts to indicate when control actions should be taken. Multiple control limits may be used to determine various levels of control points. Control charts may be developed for various aspects such as cost, schedule, resource utilization, performance, and other criteria for project evaluation.

Figure 6.12 represents a case of periodic monitoring of project progress. Cost is monitored and recorded on a monthly basis. If the cost is monitored on a more frequent

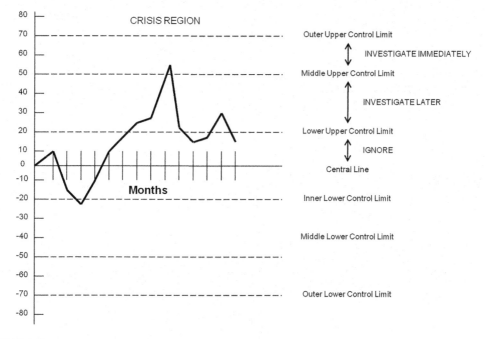

FIGURE 6.12
Control chart for project monitoring.

basis (e.g., days), then we may be able to have a more rigid control structure. Of course, one will need to decide whether the additional time needed for frequent monitoring is justified by the extra level of control provided. The control limits may be calculated with the same procedures used for X-bar and R charts in quality control or they may be based on custom project requirements. In addition to drawing control charts for cost, we can also draw control charts for other measures of performance such as task duration, quality, or resource utilization.

Figure 6.13 shows a control chart for cumulative cost. The control limits on the chart are indexed to the project percent complete. At each percent complete point, there is a control limit that the cumulative project cost is not expected to exceed. A review of the control chart shows that the cumulative cost is out of control at the 10%, 30%, 40%, 50%, 60%, and 80% completion points. The indication is that control actions should be instituted right from the 10% completion point. If no control action is taken, the cumulative cost may continue to be out of control and eventually exceed the budget limit by the time the project is finished.

The information obtained from the project monitoring capabilities of project management software can be transformed into meaningful charts that can quickly identify when control actions are needed. A control chart can provide information about resource overallocation as well as unusually slow progress of work. Figure 6.14 presents a control chart for budget tracking and control. Starting at an initial level of $B, the budget is depleted in one of the three possible modes. In case 1, the budget is depleted in such a way that a surplus

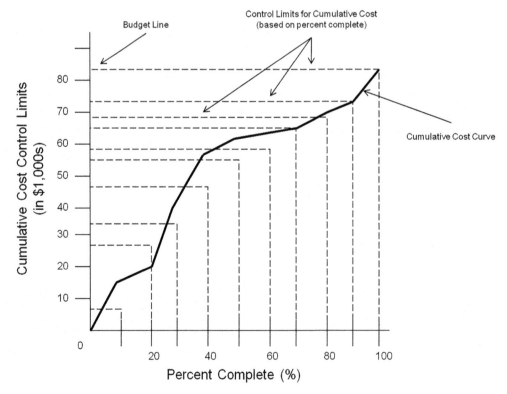

FIGURE 6.13
Control chart for cumulative cost.

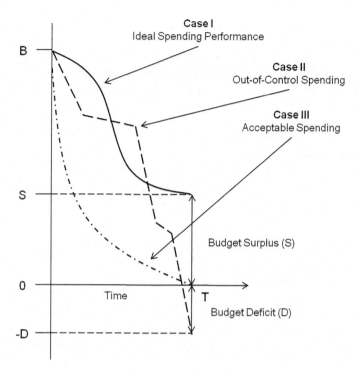

FIGURE 6.14
Chart for budget tracking and control.

is available at the end of the project cycle. This may be viewed as the ideal spending pattern. In case 2, the budget is depleted in an out-of-control pattern that leaves a deficit at the end of the project. In case 3, the budget is depleted at a rate that is proportional to the amount of work completed. In this case, the budget is zeroed out just as the project finishes. Intermediate control lines may be included in the chart so that the needed control actions can be initiated at appropriate times.

Figure 6.15 presents a chart for monitoring revenues versus expenses in a project. As explained earlier, control limits may be added to the chart to determine when control actions should be taken.

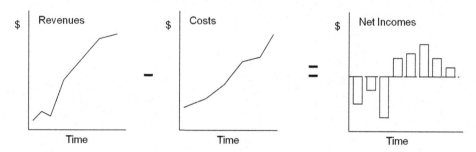

FIGURE 6.15
Review chart for revenue versus expenses.

Table 6.4 shows a report format for task progress analysis. The first three columns identify the task, its location, and its description. The activities column indicates the activities that make up the task. The next column would indicate the completed activities. Pending activities and past due activities are to be indicated in the next two columns. The last column is intended to display comments about the problems encountered. This control table helps to focus control actions on specific activities in addition to the overall project control actions.

Table 6.5 presents a format for task-based time analysis. The table has the additional feature of showing percent completion for each task. It also evaluates planned activities versus actual activities performed. Deviations from planned work should be explained.

Figure 6.16 shows a sketch of a graphical report on task progress. This bar chart analysis would generate reports showing expected completion and actual completion for task numbers, departments, or project segment. Figure 6.17 presents a chart that shows project

TABLE 6.4

Task Analysis Table

Column 1	Column 2	Column 3	Column 4	Column 5	Column 6	Column 7	Column 8
Task #	Department	Description	Activities	Completed	Pending	Past Due	Comments

TABLE 6.5

Task-Based Time Analysis Table

Column 1	Column 2	Column 3	Column 4	Column 5	Column 6	Column 7	Column 8	Column 9
Task #	Department	Description	Expected Activity	Expected % Completed	Actual Activity	Actual % Completed	Deviation	Explain

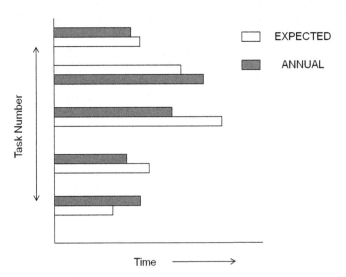

FIGURE 6.16
Graphical report on task progress.

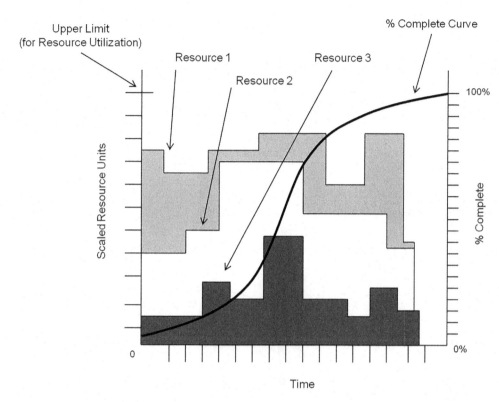

FIGURE 6.17
Resource loading versus project progress.

progress versus resource loading. This is useful for identifying how the percent comple-tion of a project is affected by the level of resource allocation.

Better performance can be achieved if more time and resources are available for a project. If lower costs and tighter schedules are desired, then performance may have to be compro-mised and vice versa. From the point of view of the project manager, the project should be at the highest point along the performance axis. Of course, this represents an extreme case of getting something for nothing. From the point of view of the project staff, the project should be at the point indicating highest performance, longest time, and most resources. This, of course, may be an unrealistic expectation since time and resources are typically in short supply. For project control, a feasible trade-off strategy must be developed.

Even though the control boundary is represented by a flat surface in Figure 6.1 earlier, it is obvious that the surface of the box will not be flat. If a multifactor mathematical model is developed for the three factors, the nature of the response surface will vary depending on the specific interactions of the factors for any given project. An example of a project performance response surface is presented in Figure 6.18. The desired trade-offs between the factors in the plot will help determine when and where control actions are required.

If we consider only two of the three constraints at a time, we can study their respective relationships better. Figure 6.19 shows some potential two-factor relationships. In the first plot, performance is modeled as the dependent variable, while cost is the indepen-dent variable. Performance increases as cost increases up to a point where performance

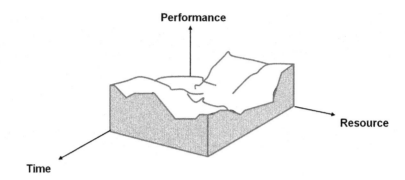

FIGURE 6.18
Project performance response surface.

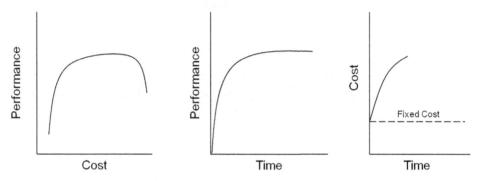

FIGURE 6.19
Trade-off relationships between project constraints.

levels off. If cost is allowed to continue to increase, performance eventually starts to drop. In the second plot, performance is modeled as being dependent on time. The more time that is allowed for a project, the higher the expected performance up to a point where performance levels off. In the third plot, cost depends on time. As project duration increases, cost increases. The increases in cost may be composed of labor cost, raw material cost, and/or cost associated with decreasing productivity. Note that there may be a fixed cost associated with a project even when a time schedule is not in effect. This is seen in the third plot.

Figure 6.20 shows an alternate time–cost trade-off relationship. In this case, the shorter the desired project duration, the higher the cost of the project. If more time is available for the project, then cost can be reduced. However, there is a limit to the possible reduction in cost. After some time, the cost function turns upward due to the increasing cost of keeping the workforce and resources tied up on the project for a long period of time. The most cost-effective duration of the project corresponds to the point where the lowest cost is shown in the figure.

The basic data analysis presented in this section can play a significant role in conveying quick information about project requirements and performance so that prompt decisions can be made. The next section presents the use of statistical analysis for project control when project parameters are subject to variabilities.

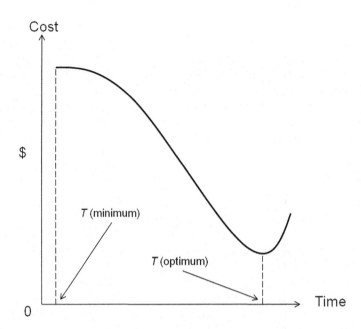

FIGURE 6.20
Time control versus cost control.

Statistical Analysis for Project Control

Statistical control of a project requires that we recognize where risk and uncertainty exist in the project. Variability is a reality in project management. Uncertainty refers to the inability to predict the future accurately. Risk deals with the probability that something will happen and the loss that will result if the thing does happen. Risk and uncertainty may affect several project parameters, including resource availability, activity durations, personnel productivity, budget, weather conditions, equipment failures, and cost. Statistical project control uses the techniques of probability and statistics to assess project performance and determine control actions. The different types of statistics relevant for project control are discussed next.

Descriptive statistics. Descriptive statistics refers to analyses that are performed to describe the nature of a process or operation. The analyses presented in the previous section fall under the category of descriptive statistics, because they are concerned with summary calculations and graphical displays of observations.

Inferential statistics. Inferential statistics refers to the process of drawing inferences about a process based on a limited observation of the process. The techniques presented in this section fall under the category of inferential statistics. Inferential statistics is of interest because it is dynamic and provides generalizations about a population by investigating only a portion of the population. The portion of the population investigated is referred to as a sample. As an example, the expected duration of a proposed task can be inferred from several previous observations of the durations of identical tasks.

Deductive statistics. Deductive statistics involves assigning properties to a specific item in a set based on the properties of a general class covering the set. For example, if it is known that 90% of projects in a given organization fail, then deduction can be used to assign a probability of 90% to the event that a specific project in the organization will fail.

Inductive statistics. Inductive statistics involves drawing general conclusions from specific facts. Inferences about populations are drawn from samples. For example, if 95% of a sample of 100 people surveyed in a 5,000-person organization favor a particular project, then induction can be used to conclude that 95% of the personnel in the organization favor the project. The different types of statistics play important roles in project control. Sampling is an important part of drawing inferences.

Sampling Techniques

A *sample space* of an experiment is the set of all possible distinct outcomes of the experiment. An *experiment* is some process that generates distinct sets of observations. The simplest and most common example is the experiment of tossing a coin to observe whether heads or tails will show up. An *outcome* is a distinct observation resulting from a single trial of an experiment. In the experiment of tossing a coin, heads and tails are the two possible outcomes. Thus, the sample space consists of only two items.

There are several examples of statistical experiments suitable for project control. A simple experiment may involve checking to see whether it rains or not on a given day. Another experiment may involve counting how many tasks fall behind schedule during a project life cycle. An experiment may involve recording how long it takes to perform a given activity in each of the several trials. The outcome of any experiment is frequently referred to as a *random outcome* because each outcome is independent and has the same chance of occurring. We cannot predict with certainty what the outcome of a particular trial of the experiment would be. An event can be a collection of outcomes.

Sample

A sample is a subset of a population that is selected for observation and statistical analysis. Inferences are drawn about the population based on the results of the analysis of the sample. The reasons for using sampling rather than complete population enumeration are as follows:

1. It is more economical to work with a sample.
2. There is a time advantage to using a sample.
3. Populations are typically too large to work with.
4. A sample is more accessible than the whole population.
5. In some cases, the sample may have to be destroyed during the analysis.

There are three primary types of samples. They differ in the manner in which their elementary units are chosen.

Convenience sample. A convenience sample refers to a sample that is selected on the basis of how convenient certain elements of the population are for observation.

Judgment sample. A judgment sample is the one that is obtained based on the discretion of someone familiar with the relevant characteristics of the population.

Random sample. A random sample refers to a sample whereby the elements of the sample are chosen at random. This is the most important type of sample for statistical analysis. In random sampling, all the items in the population have an equal chance of being selected for inclusion in the sample.

Since a sample is a collection of observations representing only a portion of the population, the way in which the sample is chosen can significantly affect the adequacy and reliability of the sample. Even after the sample is chosen, the manner in which specific observations are obtained may still affect the validity of the results. The possible bias and errors in the sampling process are discussed next.

Sampling error. A sampling error refers to the difference between a sample mean and the population mean that is due solely to the particular sample of elements that are selected for observation.

Nonsampling error. A nonsampling error refers to an error that is due solely to the manner in which an observation is made.

Sampling bias. A sampling bias refers to the tendency to favor the selection of certain sample elements having specific characteristics. For example, a sampling bias may occur if a sample of the personnel is selected from only the engineering department in a survey addressing the implementation of high-technology projects.

Stratified Sampling

Stratified sampling involves dividing the population into classes, or groups, called strata. The items contained in each stratum are expected to be homogeneous with respect to the characteristics to be studied. A random subsample is taken from each stratum. The subsamples from all the strata are then combined to form the desired overall sample. Stratified sampling is typically used for a heterogeneous population such as data on employee productivity in an organization. Under stratification, groups of employees are selected so that the individuals within each stratum are mostly homogeneous and the strata are different from one another. As another example, a survey of project managers on some important issue of personnel management may be conducted by forming strata on the basis of the types of projects they manage. There may be one stratum for technical projects, one for construction projects, and one for manufacturing projects.

A *proportionate stratified sampling* results if the units in the sample are allocated among the strata in proportion to the relative number of units in each stratum in the population. That is, an equal sampling ratio is assigned to all strata in a proportionate stratified sampling. In *disproportionate stratified sampling*, the sampling ratio for each stratum is inversely related to the level of homogeneity of the units in the stratum. The more homogeneous the stratum, the smaller its proportion included in the overall sample. The rationale for using disproportionate stratified sampling is that when the units in a stratum are more homogeneous, a smaller subsample is needed to ensure good representation. The smaller subsample helps reduce sampling cost.

Cluster Sampling

Cluster sampling involves the selection of random clusters, or groups, from the population. The desired overall sample is made up of the units in each cluster. Cluster sampling is different from stratified sampling in that differences between clusters are usually small. In addition, the units within each cluster are generally more heterogeneous. Each cluster, also known as *primary sampling unit*, is expected to be a scaled-down model that gives a good representation of the characteristics of the population.

All the units in each cluster may be included in the overall sample or a subsample of the units in each cluster may be used. If all the units of the selected clusters are included in the overall sample, the procedure is referred to as *single-stage sampling*. If a subsample is taken at random from each selected cluster and all units of each subsample are included in the overall sample, then the sampling procedure is called *two-stage sampling*. If the sampling procedure involves more than two stages of subsampling, then the procedure is referred to as *multistage sampling*. Cluster sampling is typically less expensive to implement than stratified sampling. For example, the cost of taking a random sample of 2,000 managers from different industry types may be reduced by first selecting a sample, or cluster, of 25 industries and then selecting 80 managers from each of the 25 industries. This represents a two-stage sampling that will be considerably cheaper than trying to survey 2,000 individuals in several industries in a single-stage procedure.

Once a sample has been drawn and observations of all the items in the sample are recorded, the task of data collection is completed. The next task involves organizing the raw data into a meaningful format. Frequency distribution is an effective tool for organizing data. Frequency distribution involves the arrangement of observations into classes so as to show the frequency of occurrences in each class. (Guidelines for constructing histograms are presented in the next section.)

Example 6.1

Suppose a set of data are collected about project costs in an organization. Twenty projects are selected for the study. The following observations are recorded in thousands of dollars:

$3,000	$1,100	$4,200	$800	$3000
$1,800	$2,500	$2,500	$1,700	$3,000
$2,900	$2,100	$2,300	$2,500	$1,500
$3,500	$2,600	$1,300	$2,100	$3,600

Table 6.6 shows the tabulation of the cost data as a frequency distribution. Note how the end points of the class intervals are selected such that no recorded data point falls at an end point of a class. Note also that seven class intervals seem to be the most appropriate size for this particular set of observations.

Each class internal has a spread of $500, which is an approximation obtained from the expression presented as follows:

$$W = \frac{X_{max} - X_{min}}{N}$$

$$= \frac{4,200 - 800}{7} = 485.71 \approx 500$$

TABLE 6.6

Frequency Distribution of Project Cost Data

Cost Interval ($)	Midpoint ($)	Frequency	Cumulative Frequency
750–1,250	1,000	2	2
1,250–1,750	1,500	3	5
1,750–2,250	2,000	3	8
2,250–2,750	2,500	5	13
2,750–3,250	3,000	4	17
3,250–3,750	3,500	2	19
3,750–4,250	4,000	1	20
Total		20	

TABLE 6.7

Relative Frequency Distribution of Project Cost Data

Cost Interval ($)	Midpoint ($)	Frequency	Cumulative Frequency
750–1,250	1,000	0.10	0.10
1,250–1,750	1,500	0.15	0.25
1,750–2,250	2,000	0.15	0.40
2,250–2,750	2,500	0.25	0.65
2,750–3,250	3,000	0.20	0.85
3,250–3,750	3,500	0.10	0.95
3,750–4,250	4,000	0.05	1.00
Total		1.00	

Table 6.7 shows the relative frequency distribution. The relative frequency of any class is the proportion of the total observations that fall into that class. It is obtained by dividing the frequency of the class by the total number of observations. The relative frequency of all the classes should add up to 1. From the relative frequency table, it is seen that 25% of the observed project costs fall within the range of $2,250 and $2,750. It is also noted that only 15% (0.10+0.05) of the observed project costs fall in the upper two intervals of project costs.

Figure 6.21 shows the histogram of the frequency distribution for the project cost data. Figure 6.22 presents a plot of the relative frequency of the project cost data. The plot of the cumulative relative frequency is superimposed on the relative frequency plot.

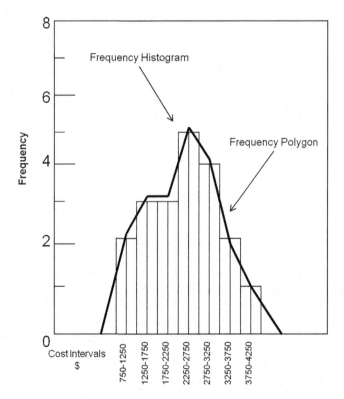

FIGURE 6.21
Histogram of project cost distribution data.

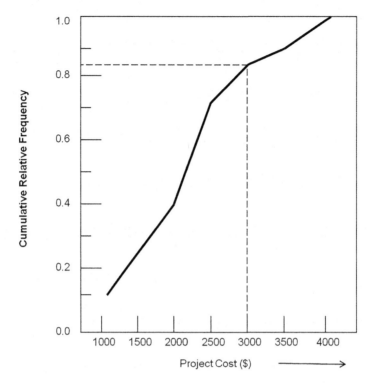

FIGURE 6.22
Plot of cumulative relative frequency.

The relative frequency of the observations in each class represents the probability that a project cost will fall within that range of costs. The corresponding cumulative relative frequency gives the probability that project cost will fall below the midpoint of that class interval. For example, 85% of project costs in this example are expected to fall below or equal to $3,000.

Diagnostic Tools

To facilitate data analysis for diagnosing a project for control purposes, we recommend using available graphical tools. The tools include flowcharts, Pareto diagrams, cause-and-effect diagrams, check sheets, scatter plots, run charts, and histograms. The tools are very effective for identifying problems that may need control actions.

Flowcharts. A flowchart is used to show the steps that a product or service follows from the beginning to the end of the process. It helps locate the value-added parts of the process steps. It also helps in locating the unnecessary steps in the process where unnecessary cost and labor exist. These unnecessary steps can be reduced or permanently eliminated.

Pareto diagram. A Pareto diagram is used to display the relative importance or size of problems to determine the order of priority for projects. It can help identify the projects to concentrate on. For example, in Figure 6.23, analysts may tend to focus on project I, since this is where the greatest dollar loss occurs.

The criticality of a project may be determined by a combination of factors. The selection of project I as the most critical project to focus on should not be made solely on the largest dollar loss alone or any other single criterion. For example, if project I involves

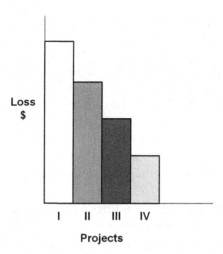

FIGURE 6.23
Relative dollar losses of quality improvement projects.

determining the number of accidents per year and project II involves determining the number of deaths per year, then project II may have priority since focusing on the number of deaths may be more critical than focusing on the number of accidents, even though the frequency of accidents is more than the frequency of deaths.

Cause-and-effect (fishbone) diagram. A fishbone diagram is used to develop a relationship between an effect and all the possible causes influencing it. It is also sometimes called a *tree* or *river* diagram. Figure 6.24 presents an example of a fishbone diagram. The diagram was originally developed for specifying the relationships between a quality characteristic and a set of factors. The diagram is now used for general applications in business and industry.

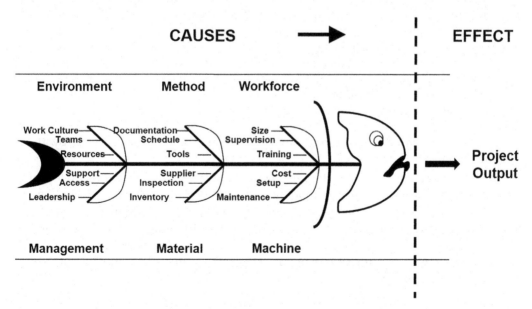

FIGURE 6.24
Fishbone diagram.

The steps for developing a fishbone diagram are as follows:

1. *Step 1*: Determine the characteristic or the response variable to be studied.
2. *Step 2*: Write the characteristic on the right-hand side of a blank sheet of paper. Start with enough room on the paper because the diagram may expand considerably during the evaluation. Enclose the characteristic in a square. Now, write the primary causes which affect the quality characteristic as big branches (or bones). Enclose the primary causes in squares.
3. *Step 3*: Write the secondary causes which affect the big branches as medium-sized branches. Write the tertiary (third-level) causes which affect the medium-sized bones as small bones.
4. *Step 4*: Assign relative importance ratings to the factors. Mark the particularly important factors that are believed to have a significant effect on the characteristic.
5. *Step 5*: Append any necessary written explanation to the diagram.
6. *Step 6*: Review the overall diagram for completeness. While it is important to expand the cause-and-effect relationships as much as possible, avoid cluttering the diagram. For a fishbone diagram to be presented to upper management, limit the contents to a few important details. At the operational level, more details will need to be provided.

Scatter plots. A scatter plot is used to study the relationships between two variables. It is sometimes called an X–Y plot. The plot gives a visual assessment of the location tendencies of data points. The appearance of a scatter plot can help identify the type of statistical analyses that may be needed for the data. For example, in regression analysis, a scatter plot can help an analyst determine the type of models to be investigated.

Run charts and check sheets. A run chart is a tool that can be used to monitor the trends in a process over time. A check sheet is a preprinted table layout that facilitates data collection. Items to be recorded are preprinted in the table. Observations are recorded by simply checking appropriate cells in the table. A check sheet helps to automatically organize data for subsequent analysis. If properly designed, a check sheet can eliminate the need for counting data points during data analysis.

Histogram. A histogram is used to display the distribution of data by organizing the data points into evenly spaced numerical groupings that show the frequency of values in each group. Histograms can be used for quickly assessing the variation and distribution affecting a project. Important guidelines for drawing histograms are as follows:

1. *Step 1*: Determine the minimum and maximum values to be covered by the histogram.
2. *Step 2*: Select a number of histogram classes between 6 and 15. Having too few or too many classes will make it impossible to identify the underlying distribution.
3. *Step 3*: Set the same interval length for the histogram classes such that every observation in the data set falls within some class. The difference between midpoints of adjacent classes should be constant and equal to the length of each interval. If N represents the number of histogram classes, determine the interval length as shown in the following:

$$W = \frac{X_{max} - X_{min}}{N},$$

where X_{min} and X_{max} are the minimum and maximum observations in the data set.

4. *Step 4*: Count the number of observations that fall within each histogram class. This can be done by using a check sheet or any other counting technique.

5. *Step 5*: Draw a bar for each histogram class such that the height of the bar represents the number of observations in the class. If desired, the heights can be converted to relative proportions in which the height of each bar represents the percentage of the data set that falls within the histogram class.

The number of classes should not be small or large such that the true nature of the underlying distribution cannot be identified. Generally, the number of classes should be between 6 and 20. The interval length of each class should be the same. The interval length should be selected such that every observation falls within some class. The difference between midpoints of adjacent classes should be constant and equal to the length of each interval.

A frequency polygon may be obtained by drawing a line to connect the midpoints at the top of the histogram bars. The polygon will show the spread and shape of the distribution of the data set. Three possible patterns of distribution may be revealed by the polygon: *symmetrical*, *positively skewed*, and *negatively skewed*. In a symmetrical distribution, the two halves of the graph are identical. In a positively skewed distribution (skewed to the right), there is a long tail stretching to the right side of the distribution. In a negatively skewed distribution (skewed to the left), there is a long tail stretching to the left side of the distribution.

Probabilistic Decision Analysis

We deal with probability in most of our day-to-day activities. It is important to understand the basic principles of probability analysis for project control. The manager of an outside construction project may reschedule available personnel on the basis of weather forecasts, which are based on probability. Probability refers to the chances of occurrence of an event out of several possible events in a sample space. This is what people often refer to as the *law of averages*. If a coin is tossed a large number of times, say several million times, the proportion of heads tends to be one-half of the total number of tosses. In that case, the number, one-half, is referred to as the probability that heads will occur on one toss of the coin. If ten items with different colors are placed in a jar and one item is pulled out of the jar at random, the probability of pulling out one specific color is one-tenth. Some general facts about probability are

1. Probabilities are real numbers between 0 and 1 inclusive that reflect an individual's belief in the chances of the occurrence of events.

2. A probability value near 0 indicates that the event in question is not expected to occur. However, it does not mean that the event will not occur.

3. A probability near 1 indicates that the event in question is expected to occur. It does not mean that the event will definitely occur.

4. A probability of one-half indicates that the event in question has equal likelihood of occurring or not occurring.

5. The sum of the probabilities of all the mutually exclusive events in a sample space is 1. This is one of the most basic facts of probability. And yet, it is the most violated rule in probability analysis by practitioners.

Normal Distribution

A *probability density function* is a mathematical expression that describes the random behaviors of events in a sample space. Probability density functions are associated with continuous sample spaces where there is an infinite number of possible events. If the number of elements in a sample space is finite or countably finite, then the behavior of events in the sample space would be described by a discrete probability distribution rather than a continuous probability density function. Countably infinite means that there is an unending sequence with as many elements as there are whole numbers. Probability distributions refer to discrete sample spaces, while probability density functions refer to continuous sample spaces.

In most practical problems, continuous random variables represent measured data, such as all possible distances a car can travel, weights, temperatures, and task duration, while discrete random variables represent counted data, such as the number of absent employees on a given day, the number of late jobs in a project, and the amount of dollars available for a particular project. Examples of discrete probability distributions are the binomial distribution, the geometric distribution, and the Poisson distribution. This section presents some of the basic properties of the normal probability density function. Other examples of probability density functions are the exponential probability density function, gamma probability density function, chi-square probability density function, and Weibull probability density function.

The normal probability density function is the most important continuous probability density function in the entire field of statistics. It is often referred to as the normal distribution, normal curve, bell-shaped curve, or Gaussian distribution. This distribution fits many of the physical events in nature, hence, its popularity and wide appeal. The normal distribution is characterized by the following formula:

$$f(x) = \frac{1}{\sqrt{2\pi}\sigma} e^{-1/2\left((x-\mu)/\sigma\right)^2}, \quad -\infty < x < \infty$$

where
  μ is the mean of the distribution
  σ is the standard deviation of the distribution
  $e = 2.71828\ldots$ (a natural constant)
  $\pi = 3.14159\ldots$ (a natural constant)

The bell-shaped appearance of the normal distribution is shown in Figure 6.25.

The values of μ and σ are the parameters that determine the specific appearance (fat, thin, long, short, narrow, or wide) of the normal distribution. Theoretically, the tails of the curve trail on to infinity. When $\mu = 0$ and $\sigma = 1$, the normal distribution is referred to as the *standard normal distribution*.

Most of the analyses involving the normal curve are done in the standardized domain. This is done by using the following transformation expression:

$$Z = \frac{X - \mu}{\sigma}$$

where
  Z is the standard normal random variable
  X is the general normal random variable with a mean of μ and standard deviation of σ

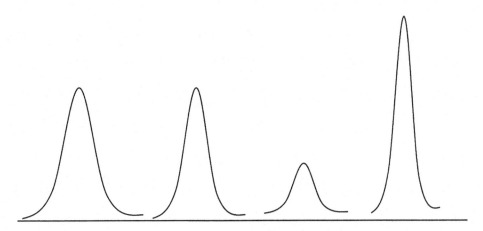

FIGURE 6.25
Shapes of the normal curve.

The variable, Z, is often referred to as the *normal deviate*. One important aspect of the normal distribution relates to the percent of observations within one, two, or three standard deviations. Approximately 68.27% of observations following a normal distribution lie within plus and minus one standard deviation from the mean. Approximately 95.45% of the observations lie within plus or minus two standard deviations from the mean, and approximately 99.73% of the observations lie within plus or minus three standard deviations from the mean. These are shown graphically in Figure 6.26.

To obtain probabilities for particular values of a random variable, it is necessary to know the probability distribution of the random variable. Because of the infinite possible combinations of means and standard deviation values, there are an infinite number of normal distributions. It is quite impractical to try and calculate probabilities directly from each one of them individually. The standard normal distribution can be applied to each and every possible normal random variable by using the transformation expression presented earlier. The standard normal distribution is of great importance in practice, because it can be used to approximate many of the other discrete and continuous random variables.

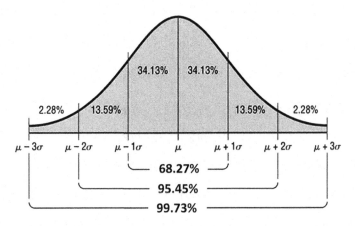

FIGURE 6.26
Areas under the normal curve.

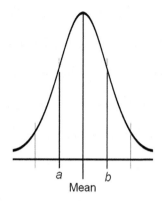

FIGURE 6.27
Probability of an interval under the normal curve.

Because the normal distribution represents a continuous random variable, it is impossible to calculate the probability of a single point on the curve. To determine probabilities, it is necessary to refer to intervals, such as the interval between points *a* and *b*. In Figure 6.27, the area under the curve from *a* to *b* represents the probability that the random variable will lie between *a* and *b*.

That probability is calculated as follows:

Given:

Normal random variable X, representing task duration

Mean of $X = 50$ days

Standard deviation of $X = 10$

Required: The probability that the task duration will lie between 45 and 62 days.

Solution: Let $X_1 = 45$ and $X_2 = 62$

Then

$$z_1 = \frac{45 - 50}{10} = -0.5$$

$$z_2 = \frac{62 - 50}{10} = 1.2$$

Therefore, we have

$$P(45 < X < 62) = P(-0.5 < Z < 0.2)$$

$$= P(Z < 1.2) - P(Z < -0.5)$$

$$= P(Z < 1.2) - [1 - P(Z < 0.5)]$$

$$= 0.884 - (1 - 0.6915) = 0.5764$$

There is a 57.64% chance that this particular task will last between 45 and 62 days. Note that the area under the curve between 45 and 62 is calculated by first finding the total area to the left of 62 (i.e., 0.8849) and then subtracting the area to the left of 45 (i.e., 0.3085). Note also that $P(Z < -0.5)$ can be computed as $1 - P(Z < 0.5)$ for the case where the normal table does not contain negative values of z.

The respective probabilities are read off the normal probability table given in Appendix A. To illustrate the use of the table, let us find the probability that Z will be less than 1.23. First, we locate the value of z equal to 1.2 in the left column of the table and then move across the row to the column under 0.03, where we read the value of 0.8907 inside the body of the table. Thus, $P(Z < 1.23) = 0.8907$. Using a *similar* process, the following additional examples are presented:

$$P(X < 55) = P\left(Z < \frac{55 - 50}{10} \right) = P(Z < 0.5) = 0.6915$$

$$P(X > 65) = P\left(Z > \frac{65 - 50}{10} \right)$$

$$= P(Z > 1.5) = 1 - P(Z < 1.5) = 1 - 0.9332 = 0.0668$$

Note that since the normal distribution table is constructed as cumulative probabilities from the left, $P(Z > 65)$ is calculated as $1 - P(Z < 65)$. Note that

$$P(Z < k) = 1.0, \quad \text{for any value } k \text{ that is greater than 3.5}$$

$$P(Z < 0) = 0.5$$

$$P(Z < k) = 0.0, \quad \text{for any value } k \text{ that is less than } -3.5$$

Decision Trees

Decision tree analysis is used to evaluate sequential decision problems. In project management, a decision tree may be useful in evaluating sequential project milestones. A decision problem under certainty has two elements: *action* and *consequence*. The decision maker's choices are the actions while the results of those actions are the consequences. For example, in critical path method (CPM) network scheduling, the choice of one task among three tasks at a specific time represents a potential action. The consequences of choosing one task over another may be characterized in terms of the slack time created in the network, the cost of performing the selected task, the resulting effect on the project completion time, or the degree to which a specified performance criterion is satisfied.

If the decision is made under uncertainty, as in program evaluation and review technique (PERT) network analysis, a third element is introduced into the decision problem. This third element is defined as *event*. Extending the CPM task selection example to a PERT analysis, the actions may be defined as select task 1, select task 2, and select task 3. The durations associated with the three possible actions can be categorized as "long task duration," "medium task duration," and "short task duration." The actual duration of each task is uncertain. Each task has some probability of exhibiting long, medium, or short durations.

The events can be identified as weather incidents: rain or no rain. The incidents of rain or no rain are uncertain. The consequences may be defined as "increased project completion time," "decreased project completion time," or "unchanged project completion time." However, these consequences are uncertain due to the probabilistic durations of the tasks and the variable choices of the decision maker. That is, the consequences are determined partly by choice and partly by chance. The consequence is dependent on which event, rain or no rain, occurs.

To simplify the decision analysis, the decision elements may be summarized by using a decision table. A decision table indicates the relationship between pairs of decision elements. The decision table for the preceding example is presented in Table 6.8.

In the table, each row corresponds to an event and each column corresponds to an action. The consequences appear as entries in the body of the table. The consequences have been coded as I (increased), D (decreased), U (unchanged). Each event–action combination has a specific consequence associated with it. In some decision problems, the consequences may not be unique. A consequence that is associated with a particular event–action pair may also be associated with another event–action pair. The actions included in the decision table are the only ones that the decision maker wishes to consider. Subcontracting or task deletion could be other possible choices for the decision maker. The actions included in the decision problem are mutually exclusive and collectively exhaustive, so that exactly one will be selected. The events are also mutually exclusive and collectively exhaustive.

The decision problem can also be represented as a decision tree as shown in Figure 6.28.

The tree representation is particularly convenient for decision problems with choices that must be made at different times over an extended period. For example, resource allocation decisions must be made several times during the life cycle of a project. The choice of actions is shown as a fork with a separate branch for each action. The events are also represented by branches in separate forks. To avoid confusion in large decision trees, the nodes for actions are represented by squares while the nodes for events are represented by circles.

The basic guideline for constructing a tree diagram is that the flow should be chronological from left to right. The actions are shown on the initial fork because the decision must be made before the actual event is known. The events are thus shown as branches in the third-stage forks. The consequence resulting from an event–action combination is shown as the end point of the corresponding path from the root of the tree.

The decision tree shows that there are six paths leading to an increase in the project duration, five paths leading to a decrease in project duration, and seven paths leading to an unchanged project duration. For a balanced tree, the total number of paths is given by the expression as follows:

$$P = \prod_{i=1}^{N} n_i$$

TABLE 6.8

Decision Table for Task Selection

| | Actions | | | | | | | | |
| | Task 1 | | | Task 2 | | | Task 3 | | |
Event	Long	Medium	Short	Long	Medium	Short	Long	Medium	Short
Rain	I	I	U	I	U	D	I	I	U
No rain	I	D	D	U	D	D	U	U	U

I = increased project duration; D = decreased project duration; U = unchanged project duration.

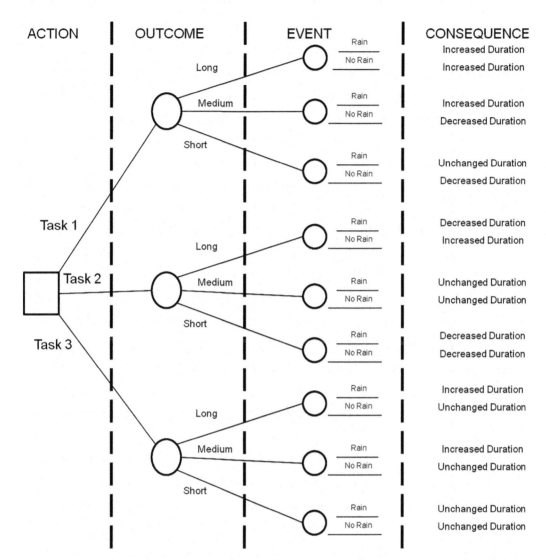

FIGURE 6.28
Decision tree for resource task selection.

where
  P is the total number of paths in the decision tree
  N is the number of decision stages in the tree
  n_i is the number of branches emanating from each node in stage i

The expression is not a general formula, because the number of outcomes for each decision node may not be the same (i.e., unbalanced tree). In Figure 6.28, the number of paths is $P = (3)(3)(2) = 18$ paths. Some of the paths lead to identical consequences even though they are distinct paths. Probability values can be incorporated into the decision structure as shown in Figure 6.29. Note that the selection of a task at the decision node is based on choice rather than probability.

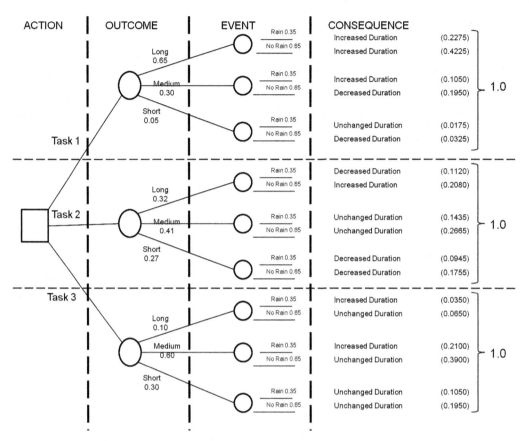

FIGURE 6.29
Probability tree diagram for task selection example.

In this example, it is assumed that the probability of having a particular task duration is independent of whether or not it rains. In some cases, the weather sensitivity of a task may influence the duration of the task. Also, the probability of rain or no rain is independent of any other element in the decision structure. If the items in the probability tree are interdependent, then the appropriate conditional probabilities would need to be computed. This will be the case if the duration of a task is influenced by whether or not it rains. In such a case, the probability tree should be redrawn as shown in Figure 6.30, which indicates that the weather event will need to be observed first before the task duration event can be determined. The conditional probability of each type of duration, given that it rains or does not rain, will need to be calculated.

The respective probabilities of the three possible consequences are shown in Figure 6.29. The probability at the end of each path is computed by multiplying the individual probabilities along the path. For example, the probability of having an increased project completion time along the first path (task 1, long duration, and rain) is calculated as

$$(0.65)(0.35) = 0.2275$$

Similarly, the probability for the second path (task 1, long duration, and no rain) is calculated as

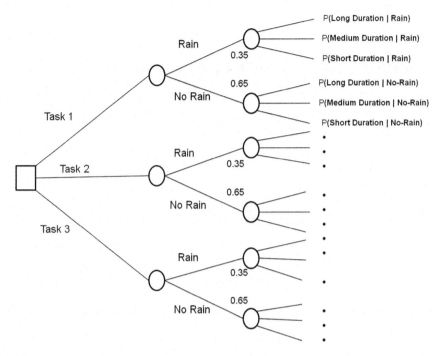

FIGURE 6.30
Probability tree for weather-dependent task durations.

$$(0.65)(0.65) = 0.4225$$

The sum of the probabilities at the end of the paths associated with each action (choice) is equal to 1 as expected. Table 6.9 presents a summary of the respective probabilities of the three consequences based on the selection of each task.

For example, the probability of having an increased project duration when task 1 is selected is calculated as

$$P(\text{increased project duration due to task 1}) = 0.2275 + 0.4225 + 0.105 = 0.755$$

Likewise, the probability of having an increased project duration when task 3 is selected is calculated as

TABLE 6.9

Probability Summary for Project Completion Time

	Selected task					
Consequence	Task 1		Task 2		Task 3	
Increased duration	0.2275 + 0.4225 + 0.105	0.755	0.112	0.112	0.035 + 0.21	0.245
Decreased duration	0.195 + 0.0325	0.2275	0.2665 + 0.0945 + 0.1755	0.5365	0.0	0.0
Unchanged duration	0.0175	0.0175	0.208 + 0.1435	0.3515	0.065 + 0.39 + 0.105 + 0.195	0.755
Sum of probabilities		1.0		1.0		1.0

$$P(\text{increased project duration due to task } 3) = 0.035 + 0.21 = 0.245$$

If the selection of tasks at the first node is probabilistic in nature, then the respective probabilities would be included in the calculation procedure. For example, Figure 6.31 shows a case where task 1 is selected 25% of the time, task 2 is selected 45% of the time, and task 3 is selected 30% of the time.

The resulting end probabilities for the three possible consequences have been revised accordingly. Note that all the probabilities at the end of all the paths add up to 1 in this case. Table 6.10 presents the summary of the probabilities of the three consequences for the case of weather-dependent task durations.

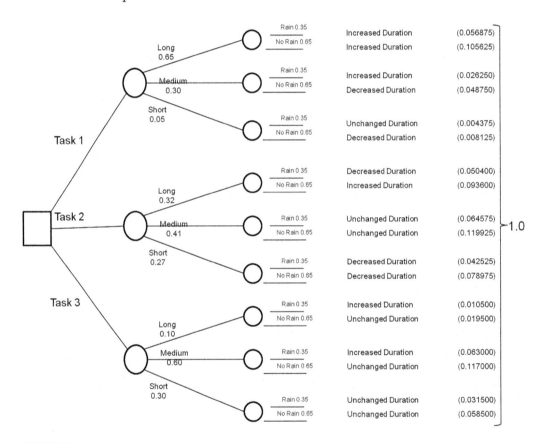

FIGURE 6.31
Modified probability tree for task selection example.

TABLE 6.10

Summary for the Case of Weather-Dependent Task Durations

Consequence	Path Probabilities	Row Total
Increased duration	0.056875 + 0.105625 + 0.02625 + 0.0504 + 0.0105 + 0.063	0.312650
Decreased duration	0.04875 + 0.119925 + 0.042525 + 0.078975	0.290175
Unchanged duration	0.004375 + 0.008125 + 0.0936 + 0.064575 + 0.0195 + 0.117 + 0.0315 + 0.0585	0.397175
	Column Sum	1.0

Example 6.2

As a project manager, Mr. Carl Q. Lator needs to decide which of the three projects to postpone due to unplanned events that took place in his department. The worth of each project depends on the state of the company when the project is completed. He has come up with the benefit matrix shown in Table 6.11.

The probability that the company will be in any of the previous stages is

$$P(\text{above average}) = 0.60$$

$$P(\text{average}) = 0.30$$

$$P(\text{below average}) = 0.10$$

Mr. Carl Q. Lator can spend extra money and time and know more about the future state of the company. For this reason, he would like to know what the value of perfect information is so that he can make a decision as to whether to go along with extra analysis or make a decision now as to which project to postpone. The expected benefit for each project is calculated as

$$\text{For Project } A: E(A) = 85(0.60) + 53(0.30) + 24(0.10) = 69.30$$

$$\text{For Project } B: E(B) = 90(0.60) + 50(0.30) + 32(0.10) = 72.20$$

$$\text{For Project } C: E(C) = 75(0.60) + 70(0.30) + 65(0.10) = 72.50$$

Mr. Carl Q. Lator would have chosen to postpone project A with the previous analysis. Therefore, the total expected benefits from projects B and C would be 72.20 + 72.50 = 144.70. Had he known that the state of the company would be above average (perfect information), then he would have postponed project C and the total benefit from projects A and B would be 175. Similarly, if the state of the company was average, then the total benefits from projects A and C would be 123, and if the state of the company was below average, the total benefit from projects B and C would be 97. Hence, the maximum expected benefit under perfect information is

$$175(0.60) + 123(0.30) + 97(0.10) = 151.60$$

Therefore, the expected value of perfect information is 151.60 − 144.70 = 6.9. At this point, Mr. Carl Q. Lator may decide not to undertake extra analysis since the added benefit may not justify it. Therefore, his final decision may be to postpone project A.

This example illustrates the use of simple probabilistic analysis to determine a decision for project control. Probabilistic and statistical analyses offer a robust approach to evaluating project performance. Measurement, evaluation, and control actions may be

TABLE 6.11

Project Benefit Matrix

	State of the Company		
	Above Average	**Average**	**Below Average**
Project A	85	53	24
Project B	90	50	32
Project C	75	70	65

influenced by probabilistic events. For example, resource allocation decision problems under uncertainty can be handled by appropriate decision tree models. With the statistical approach, the overall function of project control can be improved.

Project Control through Rescheduling

This section presents project control through rescheduling based on project progress. CPM and PERT methods are generally used in the planning phase to enable managers to get an overall picture of how long the project will take, which set of tasks are critical to the completion of the project at the target date, and what the estimated costs will be to run the project. They can also be used to generate actual schedules and to monitor the progress of the project. Actual times and costs can be compared to the scheduled times and costs and actions that need to be taken if large deviations exist between the actual versus scheduled. In this section, we discuss how CPM and PERT can help managers in making decisions related to rescheduling.

As the project progresses, it may be that some activities take a longer time or more resources than expected. If only a few activities are affected, then the effect on the total project completion time or the anticipated cost may be projected without having to recalculate early start times, early finish times, or new resource level requirements, and so on. However, if there have been several changes, then one needs to reevaluate the remainder of the project and maybe reallocate resources or reschedule activities to minimize deviations from the targeted realizations.

Consider the project in Figure 6.32. Suppose a week after the project has started an evaluation of the progress is made. It is found that activities *a, b, c, d,* and *e* have been completed. Activity *i* started 2 days earlier and activities *f* and *h* have not yet started. Also the

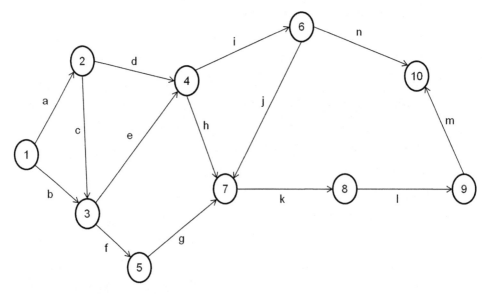

FIGURE 6.32
Original project network for rescheduling example.

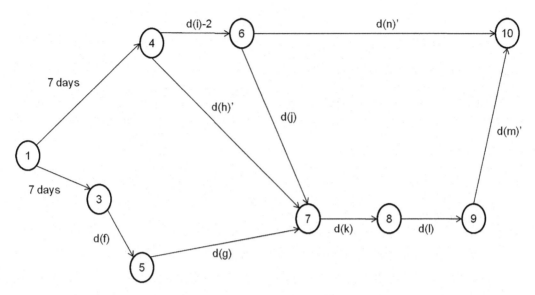

FIGURE 6.33
Modified network for rescheduling.

durations of activities h, m, and n have been modified to $d'(h)$, $d'(m)$, and $d'(n)$. Nodes 2, 3, and 4 were realized at times t_2, t_3, and t_4, respectively.

Figure 6.33 is the modified network that contains the start node, the set of nodes realized with unfinished immediate successors, and the set of nodes not yet realized. The start node is connected to each of the realized nodes with an arc whose duration is equal to the project elapsed time and does not require any resources. The durations of activities h, m, and n are updated. The duration of activity i is decreased by two days, since two days have elapsed since the start of the activity. The modified network can then be used to reschedule the activities with the available methods described earlier.

Experimental Analysis for Project Control

Many project analysis and control activities require data collection, experimentations, and simulation, which necessitate interaction or project personnel with support staff in other units of the organization. This often creates conflicts, power play, operational tension, and difficulties for project teams. Adversarial relationships can be avoided by following project management guidelines for the data collection function itself.

Personnel Interactions for Experimentations

From a systems perspective, it is very important to interact positively with support staff and laboratory technicians in executing project control activities. Technical professionals involved in experimentation spend much of their time in the laboratory or with specialized equipment, often without the direct presence of the project manager. With little or no interpersonal skills, they often get into tense situations with the laboratory technicians.

This is frequently due to the mutual disrespect that both groups have for each other. For example, engineers may view lab technicians as low-tech individuals not familiar with the underlying theories of their work, while the technicians may view engineers as theoreticians with no pragmatic equipment skills.

Laboratory technicians jealously guard their equipment and often restrict access by others. This can seriously impede project experimentation and analysis. Project managers often end up playing both sides of the field, acting as cautious mediators. On the one hand, they support the right of the engineers to have access to lab facilities. On the other hand, they do not want to step on the toes of the lab technicians. The engineers blame the lab for failed or delayed projects. The lab blames engineers for disrupting lab work flow and schedules. The following guidelines should help minimize conflicts in the laboratory:

1. Avoid arrogant approach to other personnel.
2. Recognize that lab technicians are experts at what they do.
3. Respect lab attendants as professionals in their own fields.
4. Schedule experiments in advance with lab technicians.
5. Do not expect the laboratory staff to answer all your questions.
6. Recognize that lab staff are limited in what they can do for you.
7. Develop a friendly (out-of-lab) rapport with laboratory staff.

The Triple C model presented in an earlier chapter can be used to resolve laboratory-based project conflicts. Such conflicts may include technical conflict, power conflict, schedule conflict, cost conflict, expectation conflict, management conflict, priority conflict, resource conflict, and personality conflict. All of these can be alleviated with the Triple C approach through communication, cooperation, and coordination to facilitate cross-system harmony and successful project execution.

Need for Project Experimentation

To solve a problem, you sometimes need more information than you have available. You may need to carry out an experimental program to generate the necessary data or information needed for project decisions. It is important to carry out project experiments efficiently. For example, the objective in developing an experimental model is to make the model as parsimonious as possible. Important questions to address when developing project experimentation strategies include the following:

1. Why is the experiment needed?
2. Can the information to be generated from the experiment be obtained through other means?
3. Are there adequate resources to initiate and carry the experiment through completion?
4. Is the experimental facility available and cooperative?
5. What are the safety considerations for the experiment?
6. Are human and/or animal subjects to be involved in the experiment? If so, what are the institutional safety and approval requirements for such subjects?

Experimental Procedure

1. *Examine the need for the experiment.* Is this a retrospective experiment or prospective experiment? A careful examination of the need and nature of the experiment will effectively guide the rest of the experimental work.

2. *Define objectives for the experiment.* Develop and prioritize the goals you want the experiment to accomplish. Have a focus! Determine the problem to be solved. Specify the experimental criterion (i.e., dependent variable or response variable). The nature of the criterion helps to determine what statistical tests are applicable.

   Is the response measurable?

   How accurate can it be measured?

   What independent variables are involved?

   Are they measurable?

   Can their levels be manipulated?

   Over what range can they be manipulated?

   Are the levels fixed or random?

3. *Identify the variables of interest.* Most experiments will have many variables, some of which will be of main interest while others will be mere concomitant variables. You will need to identify which are of major interest in your proposed experiment.

   Is the variable you want to measure really measurable within your experimental scope and capabilities?

   What are the control variables?

   What are the dependent variables?

   What are the independent variables?
   The three ways to handle independent variables are

   Rigidly controlled (extrapolation not recommended)

   Manipulated (variable is set at desired levels)

   Randomized (levels of variable selected at random)

4. *Identify data categories and classifications.* The type of data involved in your experiment will dictate what you can or cannot do. Different data characteristics may be encountered in an experimental setting. How do you handle or record each type? What are the really important measurements to make and how to make them? What are the expected ranges of measurements? What measurement scales will be used?

   Data can be classified based on their variability and volatility characteristics. Examples of the relevant classifications are transient data, recurring data, static data, and dynamic data. *Transient data* are defined as a volatile set of data that are encountered once during an expert system consultation and is not needed again. Transient data need not be stored in a permanent database unless it may be needed for future analysis or uses.

   Recurring data refer to data that are encountered frequently enough to necessitate storage on a permanent basis. Recurring data may be further categorized into *static* and *dynamic data*. Recurring data that are static will retain their original parameters and values each time they are encountered during an experiment. Recurring data that are dynamic have the potential for taking on different parameters and values each time they are encountered.

For proper data analysis, data should be recorded in such a way that the structure and format are easy to use and understood. This will enable the analyst to identify possible stratifications and populations that may exist in the data.

5. *Design the experiment.* Design of experiment refers to the determination of how observations are taken. Experiments must be properly designed to obtain the maximum amount of information with the minimum expenditure of time and resources. A successfully designed experiment is a series of organized trials that enable you to obtain the most experimental information with the least amount of effort. Pertinent questions for designing experiments are

What is the minimum number of experiments needed to achieve desired experimental confidence?

What are the protocols for experimental replications?

What are the types of errors to avoid?

Type I error is one in which a conclusion is made that a variable has an effect on the experimental outcome, when in fact it really does not.

Type II error occurs when the analyst fails to discover a real effect that exists. Questions addressed by design of experiment are as follows:

How many observations are needed?

Where should the observations be taken?

In what order should the observations be taken?

What variables should be controlled?

What is the randomization procedure?

For example, in an analysis of a family's financial welfare, if a financial analyst blames one spouse's spending habits for the family's poor financial status when, in fact, the poor financial status is due to the low income of the family, then the analyst has committed Type I error. In an industrial setting, Type I error is often referred to as the producer's risk while Type II error is referred to as the consumer's risk. In this sense, the former is viewed as the probability of rejecting a good product from a producer while the latter is the probability of a consumer accepting a bad product.

6. *Carry out the experiment.* The minimum number of experiments that must be performed depends on the number of important independent variables in the experiment and how precisely we can measure the results. If there is some error associated with measuring the outcome of an experiment, then it is prudent to repeat the experimental trials. If an experiment is very precise, it is said to be reproducible, and consequently, with little error or variation from trial to trial. The less reproducible the experiment and the smaller the error we want to detect, the more the data that must be collected.

Reproducibility (operator variation) is the variation in measurements obtained when several operators use one instrument to measure an identical characteristic on the same part.

Repeatability (instrument variation) is the variation in measurements obtained with one instrument when one operator uses it several times to measure an identical characteristic on the same part.

7. *Analyze the results*. Use analytical techniques to analyze the results of the experiment. There are numerous software tools available for experimental data analysis. Software tools permit different views and presentations of the experimental data. Mathematical, graphical, and logarithmic representation of experimental data are some of the analytical approaches you should consider for project experiments.

8. *Write a report on the experiment*. The best way to keep up with experimental results is to document them as they occur. Write your immediate reactions and observations to specific results as they are obtained. These will help you recollect and organize your report later. You should date each item to help with chronological ordering if needed.

9. *Recommend how the results can be used*. Your experimental results are of no use if they cannot be used independently. Describe the intrinsic details of the experimental results as they may be used for decision-making purposes. Operational decisions are based on the interpretation of experimental results.

10. *Follow up on the experiment as appropriate*. Your experiment should not be an end in itself. Consider follow-up experiments or activities that may enhance the utility of the experimental results. Investigate how the experimental work can lead to other nonexperimental type of research.

Types of Experimentation

Variable manipulation and randomization are the basis for experimentation. The aim is often to infer cause-and-effect relationships. *Ex-post-facto experimental research* does not involve experimentation. The variables have already acted or interacted to produce intrinsic results. The researcher only measures, analyzes, and interprets what has occurred. Some important concepts to know are

1. *Regression*: Mathematical equation used for predictive purposes. This is the mathematical model describing the experiment. It expresses the response variable as a function of independent variables.

2. *Correlation*: Determination of strengths of relationships among variables.

3. *Double-blind experiment*: The control group does not know that it is the control group, and the analyst does not know which one is the control group.

4. *Research hypothesis*: A research hypothesis indicates what the experimenter expects to find in the data.

Hypothesis Testing

Statistical inference is usually divided into two categories: *hypothesis testing* and *estimation*. Hypothesis testing involves rejecting or accepting statements about process parameters. Estimation involves estimating the values of process parameters. There are two types of hypotheses: the *null hypothesis* and the *alternate hypothesis*. The null hypothesis is denoted as H_0 while the alternate hypothesis is denoted as H_1.

The null hypothesis is formed primarily to determine whether it can be rejected. It represents a probable statement that is viewed as being correct until it is statistically rejected or accepted. The null hypothesis often involves a statement that a parameter is equal to a specified value. The idea of "no action needed" or "no difference exists" is often conveyed by the null hypothesis. Hence, the name *null hypothesis*, where *null* implies "no action" or "no difference" or "do-nothing."

The implication of rejecting the null hypothesis is that no further action is required either to identify the cause of a difference or to explore the process further. As an example, the statement *innocent until proven guilty* is a judicial null hypothesis that can be stated as follows:

1. H_0: The suspect is innocent
2. H_1: The suspect is guilty

Rejecting or accepting the null hypothesis in the aforementioned example does not confirm or repudiate the suspect's innocence beyond reasonable doubt. Rejection or acceptance is based on available evidence, which may not be enough to infer the truth. Similarly, in a statistical evaluation of a process, rejecting or accepting the null hypothesis does not mean that we have arrived at the final conclusion about the process. The tentative conclusion may be limited by the quality of data (evidence) available.

One-Tailed versus Two-Tailed Hypothesis Testing

A test of hypothesis may be one-tailed or two-tailed based on the direction of the statement contained in the alternate hypothesis. An example of a two-tailed test is as follows:

1. H_0: Process average = 200
2. H_1: Process average $\neq$ 200

In the previous example, H_0 will be rejected if the process average is above or below 200. The one-tailed tests for the example are explained as follows:

1. H_0: Process average = 200
2. H_1: Process average > 200

In this case, H_0 will be rejected only if the process average is above 200.

1. H_0: Process average = 200
2. H_1: Process average < 200

In this case, H_0 will be rejected only if the process average is below 200.

Hypothesis testing can be done to evaluate a single process based on a sample from the process. This is referred to as a one-sample study. If the study is done to compare two processes based on samples drawn from the processes, then the study is referred to as a two-sample study.

The significance level in hypothesis testing is expressed in terms of α. For example, an alpha value of 0.001 ($\alpha = 0.001$) implies taking one chance in 1,000 of rejecting H_0 when it is true—that is, probability of rejecting H_0 when it is true (Type I error risk level). The beta value (β) refers to the probability of accepting H_0 when it is false (Type II error). The power of the test ($1 - \beta$) refers to the probability of rejecting H_0 when it is false. In practical terms, Type II error is more damaging because it involves accepting a false statement. Power of the test may be increased by

1. Increasing the sample size.
2. Increasing the alpha level, α. Increasing α means that the significance level, $1 - \alpha$, will be decreased. Thus, the test will be more effective in detecting differences. The higher the desired significance level, the more difficult it will be for the test to be powerful (or effective).

Increasing (α) means that the significance level ($1 - \alpha$) will be decreased. Thus, the test will be more effective in detecting differences. The tighter the *tolerance*, the more difficult it is for the test to be effective or powerful. *Operating characteristic curve* is a plot of beta versus the population mean (β versus μ). A *power curve* is a plot of power versus population mean ($1 - \beta$ versus μ). Complete randomization averages out time-dependent or factor-dependent effects on an experiment. The experimental process is summarized as follows:

1. Experiment
   a. Statement of problem
   b. Choice of response variable
   c. Selection of factors to be varied
   d. Choice of factor levels
      i. Quantitative versus qualitative
      ii. Fixed versus random
2. Design
   a. Number of observations to be taken
   b. Order of experimentation
   c. Method of randomization to be used
   d. Mathematical model to describe the experiment
   e. Specification of hypothesis to be tested
3. Analysis
   a. Data collection and processing
   b. Computation of test statistics
   c. Interpretation of results

Type I error refers to the rejection of the null hypothesis when it is true. Type I error is normally expressed in terms of a significance level denoted as α. That is,

$$\alpha = P\left(\text{Type I error}\right)$$

$$= P(\text{rejecting } H_0 | H_0 \text{ is true})$$

The significance level for the test of hypothesis is often denoted as

$$\text{Significance level} = 1 - \alpha.$$

As an example, $\alpha = 0.0001$ implies taking once chance in 1,000 of rejecting the null hypothesis when it is true.

Type II error refers to the acceptance of the null hypothesis when it is false. This is denoted as

$$\beta = P(\text{Type II error})$$

$$= P(\text{accepting } H_0 | H_0 \text{ is false})$$

In terms of process improvement studies, there will be an opportunity cost associated with accepting the null hypothesis when it is false. The probability of rejecting the null hypothesis when it is false is referred to as the *power* of the hypothesis test, and it is denoted as

$$\text{Power} = 1 - \beta$$

$$= P(\text{rejecting } H_0 | H_0 \text{ is false})$$

This may be viewed as the discriminating capability of the test. The *p*-value of a statistical test is the smallest α level for which H_0 can be rejected. The objective is to minimize the *p*-value. That is, minimize the probability of Type I error.

Producer's Risk versus the Consumer's Risk

Type II error is often referred to as the *consumer's risk* while Type I error is referred to as the *producer's risk*. Suppose we have a batch or products from a process, and we want to ascertain the quality level of the batch through a test of hypothesis. Rejecting the batch when it is good implies a risk to the producer since a rejected batch never makes it to the market. On the other hand, accepting the batch when it is bad implies a risk to the consumer since there is the potential that a consumer will end up with a bad product. This illustration is summarized in Table 6.12.

Control through Termination

Project termination is an important aspect of project control. Termination should be viewed as a control function since some projects can drag on unnecessarily if control is not instituted. There are several reasons for terminating projects. Some projects are terminated under cordial, arranged, and expected circumstances while others are terminated under unpleasant circumstances that call for managerial control. If necessary, a project audit should be conducted to ascertain the need to terminate a project. Some of the common reasons include the following:

TABLE 6.12

Producer's Risk Versus Consumer's Risk

	Good Batch	Bad Batch
Accept batch	Correct decision $(1 - \alpha)$	Type II Error β (Consumer's risk)
Reject batch	Type I error α (Producer's risk)	Correct decision $(1 - \beta)$

1. Cost overruns
2. Alternate technology
3. Missed deadline
4. Product obsolescence
5. Environmental concern
6. Government requirement
7. Excessive delay penalties
8. Technically impossible goals
9. Lack of project justification
10. Poor performance beyond remedy
11. Alternate objective to the initial plan
12. Project objective accomplished
13. Poor unachievable project plan
14. Lack of required personnel or other resources

Even after the reasons for terminating the project have been identified, actual termination may not be easy to implement, especially for long-range and large projects that have spread their tentacles throughout an organization. Problems of morale may develop. Some workers may have grown accustomed to the extra attention, recognition, or advancement opportunities associated with the project. They may not see the wisdom of terminating the project. The Triple C approach should be used in setting the stage for the termination of a project at the appropriate time. The termination process should cover the following items:

1. Communicating with the personnel on the need for termination
2. Retraining workers for new functions
3. Reassigning workers to other functions
4. Assuring the cooperation of those involved
5. Returning workers to their previous functions
6. Coordinating the required actions for termination
7. Withdrawing funding from the project (pulling the plug)

If the termination is handled properly, workers will be less agonized by the loss and there will be a smooth transition to other projects.

Project Control Verification and Validation

Project system control must be verified and validated before being implemented in a functional setting. Without proper verification and validation, disappointing results can occur. An important reason for performing careful verification is based on the fact that when a system malfunctions, the source of the problem may not be as obvious as in the case of conventional programs. Thus, the problem may go undetected until serious harm has been done to the project or the organization.

What Is Verification?

Verification involves the determination of whether or not the system is functioning as intended. This may involve program debugging, error analysis, input acceptance check, output verification, reasonableness of operation, run-time control, and result documentation.

What Is Validation?

Validation concerns a diagnosis of how closely a system's solution matches expected solution. If the system is valid, then the decisions, conclusions, or recommendations it offers can form the basis for setting actual operating conditions. The validation should be done by using different problem scenarios to simulate actual system operations.

Sometimes, it may be impossible to validate a system for all the anticipated problems, because the data for such problems may not yet be available. In such a case, the closest possible representation of the expected scenarios should be utilized for the validation.

What to Validate

The crucial characteristic components of a system should be identified and used as the basis for validation. For example, if the project involves the development of a software tool, then the knowledge base component of the software is the area which will require the most thorough evaluation since it contains the problem-solving strategies of the software.

How Much to Validate

The methodology that one uses to determine how much validation to perform on a system is dependent on the number of representative cases that are available for evaluation. For example, evaluating a medical knowledge base that diagnoses rare diseases will be much more difficult than evaluating a knowledge base that addresses a common problem of workshop layout. Based on the availability of representative cases, special techniques may need to be used to perform validation (i.e., sensitivity analysis, what-if analysis).

The degree of validation to be performed on a system is dependent on the degree of significance placed on the system. This is based on the context in which the system will be used. In some cases, a system may be viewed as a complete replacement for a human operator. This results in total dependence on the system and a need for greater validation. In other cases, a system may be developed as a complementary tool in problem-solving. This type of use does not require a rigorous validation.

When to Validate

The appropriate time to perform validation is a key decision in any project. Since errors can occur anywhere in the development process from data collection to methodology development, validating a system in stages is very important. Validating a system in stages facilitates catching errors before they become compounded. For very small systems, validation can be performed in one single stage at the completion of the system.

Verification and Validation Stages

In a large system, validation should occur at each stage of the development cycle. Small and medium systems may be validated at only a few selected intervals. Recommended stages for validation are as follows:

1. *Conceptualization stage with overall goal definition*: State what the measures of the project's success will be and how failure or success will be evaluated.

2. *First version prototype showing feasibility*: Demonstrate feasibility of the system and perform preliminary evaluation with a few special test cases.

3. *System refinement*: Evaluate with informal test cases and get feedback from experts, end users, and other stakeholders.

4. *Evaluation of performance*: Perform formalized evaluation using randomly selected data inputs.

5. *Evaluation of acceptability to users*: Evaluate the system in its intended users' surroundings. Verify that the system has good human factors interface (i.e., input/output devices and ease of use).

6. *Evaluation of functionality for extended period in a prototype environment*: Field test and verify the system. Observe performance of the system and reactions of the users.

7. *Pre-implementation evaluation*: Evaluate the overall system before deployment in an operating environment.

Factors Involved in Validation

Several major factors should be examined carefully in the verification and validation stages of a system. The objective is to verify and validate that, for any correct input to the system, a correct output can be obtained. Factors of interest in system validation include

1. *Completeness*: This refers to the thoroughness of the system and checks if the system can address all desired problems within its problem domain.

2. *Efficiency*: Efficiency checks how well the system makes use of the available knowledge, data, hardware, software, and time in solving problems within its specified domain.

3. *Validity*: This involves the correctness of the system outputs. Validity may be viewed as the ability of the system to provide accurate results for relevant data inputs.

4. *Maintainability*: This involves how well the integrity of the system can be preserved even when operating conditions change.

5. *Consistency*: Consistency requires that the system provide similar results to similar problem scenarios.

6. *Precision*: This refers to the level of certainty or reliability associated with the consultations provided by the system. Precision is often application dependent. For example, precision in a medical diagnosis may have more importance than precision in other domains of diagnosis. Compliance with any prevailing rules and regulations is an important component of the precision of a technical system.

7. *Soundness*: Soundness refers to the quality of the scientific and technical basis for the methodology of the system.

8. *Usability*: This involves an evaluation of how the system might meet users' needs. Questions to be asked include the following: Is the system usable by the end user? Are questions worded in an easily understood format? Is help available? Is the system able to explain its reasoning process to the user? Is the system compatible with the delivery environment?

9. *Justification*: A key factor of validation involves justification. A system should be justified in terms of cost requirements, operating characteristics, maintainability, and responsiveness to user requests.

10. *Reliability*: Under reliability evaluation, the system is expected to perform satisfactorily whenever it is used. It should not be subject to erratic performance and results. Several test runs are typically needed to ascertain the reliability of a system.

11. *Accommodating*: To be accommodating, the system has to be very forgiving for minor data entry errors by the user. Appropriate prompts should be incorporated into the user interface to inform users of incorrect data inputs and allow corrections of inputs.

12. *Clarity*: Clarity refers to how well the system presents its prompts to avoid ambiguities in the input/output processes. If the system possesses a high level of clarity, there will be assurance that it will be used as intended by the users.

13. *Quality*: The quality of a system refers to the subjective perception of the user of the system. Quality is often defined as a measure of the user's satisfaction. It refers to the comprehensive combination of the characteristics of a system that determines the system's ability to satisfy specific needs.

14. *Other "ilities"*: Modularity, reconfigurability, interoperability, utility, etc.

How to Evaluate the System

To correctly validate a system based on empirical analysis, the correct results for test cases must be known and accessible. With known results, an absolute measure of the effectiveness of the system may be estimated as the proportion of correct to incorrect results produced by the system. If standard results are not available, then a relative evaluation of the system may be performed on the basis of the performance of other systems designed to perform similar functions. Provided in the following text are some guidelines on how to perform the evaluation process:

1. Set realistic standards for the performance of the system.

2. Define the minimum acceptable standard required for the system to be considered successful.

3. Use performance standards that are comparable to those used in evaluating comparable systems.

4. Use controlled experiments whereby the evaluators are not biased by the sources of the results being evaluated.

5. Distinguish between "false positive" and "true positive" results produced by the expert system. In a false positive result, the system would diagnose as "true" what is not really "true." In true positive results, the system would diagnose as "true" what is really "true."

6. In cases of incorrect results, identify which correct solutions are closest to being reached. This will be valuable in performing a refinement of the system later on.

Sensitivity Analysis for Project Control

To improve the precision of a project control system, sensitivity analysis can be performed by the project team. Sensitivity analysis establishes the variability in the conclusions of the system as a function of the variability of data. That is, we would identify the differences in results that are caused by different levels of changes in the input to the system. If minor changes in the inputs lead to large differences in the result, then the system is said to be sensitive to changes in inputs. One effective method of using sensitivity analysis to improve precision is to display as a histogram output values against possible answers for one given input. Sensitive points will be displayed as significant changes in the histogram. These visual identifications help identify potential trouble spots in the system.

Exercises

6.1 Give one definition of productivity that relates to managerial control.

6.2 Some common impediments to control.

6.3 Prepare a taxonomy of what should be included in a PMIS as measures of control.

6.4 How is the WBS valuable in project monitoring and control?

6.5 How can schedule control be tied in with cost control?

6.6 For each of the causes of control problems listed in this chapter, discuss what corrective actions should be taken.

6.7 List some additional reasons for terminating a project as a measure of control.

6.8 Given the following data for three projects, perform the complete data analysis as was done for Table 6.2. How would you perform the average cross-check in the cell marked "XXX"?

Project	January	February	March	April	Row Total	Row Average
A	3,000	3,200	3,400	2,800		
B	1,200	1,900	2,500	2,400		
C	4,500	3,400	4,600	4,200		
Column total						
Column average						XXX

6.9 For each of the following measurement scales—nominal scale, ratio scale, ordinal scale, and interval scale—list at least five factors or data types associated with project management that can be measured on the scale.

6.10 The employees selected to work on a project are surveyed to select the type of organization structure suitable for the project. Which measure of central tendency would be the most appropriate to determine the preference by the greatest number of employees?

6.11 Use the raw data presented in Exercise 6.8 to verify that $\sum(X - \bar{X}) = 0$.

6.12 Using a software tool, such as a spreadsheet program, compute the average deviation, standard deviation, and variance for the raw data presented in Exercise 6.8.

6.13 What types of data would you recommend to be collected for a project involving the construction of a new soccer playground in a small community?

6.14 The duration of a certain task is known to be normally distributed with a mean of 7 days and a standard deviation of 3 days. Find the following:

a. The probability that the task can be completed in exactly 7 days

b. The probability that the task can be completed in 7 days or less

c. The probability that the task will be completed in more than 6 days

6.15 Alctrex Construction Company is bidding against Betatrex for a building project. Due to past performance of both companies, Alctrex knows that if the company bids a lower or an equal amount, it will win the bid. It will cost Alctrex $9,500 to complete the project. Betatrex's bid is a random variable B with the following probabilities:

$$P(B = \$9,500) = 0.45$$

$$P(B = \$10,500) = 0.35$$

$$P(B = \$12,000) = 0.20$$

Suppose that Alctrex is thinking of bidding between $9,500 and $12,000 in increments of $500. Determine the profit matrix for Alctrex. What is the best decision for the company? What is the value of perfect information?

6.16 Consider the following payoff matrix for a decision-making problem:

	States	
	θ_1	θ_2
Alternative A	100,000	−40,000
Alternative B	50,000	−10,000
Alternative C	0	0

Determine the best alternative as a function of $P(\theta_1) = p$, where $0 \leq p \leq 1$. What is the best alternative for $p = 0.3$? What is the value of perfect information for $p = 0.3$?

6.17 Draw a fishbone diagram for evaluating the causes of deficiencies for each of the following project parameters:

Cost

Schedule

Performance

6.18 The *birthday problem* is a popular problem in probability. The problem involves finding the probability that at least two people have the same birthday in a group of N individuals. An activity scheduling formulation of the problem can be stated as follows: N activities must be scheduled at random during a given scheduling cycle consisting of 200 days. Each activity takes exactly 1 day to complete. The days all have equal likelihood of being selected for any of the activities. Find the probability of a schedule conflict. That is, the probability that two or more activities will be scheduled on the same day ($N \leq 200$). Solve for $N = 32$.

6.19 Suppose N independent candidates are to be scheduled for interviews during 1 year (365 days). Each interview takes exactly 1 day. There is only one interviewer available and only one candidate can be interviewed at a time. Each candidate is requested to specify a preferred date for his or her interview. It is assumed that the candidates pick their interview days at random and independent of one another. Find the probability of having a conflict in scheduling the candidates.

6.20 Develop a computer simulation model to solve the problem in Exercise 6.18. Run the simulation for $N = 1, 2, 3, \ldots, 200$. Plot the probabilities versus the values of N. Discuss your findings.

7

Modeling for Project Optimization

Project Modeling

Schedule optimization is often the major focus in project management. While heuristic scheduling is simple to implement, it does have some limitations. The limitations of heuristic scheduling include subjectivity, arbitrariness, and simplistic assumptions. In addition, heuristic scheduling does not handle uncertainty very well. On the other hand, mathematical scheduling is difficult to apply for practical problems. However, the increasing access to low-cost high-speed computers has facilitated the increased use of mathematical scheduling approaches that yield optimal project schedules. The advantages of mathematical scheduling include the following facts:

1. It provides optimal solutions.
2. It can be formulated to include realistic factors influencing a project.
3. Its formulation can be validated.
4. It has proven solution methodologies.

With the increasing availability of personal computers and software tools, there is very little need to solve optimization problems by hand nowadays. Computerized algorithms are now available to solve almost any kind of optimization problem. What is more important for the project analyst is to be aware of the optimization models available, the solution techniques available, and how to develop models for specific project optimization problems? It is crucial to know which model is appropriate for which problem and to know how to implement optimized solutions in practical settings. The presentation in this chapter concentrates on the processes for developing models for project optimization, as presented by Badiru and Pulat (1995).

General Project Scheduling Formulation

Several mathematical models can be developed for project scheduling problems, depending on the specific objective of interest and the prevailing constraints. One general formulation is

$$\text{Minimize:} \left\{ \underset{\forall i}{\text{Max}} \left\{ s_i + d_i \right\} \right\}$$

$$\text{Subject to: } s_i \geq s_j + d_j \quad \text{for all } i; j \in P_i$$

$$R_k \geq \sum_{i \in A_t} r_{ik} \quad \text{for all } t; \text{for all } k$$

$$s_i \geq 0 \quad \text{for all } i$$

$$r_{ik} \geq 0 \quad \text{for all } i; \text{for all } k$$

where
s_i is the start time of activity i
d_i is the duration of activity i
P_i is the set of activities that must precede activity i
R_k is the availability level of resource type k over the project horizon
A_t is the set of activities ongoing at time t
r_{ik} is the number of units of resource type k required by activity i

The objective of the aforementioned model is to minimize the completion time of the last activity in the project. Since the completion time of the last activity determines the project duration, the project duration is indirectly minimized. The first constraint set ensures that all predecessors of activity i are completed before activity i may start. The second constraint set ensures that resource allocation does not exceed resource availability. The general model may be modified or extended to consider other project parameters. Examples of other factors that may be incorporated into the scheduling formulation include cost, project deadline, activity contingency, mutual exclusivity of activities, activity crashing requirements, and activity subdivision.

An *objective function* is a mathematical representation of the goal of an organization. It is stated in terms of maximizing or minimizing some quantity of interest. In a project environment, the objective function may involve any of the following:

1. Minimize project duration.
2. Minimize project cost.
3. Minimize number of late jobs.
4. Minimize idle resource time.
5. Maximize project revenue.
6. Maximize net present worth.

Linear Programming Formulation

Linear programming (LP) is a mathematical technique for maximizing or minimizing some quantity, such as profit, cost, or time to complete a project. It is one of the most widely used quantitative techniques. It is a mathematical technique for finding the optimum solution to a linear objective function of two or more quantitative decision variables subject to a set of linear constraints. The technique is applicable to a wide range of decision-making problems. Its wide applicability is due to the fact that its formulation is not tied to any particular class of problems, as the critical path method (CPM) and program evaluation and review technique (PERT) techniques are. Numerous research and application studies of LP are available in the literature.

The objective of an LP model is to optimize an objective function by finding values for a set of decision variable subject to a set of constraints. We can define the optimization problem mathematically as

$$\text{Optimize: } z = c_1 x_1 + c_2 x_2 + \cdots + c_n x_n$$

$$a_{11} x_1 + a_{12} x_2 + \cdots + a_{1n} x_n \left\{ \leq, =, \geq \right\} b_1$$

$$a_{21} x_1 + a_{22} x_2 + \cdots + a_{2n} x_n \left\{ \leq, =, \geq \right\} b_2$$

Subject to:

$$\cdots$$
$$\cdots$$

$$a_{m1} x_1 + a_{m2} x_2 + \cdots + a_{mn} x_n \left\{ \leq, =, \geq \right\} b_m$$

$$x_1, x_2, \ldots, x_n \geq 0$$

where
   Optimize is replaced by *maximize* or *minimize* depending on the objective
   z is the value of the objective function for specified values of decision variables
   $x_1, x_2, \ldots, x_n$ are the n decision variables
   $c_1, c_2, \ldots, c_n$ are the objective function coefficients
   $b_1, b_2, \ldots, b_m$ are the limiting values of the resources (*right-hand* side)
   $a_{11}, a_{12}, \ldots, a_{mn}$ are the constraint coefficients (per-unit usage rates)

The word *programming* in LP does not refer to computer programming, as some people think. Rather, it refers to choosing a *program of action*. The word *linear* refers to the *linear relationships* among the variables in the model. The characteristics of LP formulation are explained next.

Quantitative decision variables. A decision variable is a factor that can be manipulated by the decision maker. Examples are number of resource units assigned to a task, number of product types in a product mix, and number of units of a product to produce. Each decision variable must be defined numerically in some unit of measurement.

Linear objective function. The objective function relates to the measure of performance to be minimized or maximized. There is a linear relationship among the variables that make up the objective function. The coefficient of each variable in the objective function indicates its per-unit contribution (positive or negative) toward the value of the objective function.

Linear constraints. Every decision problem is subject to some specific limitations or constraints. The constraints specify the restrictions on how the decision maker may manipulate the decision variables. Examples of decision constraints are capacity limitations, maximum number of resource units available, demand and supply requirements, and number of work hours per day. The relationships among the variables in constraint must be expressed as linear functions represented as equations or inequalities.

Nonnegativity constraint. The nonnegativity constraint is common to all LP problems. This requires that all decision variables are restricted to nonnegative values.

The general procedure for using an LP model to solve a decision problem involves an LP formulation of the problem and a selection of a solution approach. The procedure is summarized as follows:

1. Determine the decision variables in the problem.
2. Determine the objective of the problem.
3. Formulate the objective function as an algebraic expression.
4. Determine the real-world restrictions on the problem scenario.
5. Write each of the restrictions as an algebraic constraint. Make sure that units match throughout the constraints. Otherwise, the terms cannot be added.
6. Select a solution approach. The *graphical method* and the *simplex technique* are the two most popular approaches. The graphical method is easy to apply when the LP model contains just two decision variables. Several computer software packages are available for solving LP problems. Examples are LINDO, LINGO, @RISK, ILOG, Linear Optimizer, LP88, MathPro, What-if Solver, and Turbo-Simplex.

An important aspect of using LP models is the interpretation of the results to make decisions. An LP solution that is optimal analytically may not be practical in a real-world decision scenario. The decision maker must incorporate his or her own subjective judgment when implementing LP solutions. Final decisions are often based on a combination of quantitative and qualitative factors. The examples presented in this chapter illustrate the application of optimization models to project planning and scheduling problems.

Activity Planning Example

Activity planning is a major function in project management. LP can be used to determine the optimal allocation of time and resources to the activities in a project. Suppose a program planner is faced with the problem of planning a 5-day development program for a group of managers in a manufacturing organization. The program includes some combination of four activities: a seminar, laboratory work, case studies, and management games. It is estimated that each day spent on an activity will result in productivity improvement for the organization. The productivity improvement will generate annual cost savings as shown in Table 7.1. The program will last for 5 days, and there is no time lost between

TABLE 7.1

Data for Activity Planning Problem

Activity	Cost Savings ($/Year)	% Active	% Passive	Cost ($/Day)
Seminar	3,200,000	10	90	400
Laboratory work	2,000,000	40	60	200
Case studies	400,000	100	0	75
Management games	2,000,000	60	40	100

activities. To balance the program, the planner must make sure that not more than 3 days is spent on active or passive elements of the program. The active and passive percentages of each activity are also shown in the table. The company wishes to spend at least half a day on each of the four activities. A total budget of \$1,500 is available. The cost of each activity is shown in the tabulated data.

The program planner must determine how many days to spend on each of the four activities. The following variables are defined for the problem:

x_1 represents number of days spent on a seminar

x_2 represents number of days of laboratory work

x_3 represents number of days for case studies

x_4 represents number of days with management games

The objective is to maximize the estimated annual cost savings. That is,

$$\text{Maximize: } f = 3{,}200x_1 + 2{,}000x_2 + 400x_3 + 2{,}000x_4$$

Subject to the following constraints:

1. The program lasts exactly 5 days.

$$x_1 + x_2 + x_3 + x_4 = 5$$

2. Not more than 3 days can be spent on active elements:

$$0.10x_1 + 0.40x_2 + x_3 + 0.60x_4 \leq 3$$

3. Not more than 3 days can be spent on passive elements:

$$0.90x_1 + 0.60x_2 + 0.40x_4 \leq 3$$

4. At least 0.5 day must be spent on each of the four activities:

$$x_1 \geq 0.50$$

$$x_2 \geq 0.50$$

$$x_3 \geq 0.50$$

$$x_4 \geq 0.50$$

5. The budget is limited to \$1,500:

$$400x_1 + 200x_2 + 75x_3 + 100x_4 \leq 1{,}500$$

The complete LP model for the example is presented as follows:

$$\text{Maximize: } x_1 + x_2 + x_3 + x_4 = 5$$

TABLE 7.2

LP Solution to the Activity Planning Example

Activity	Cost Savings ($/Year)	Number of Days	Annual Cost Savings ($)
Seminar	3,200,000	2.20	7,040,000
Laboratory work	2,000,000	0.50	1,000,000
Case studies	400,000	0.50	200,000
Management games	2,000,000	1.80	3,600,000
Total		5	11,840,000

$$0.1x_1 + 0.4x_2 + x_3 + 0.6x_4 \leq 3$$

$$0.9x_1 + 0.6x_2 + 0x_3 + 0.4x_4 \leq 3$$

$$x_1 \geq 0.5$$

$$x_2 \geq 0.5$$

Subject to:

$$x_3 \geq 0.5$$

$$x_4 \geq 0.5$$

$$400x_1 + 200x_2 + 75x_3 + 100x_4 \leq 1500$$

$$x_1, x_2, x_3, x_4 \geq 3$$

The optimal solution to the problem is shown in Table 7.2. Most of the conference time must be allocated to the seminar (2.20 days).

The expected annual cost savings due to this activity is $7,040,000. That is, 2.20 days × $3,200,000/year/day. Management games are the second most important activity. A total of 1.8 days for management games will yield annual cost savings of $3,600,000. Fifty percent of the remaining time (0.5 day) should be devoted to laboratory work, which will result in annual cost savings of $1,000,000. Case studies also require half a day, with a resulting annual savings of $200,000. The total annual savings, if the LP solution is implemented, is $11,840,000. Thus, an investment of $1,500 in management training for the personnel can generate an annual savings of $11,840,000, a huge rate of return on investment!

Resource Combination Example

This example illustrates the use of LP for energy resource allocation. Suppose an industrial establishment uses energy for heating, cooling, and lighting. The required amount of energy is presently being obtained from conventional electric power and natural gas. In recent years, there have been frequent shortages of gas, and there is a pressing need to reduce the consumption of conventional electric power. The director of the energy management department is considering solar energy system as an alternate source of energy. The objective is to find an optimal mix of three different sources of energy to meet the plant's energy requirements. The three energy sources are

1. Natural gas
2. Conventional electric power
3. Solar power

It is required that the energy mix yields the lowest possible total annual cost of energy for the plant. Suppose a forecasting analysis indicates that the minimum kWh (kilowatt hour) needed per year for heating, cooling, and lighting, are 1,800,000, 1,200,000, and 900,000 kWh, respectively. The solar energy system is expected to supply at least 1,075,000 kWh annually. The annual use of conventional electric power must be at least 1,900,000 kWh due to a prevailing contractual agreement for energy supply. The annual consumption of the contracted supply of gas must be at least 950,000 kWh. The cubic foot unit for natural gas has been converted to kWh (1 ft³ of gas = 0.3024).

The respective rates of $6, $3, and $2 per kWh are applicable to the three sources of energy. The minimum individual annual savings desired are $600,000 from solar power, $800,000 from conventional electric power, and $375,000 from natural gas. The savings are associated with the operating and maintenance costs. The energy cost per kWh is $0.30 for conventional electric power, $0.20 for natural gas, and $0.40 for solar power. The initial cost of the solar energy system has been spread over its useful life of 10 years with appropriate cost adjustments to obtain the rate per kWh. The problem data is summarized in Table 7.3. If we let x_{ij} be the kWh used from source i for purpose j, then we would have the data organized as shown in Table 7.4.

The optimization problem involves the minimization of the total cost function, Z. The mathematical formulation of the problem is presented as follows:

$$\text{Minimize: } Z = 0.4 \sum_{j=1}^{3} x_{1j} + 0.3 \sum_{j=1}^{3} x_{2j} + 0.2 \sum_{j=1}^{3} x_{3j}$$

TABLE 7.3

Energy Resource Combination Data

Energy Source	Supply (1,000s kWh)	Savings (1,000s $)	Unit Savings ($/kWh)	Unit Cost ($/kWh)
Solar power	1,075	600	6	0.40
Electric power	1,900	800	3	0.30
Natural gas	950	375	2	0.20

TABLE 7.4

Tabulation of Data for an LP Model

Energy Source	Heating	Type of Use Cooling	Lighting	Constraint
Solar power	X_{11}	X_{12}	X_{13}	≥1,075K
Electric power	X_{21}	X_{22}	X_{23}	≥1,900K
Natural gas	X_{31}	X_{32}	X_{33}	≥950K
Constraint	≥1,800	≥1,200	≥ 900	

TABLE 7.5

LP Solution to the Resource Combination Example

Energy Source	Type of Use		
	Heating	Cooling	Lighting
Solar power	1,075	0	0
Electric power	750	250	900
Natural gas	0	950	0

$$x_{11} + x_{21} + x_{31} \geq 1800$$

$$x_{12} + x_{22} + x_{32} \geq 1200$$

$$x_{13} + x_{23} + x_{33} \geq 900$$

$$6(x_{11} + x_{21} + x_{31}) \geq 600$$

$$3(x_{21} + x_{22} + x_{23}) \geq 800$$

Subject to:

$$2(x_{31} + x_{32} + x_{33}) \geq 375$$

$$x_{11} + x_{21} + x_{31} \geq 1075$$

$$x_{21} + x_{22} + x_{23} \geq 1900$$

$$x_{31} + x_{32} + x_{33} \geq 950$$

$$x_{ij} \geq 0, \quad i,j = 1,2,3$$

Using the LINDO LP computer package, the solution presented in Table 7.5 was obtained. The table shows that solar power should not be used for cooling and lighting if the lowest cost is to be realized. The use of conventional electric power should be spread over the three categories of use. The solution indicates that natural gas should be used for cooling purposes. In pragmatic terms, this LP solution may have to be modified before being implemented on the basis of the prevailing operating scenarios and the technical aspects of the units involved.

Resource Requirements Analysis

Activity–resource assignment combinations provide opportunities for finding the best allocation of resources to meet project goals within the prevailing constraints in the project environment (Badiru, 1993). Suppose a manufacturing project requires that a certain number of workers be assigned to a workstation. The workers produce identical units of the same product. The objective is to determine the number of workers to assign to the workstation in order to minimize the total production cost per shift. Each shift is 8 h long. Each worker can be assigned a variable number of hours and/or variable production rates to work during a shift. Four different production rates are possible: *slow rate, normal rate, fast rate,* and *high-pressure rate.* Each worker is capable of working at any of the production rates during a shift. The total number of work hours available per shift is determined by multiplying the number of workers assigned by the 8h available in a shift.

There are variable costs and percent defective associated with each production rate. The variable cost and the percent defective increase as the production rate increases. At least 450 units of the product must be produced during each shift. It is assumed that the workers' performance levels are identical. The production rates (r_i), the respective costs (c_i), and percent defective (d_i) are presented as follows:

Operating Rate 1 (Slow)
  $r_1 = 10$ units/h
  $c_1 = \$5/h$
  $d_1 = 5\%$
Operating Rate 2 (Normal)
  $r_2 = 18$ units/h
  $c_2 = \$10/h$
  $d_2 = 5\%$
Operating Rate 3 (Fast)
  $r_3 = 30$ units/h
  $c_3 = \$15/h$
  $d_3 = 12\%$
Operating Rate 4 (High Pressure)
  $r_4 = 40$ units/h
  $c_4 = \$25/h$
  $d_4 = 15\%$

LP Formulation

Let x_i represent the number of hours worked at production rate i.

Let n represent the number of workers assigned.

Let u_i represent the number of good units produced at operation rate i.

$$u_1 = (10 \text{ units/h}) \cdot (1 - 0.05) = 9.50 \text{ units/h}$$

$$u_2 = (18 \text{ units/h}) \cdot (1 - 0.08) = 16.56 \text{ units/h}$$

$$u_3 = (30 \text{ units/h}) \cdot (1 - 0.12) = 24.40 \text{ units/h}$$

$$u_4 = (40 \text{ units/h}) \cdot (1 - 0.15) = 34.00 \text{ units/h}$$

Minimize: $z = 5x_1 + 10x_2 + 15x_3 + 25x_4$

$$x_1 + x_2 + x_3 + x_4 \le 8n$$

Subject to: $9.50x_1 + 16.56x_2 + 25.40x_3 + 34.00x_4 \le 450$

$$x_1, x_2, x_3, x_4 \ge 0$$

The solution will be obtained by solving the LP model for different values of n. A plot of the minimum costs versus values of n can then be used to determine the optimum assignment policy. The complete solution is left as an exercise at the end of this chapter.

Integer Programming Approach for Resource Scheduling

Integer programming is a restricted model of LP that permits only solutions with integer values of the decision variables. Suppose we are interested in minimizing the project completion time while observing resource limitations and job precedence relationships. The basic assumption is that once a job starts, it has to be completed without interruption. We can construct several different integer programming models for the problem, all of which will give the same optimal solution whose execution times will differ considerably. An efficient integer programming model for the problem should use as few integer variables as possible.

Define variables as

$$x_{ij}: 1 \text{ if job } i \text{ starts in period } j; \quad 0, \text{ otherwise}$$

$$t_p: \text{ completion time of the project}$$

Only x_{ij}'s are restricted integers. For each job i, one can determine the early start and latest start times, ES_i and LS_i, respectively. Therefore, assuming that there are n jobs in the project, we have $1 \le i \le n$ for each i, then we have $ES_i \le j \le LS_i$. Let t_i denote the duration of job i. Resource availability constraints can be smartly handled by defining a vector V_{ij} that has 0's everywhere except positions $j, j+1, \ldots, j + t_i - 1$, where it has 1's. It indicates the time period where job i uses the resource, assuming that $x_{ij} = 1$. Let r_i and R_j be the resource required by job i and the resource available on day j, respectively. Let R be a row vector containing R_j.

Then, the integer programming model for the scheduling problem with limited resource can be defined as

Minimize t_p

Subject to:

$$\sum_{j=ES_i}^{LS_i} x_{ij} = 1 \quad \forall i = 1, \ldots, n \tag{7.1}$$

$$-\sum_{j=ES_i}^{LS_i} j x_{ij} + \sum_{j=ES_k}^{LS_k} j x_{kj} \le t_i \quad \forall k \in S(i) \tag{7.2}$$

$$\forall i = 1, \ldots, n$$

$$t_p - \sum_{j=ES_i}^{LS_i} jx_{ij} \geq t_i - 1 \quad \forall i \text{ with } S(i) = \emptyset \tag{7.3}$$

$$\sum_{i=1}^{n} \sum_{j=ES_i}^{LS_i} x_{ij} r_i V_{ij} \leq R$$

$$x_{ij} = 0, 1 \qquad \forall i = ES_i, \dots, LS_i \tag{7.4}$$
$$\forall i = 1, \dots, n$$

where $S(i)$ is the set of immediate successor jobs of job i.

Equation 7.1 indicates that each job must start on the same day. Equation 7.2 collectively makes sure that a job cannot start until all of the predecessor jobs are completed. Equation 7.3 determines the project completion time t_p. The project is completed after all the jobs without any successors are completed. The last set of equations makes sure that daily resource requirements are met. The indicator variable x_{ij} is restricted to values 0 and 1.

The previous integer programming model can be solved using LINDO computer code and declaring x_{ij}'s as binary variables. The code uses the branch and bound method of integer programming to solve the problem. As an example, consider the project shown in Figure 7.1. Assume that the daily resource availability is 10 units. Figure 7.2 contains a V_{ij} vector chart for the example. The lines indicate the positions of 1's in the vector. It took a 386/33 MHz machine about 3 min to reach the optimal solution using the LINDO package. An interesting observation is that the optimal solution was found after examining 116 branches of the branch and bound tree. However, it was only after examining 1,609 branches that the method assured optimality of the solution. The optimal solution is illustrated in Figure 7.3.

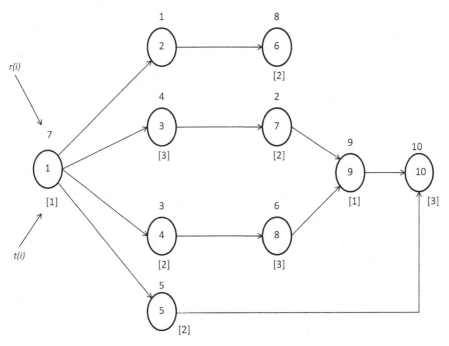

FIGURE 7.1
Example network for integer programming model.

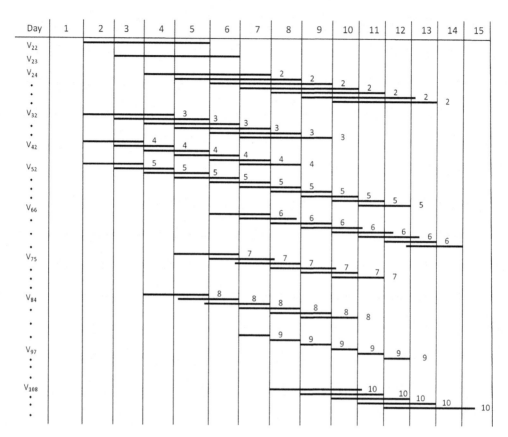

FIGURE 7.2
Vector chart for example.

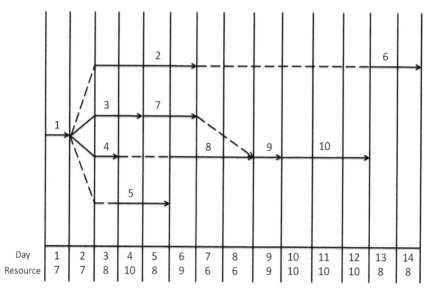

FIGURE 7.3
Optimal solution for integer programming.

Time–Cost Trade-Off Model

For projects involving several activities and complex precedence structure, optimization techniques provide efficient solutions for the time–cost trade-off problem. In fact, one can generate the complete time–cost trade-off curve with such techniques. We will provide the LP model for the time–cost trade-off problem and solve it using a network flow approach. Let a_{ij} be the cost of crashing activity (i, j) by one time unit. The duration of activity (i, j) is denoted by y_{ij}, which is bounded by the crash duration, d_{ij}, and the normal duration, u_{ij}, as shown in Figure 7.4.

If t_i represents realization time of node i, then the precedence relationship implies that

$$t_i = y_{ij} \le t_j \quad \text{for all } (i, j) \in A$$

where A is the arc(activity) set. The preceding inequality states that node j can be realized before y_{ij} time units have elapsed after the realization time of node i.

Given a project completion time T, the optimal activity durations can be found by solving the following LP model.

$$\text{Minimize: } z = \sum_{(i, j) \in A} \left(b_{ij} - a_{ij} y_{ij} \right) \tag{7.5}$$

Subject to:

$$t_i + y_{ij} - t_j \le 0n \quad \forall (i, j) \in A \tag{7.6}$$

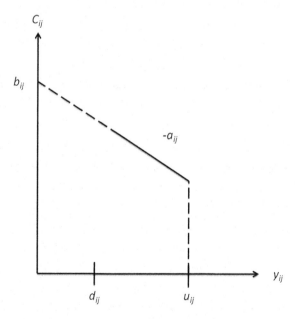

FIGURE 7.4
Linear cost function for example.

$$t_n = T \tag{7.7}$$

$$d_{ij} \le y_{ij} \le u_{ij}, \quad \forall(i,j) \in A \tag{7.8}$$

where t_n denotes the realization time of the end node. By simply decreasing T and solving the previous LP problem, the complete time–cost trade-off curve can be generated. Clearly, $T_{min} \le T \le T_{max}$ where T_{min} and T_{max} are the minimum and maximum achievable project completion times. We will explain a network flow procedure to solve the time–cost trade-off problem without getting into the theoretical details. Minimizing the objective function of Equation 7.5 is equivalent to maximizing $a_{ij} \, y_{ij}$, since b_{ij} is a constant. Given a project network $G = (N, A)$ where N is the node set and A is the arc set, the procedure starts with $y_{ij} = u_{ij}$ for all $(i, j) \infty A$. Node realization (early start) times are calculated using $t_1 = 0$ and the precedence constraint (7.6). At this point, $t_n = T_{max}$. The activities which satisfy (7.6) as equality are called the critical activities and form the critical subgraph, $G' = (N, A')$. The cheapest set of critical arcs are determined using a network flow approach and crashed until either an arc of the set reaches its crash duration or a noncritical activity becomes critical. The crashing decreases t_n and requires an update of G'. The new set of cheapest critical activities is next determined and t_n is crashed further. The procedure continues until there exists a critical path from node 1 to node n with all arcs at their crash durations in which case $t_n = T_{min}$.

We will now explain the maximum flow procedure that can be used to determine the cheapest set of critical arcs to crash.

Maximum Flow Procedure

For $t_n = T_{max}$ set $f_{ij} = 0$, $c_{ij} = a_{ij}$ for all $(i, j) \in A$. Otherwise, take the current f_{ij} and y_{ij} values.

Step 1: Label node 1 with $(\infty, 0)$. In general, label node j from a labeled node i with (q_j, i), where

$$q_j = \begin{cases} c_{ij} - f_{ij} & \text{if } c_{ij} > 0 \text{ and } (i,j) \in A \\ f_{ij} & \text{if } f_{ij} > 0 \text{ and } (j,i) \in A \end{cases}$$

Continue labeling until either node n is labeled or labeling terminates with node n unlabeled (nonbreakthrough). Go to step 2 with $k = n$ if $q_n > 0$ and is finite. Otherwise, stop. The maximum flow has been reached.

Step 2: Node k has label (q_k, i). If $(i, k) \in A$, then $f_{ik} = f_{ik} + q_n$.

Otherwise, $(k, i) \in A$, set $f_{ik} = f_{ik} - q_n$. Set $k = i$ and repeat this step until $k = 1$

The aforementioned procedure locates flow augmenting paths and augments flow of q_n units along the path. The maximum flow is determined when no more flow augmenting paths exist in the network. This is indicated by a nonbreakthrough. Then, the set of labeled nodes form the set X and the set of unlabeled nodes for $\bar{X}$. The cut-set C is defined by

$$C = \left\{(i,j) \in A \mid i \in X, j \in \bar{X} \text{ or } i \in \bar{X}, j \in X\right\}$$

The set of arcs in C define the minimum cut-set separating nodes 1 and n. If u is the value of the maximum flow, then the maximum flow–minimum cut theorem indicates that

$$v = \sum_{i \in x, j \in \overline{X}} c_{ij}$$

In the time–cost trade-off procedure, the arc capacities are determined using a_{ij} values. The maximum flow value corresponds to the increase in cost per-unit decrease in T value. The set of activities to be crashed is given by the set of critical arcs in C oriented from X to $\overline{X}$. As it will be pointed out later, if there exists a critical arc $(i, j) \in C$ with $i \in \overline{X}$ and $j \in X$, the duration of this arc will be lengthened when t_i is reduced.

Time–Cost Trade-Off Procedure

1. *Step 1*: Determination of the critical subnetwork

$$\text{Set } y_{ij} = u_{ij} \quad \text{for all } (i, j) \in A$$
$$\text{Set } t_1 = 0$$
$$\text{Find } t_j = \text{Max}_{(i,j)} \in A\left(t_i + y_{ij}\right) \quad \text{for all } j \in N$$
$$\text{Define } A' = \left\{(i, j) \mid t_j = t_i + y_{ij}\right\}$$

2. *Step 2*: Preparation of the flow network

$$\text{Set flow values, } f_{ij} = 0 \quad \text{for all } (i, j) \in A'$$
$$\text{Set } c_{ij} = 0 \quad \text{for all } (i, j) \in A - A'$$
$$\text{For all } (i, j) \in A', \text{ define}$$

$$c_{ij} = \begin{cases} a_{ij}, & \text{if } y_{ij} = u_{ij} \\ 0, & \text{if } d_{ij} < y_{ij} < u_{ij} \\ \infty, & \text{if } y_{ij} = d_{ij} \end{cases}$$

3. *Step 3*: Use the maximum flow procedure (discussed earlier) to determine the maximum flow. Define

$$C = C_{11} \cup C_{12} \cup C_{21} \cup C_{22}$$

where

$$C_{11} = \left\{(i, j) \in C \cap A' \mid i \in X, j \in \overline{X}\right\}$$

$$C_{12} = \left\{(i, j) \in C \cap A' \mid i \in \overline{X}, j \in X\right\}$$

$$C_{21} = \left\{(i, j) \in C \cap (A - A') \mid i \in X, j \in \overline{X}\right\}$$

$$C_{22} = \left\{(i, j) \in C \cap (A - A') \mid i \in X, j \in X\right\}$$

Calculate

$$\Delta_1 = \min\left(\min_{(i,j)\in C_{11}}\{y_{ij} - d_{ij}\}, \min_{(i,j)\in C_{12}}\{u_{ij} - y_{ij}\}\right)$$

and

$$\Delta_2 = \min_{(i,j)\in C_{21}}\{sij\}, \quad \text{where } s_{ij} = t_j - \left(t_i + y_{ij}\right)$$

where s_{ij} represents the amount of slack for activity (i, j). Let $\Delta = \min(\Delta_1\Delta_2)$.

4. *Step 4*: Let X, $\bar{X}$ denote the set of labeled and unlabeled nodes leading to the non-breakthrough condition, respectively. Update node realization times and activity durations as follows:

$$\text{Set } t_j = t_j - \Delta \quad \text{for all } j \in \bar{X}$$

$$\text{Set } y_{ij} = y_{ij} - \Delta \quad \text{for all } (i, j)C_{11}$$

$$\text{Set } y_{ij} = y_{ij} + \Delta \quad \text{for all } (i, j)C_{12}$$

Update $\mathbf{A}'$ and c_{ij} for all $(i, j) \in \mathbf{A}'$ according to the c_{ij} scale equation presented earlier. Return to step 3 with the updated flow network.

The procedure terminates when step 3 labels node n with $q_n = \infty$. At this point, $T = T_{\min}$.

An application of the procedure to the sample network from Chapter 4 is as follows. Figure 7.5 is an activity-on-node (AOA) representation of Figure 4.2. The critical arcs are shown in Figure 7.6 by heavy arrows.

From Figure 7.6, we have $A' = \{(1,2),(1,3),(3,4),(4,5)\}$. Figure 7.7 shows the flow network where $c_{ij} = a_{ij}$ for $(i, j) \in \mathbf{A}'$ with $y_{ij} = u_{ij}$. For arc(1, 2), $c_{12} = \infty$ since $y_{12} = u_{12} = d_{12}$. For $(i, j) \in A - \mathbf{A}'$, $c_{ij} = 0$.

The only path through which the flow can be augmented is 1–3–4–5 with a maximum flow of 25 units. Figure 7.8 performs labeling to detect this path. Figure 7.9 shows the updated flow values. The labeling in Figure 7.10 results in a nonbreakthrough, with $X = \{\text{labeled nodes}\} = \{1, 2, 3, 4\}, = \bar{X}\{5\}$ and $C = \{(2, 5),(4, 5)\}$.

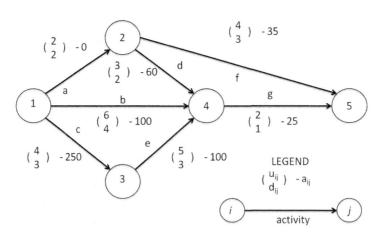

FIGURE 7.5
AOA representation of Figure 4.2.

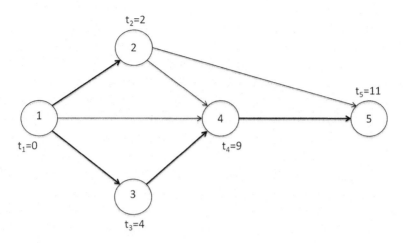

FIGURE 7.6
Critical subnetwork for $T = 11$.

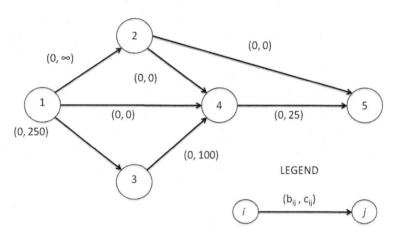

FIGURE 7.7
The flow network.

Next, Δ is calculated as follows:

$$\Delta_2 = t_5 - t_2 - y_{25} = 11 - 2 - 4 = 5$$

$$\Delta_1 = y_{45} - d_{45} = 2 - 1 = 1$$

Therefore, $\Delta = \min(\Delta_1, \Delta_2) = \Delta_1 = 1$. Reducing t_j by one unit for all $j \in \bar{X}$, one gets $t_5 = 10$. The duration of (4, 5) reaches its crash limit. The cost of crashing the project duration from 11 to 10 is $(a_{45}) \cdot (11-10) = \$25$. The critical path remains the same as shown in Figure 7.11. The flow network of Figure 7.12 reflects the current flow values and the updated c_{ij} values. $c_{45} = \infty$ since $y_{45} = d_{45}$. The only existing augmenting path is 1–3–4–5 through which 75 units of additional flow are augmented. Further labeling results in nonbreakthrough with $X = \{1, 2, 3\}$, $\bar{X} = \{4, 5\}$, and $C = \{(1, 4), (2, 5), (2, 4), (3, 4)\}$. Figure 7.13 summarizes the steps of the maximum flow procedure.

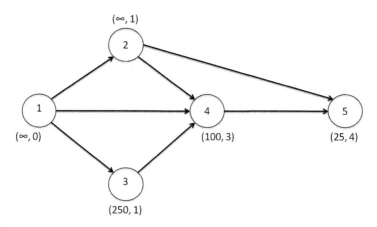

FIGURE 7.8
Maximum flow iteration 1.

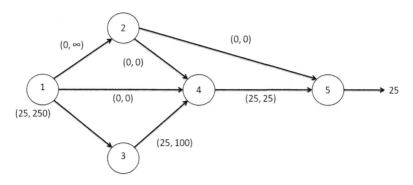

FIGURE 7.9
Maximum flow iteration 2.

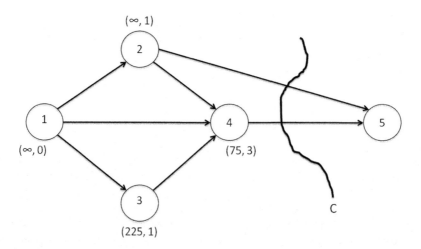

FIGURE 7.10
Maximum flow iteration 3.

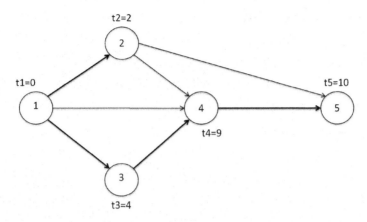

FIGURE 7.11
The critical path subnetwork for $T = 10$.

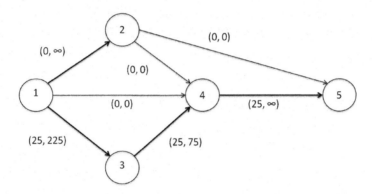

FIGURE 7.12
The flow network for computational example.

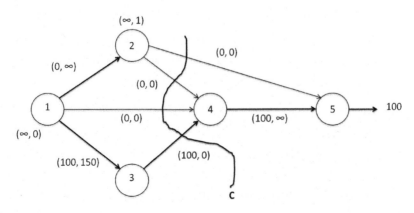

FIGURE 7.13
The maximum flow solution.

The Δ values are calculated as shown in the following:

$$\Delta_2 = \min\left\{(t_4 - t_1 - y_{14}),(t_5 - t_2 - y_{25}),(t_4 - t_2 - y_{24})\right\}$$

$$= \min\left\{(9-0-6),(10-2-4),(9-2-3)\right\} = 3$$

$$\Delta_1 = y_{34} - d_{34} = 5 - 3 = 2$$

$$\Delta = \min(\Delta_1, \Delta_2) = 2$$

Therefore, the project duration can be reduced another 2 units by crashing the duration of activity (3, 4) to its crash duration. The slacks on arcs (1, 4), (2, 5), and (2, 4) will be reduced by 2 units. Figure 7.14 contains the critical subnetwork and the new node realization times.

T is reduced from 10 to 8 at a cost of $a_{34}(\Delta)$ – \$200. The updated flow network shown in Figure 7.15 indicates that a flow of 150 units can be augmented through the path 1–3–4–5. Further labeling results in a nonbreakthrough as shown in Figure 7.16. $\Delta = 1$, which results in $y_{13} = d_{13}$ and activity (1, 4) joining the critical subgraph. Figure 7.17 and Figure 7.18 give the expanded critical subgraph and the corresponding flow network, respectively. Labeling results in node 5 being labeled with infinite symbol, which indicates that a critical path has reached its crash limit. Therefore, the procedure stops with $T = T_{\min} = 7$.

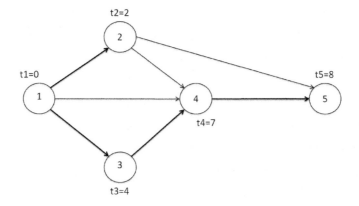

FIGURE 7.14
The maximum flow solution continued.

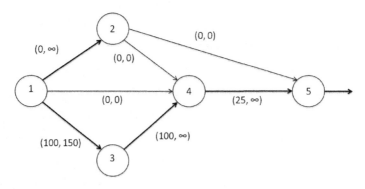

FIGURE 7.15
Network flow graph for time–cost trade-off example.

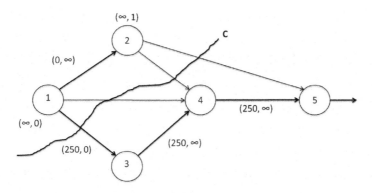

FIGURE 7.16
Maximum flow solution for time–cost trade-off example.

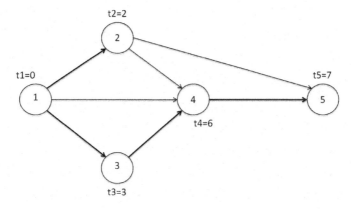

FIGURE 7.17
Expanded critical subnetwork for time–cost trade-off example.

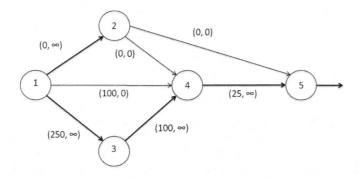

FIGURE 7.18
Network flow graph for time–cost trade-off example.

The complete time–cost trade-off curve is given by Figure 7.19. The figure only illustrates the crashing cost as a function of project duration. In general, one needs to consider the indirect cost that decreases as the project duration is decreased. The optimal project duration, T^*, is the one which minimizes the total crashing cost and the indirect cost. This is illustrated in Figure 7.20.

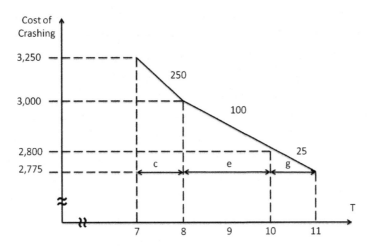

FIGURE 7.19
Time–cost trade-off curve.

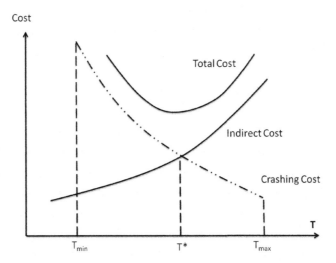

FIGURE 7.20
The total project cost curve.

If the crash duration of activity c was 2 time units, then we would have $c_{13} = 0$ in Figure 7.18, which would result in flow augmentation along 1–4–5 of 100 units. The maximum flow value of 350 units will indicate that a further decrease in project duration will cost \$350 per time unit with activities b and c being the cheapest set of activities to crash.

Sensitivity Analysis for Time–Cost Trade-Off

In this section, we discuss the sensitivity analysis on the cost parameters for the time–cost trade-off problem. The cost of crashing an activity time, say for activity k by one unit, a_k, is generally estimated and is hence subject to error. Therefore, the project manager would like to know how sensitive the solution is to the estimated cost parameters. Consider the critical path subnetwork in Figure 7.21.

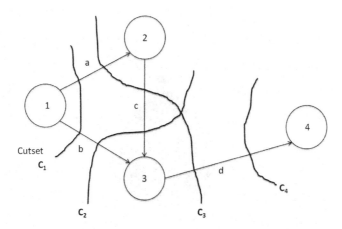

FIGURE 7.21
Critical path subnetwork for sensitivity analysis.

Four cut-sets separate node 1 from node 4. Suppose the original costs were $a_a = 12$, $a_b = 6$, $a_c = 12$, and $a_d = 16$. The cheapest cut-set is C_1 with the total cost of crashing being equal to \$15. Suppose the manager has found out that the crashing costs for activities a and b were underestimated by 20%. Therefore, $a'_a = 12$, $a'_b = 6$, $a_c = 12$, and $a_d = 16$. C_1 is no longer the cheapest cut-set. Cut-set C_4 has a crashing cost of \$16 per time unit and hence is the cheapest cut-set.

Clearly, if an activity has not been critical for $T_{min} \leq T \leq T_{max}$, then its cost will not be relevant for the time–cost trade-off problem. However, if an activity (i, j) has been added to the critical subnetwork at project completion time T_m, then we need to reevaluate the crashing decisions concerning $T_m \leq T \leq T_{max}$. Suppose the cost of activity (i, j) was overestimated, that is, the modified cost $a'_{ij} < a'_{ij}$. This may affect the decisions over the period $[T_m, T_{max}]$, even though activity (i, j) was actually crashed when $T = T'$ where $T_m \leq T' \leq T_{max}$. However, if $a'_{ij} > a'_{ij}$, then one needs to only reevaluate the decisions over the period $[T', T_{max}]$ even though activity (i, j) became critical at T_m. For example, given in Figure 7.5, if $a_c = 275$, then the reevaluations should start at Figure 7.15 with the modification C_{13}(residual capacity) = 175 instead of 150.

Knapsack Problem

The *knapsack problem* is a famous operations research problem dealing with resource allocation. The general nature of the problem is as follows: Suppose that n items are being considered for inclusion in a travel supply bag (knapsack). Each item has a certain per-unit value to the traveler. Each item also has a per-unit weight that contributes to the overall weight of the knapsack. There is a limitation on the total weight that the traveler can carry. Figure 7.22 shows a representation of the problem. The objective is to maximize the total values of the knapsack, subject to the total weight limitation.

The mathematical formulation is for $j = 1, \ldots, n$, let $c_j > 0$ be the value per-unit for item j. Let $w_j > 0$ be the weight per-unit of item j. If the total weight limitation is W, then the problem of maximizing the total value of the knapsack is

$$\text{Maximize: } z = \sum_{j=1}^{n} c_j x_j$$

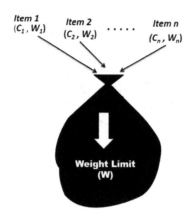

FIGURE 7.22
The knapsack loading problem.

$$\text{Subject to:} \qquad \sum_{j=1}^{n} w_j x_j \leq W$$

$$x_j \geq 0 \text{ integer}, \quad j = 1, \ldots, n$$

where x_j is the number of units of item j included in the knapsack. All data in the knapsack formulation are assumed to be integers. The knapsack problem can be solved by any of the available general integer LP (ILP) algorithms. However, because it has only one constraint, more efficient solution algorithms are possible. For simple models, the problem can be solved by inspection. For some special problems, such as project scheduling, it may be necessary to solve a large number of knapsack problems sequentially. Even though each knapsack formulation and solution may be simple, the overall problem structure may still be very complex.

Knapsack Formulation for Scheduling

The knapsack problem formulation can be applied to a project scheduling problem that is subject to resource constraints. In this case, each activity to be scheduled at a specific instant is modeled as an item to be included in the knapsack. The composition of the activities in a scheduling window is viewed as the knapsack. Figure 7.23 presents a representation of a scheduling window with its composition of scheduled activities at a given instant.

One important aspect of the knapsack formulation for activity scheduling is that only one unit of each activity (item) can be included in the schedule at any given scheduling time. This is because the same activity cannot be scheduled more than once at the same time. However, in certain applications, such as concurrent engineering, it may be desired to have multiple executions of the same activity concurrently. For such cases, each execution of the activity may be modeled as a separate entity with one additional constraint to force concurrent execution. The formulation of the knapsack problem for activity scheduling is done at each and every scheduling time t.

The objective is to schedule as many activities of high priority as possible while satisfying activity precedence relationships without exceeding the resource availability limitations.

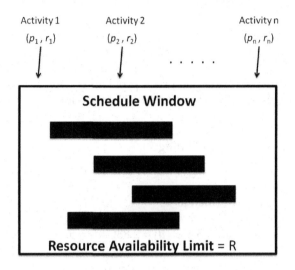

Activity 1   Activity 2   Activity n
(p_1, r_1) (p_2, r_2) (p_n, r_n)

Schedule Window

Resource Availability Limit = R

FIGURE 7.23
The knapsack formulation of project scheduling.

The activity precedence requirements are used to determine the activities that are eligible for scheduling at a given scheduling time.

For $j = 1, \ldots, N$ activities, let $p_j > 0$ be the priority value for activity j. The priority value is used to prioritize activities for resource allocation when activities compete for limited resources. It is assumed that the priority values for the activities do not change during the scheduling process. This is referred to as a *fixed prioritization* of activities. If the activity priority values can change depending on the state and time of the scheduling problem, then the activity prioritization is referred to as a *variable prioritization* of activities. Let $r_{ij} > 0$ be the number of units of resource type i (weight) required by activity j. It is also assumed that the resource requirements do not change during the scheduling process. Let R_{it} be the limit on the units of resource type i available at time t. The formulation of the scheduling problem for time t is as follows:

$$\text{Maximize: } z_t = \sum_{j \in S_t} p_j x_{jt}$$

$$\text{Subject to: } z_t = \sum_{j \in S_t} r_{ij} x_{jt} \leq R_{it}, \quad i = 1, \ldots, k$$

$$x_{jt} = 0 \text{ or } 1, \quad j \in S_t$$

where
    z_t represents the overall performance measure of the schedule generated at time t
    p_j represents the priority measure (value) for activity j
    t represents the current time of scheduling
    x_{jt} represents the indicator variable specifying whether or not activity j is scheduled at
        time t
    S_t represents the set of activities eligible for scheduling at time t

k represents the number of different resource types

r_{ij} represents the units of resource type i required by activity j

R_{it} represents the units of resource type i available at time t

The next scheduling time, t, for the knapsack problem is determined as the minimum of the finishing times of the scheduled and unfinished activities.

Example of Knapsack Activity Scheduling

Suppose we have a project consisting of seven activities: activities A, B, C, D, E, F, and G, which are labeled 1–7.

The activity durations are specified as d_j:

$$d_1 = 2.17, \quad d_2 = 6, \quad d_3 = 3.83, \quad d_4 = 2.83, \quad d_5 = 5.17, \quad d_6 = 4, \quad d_7 = 2$$

The scaled priority values of the activities are specified as p_j:

$$p_1 = 55.4, \quad p_2 = 100, \quad p_3 = 72.6, \quad p_4 = 54, \quad p_5 = 88, \quad p_6 = 66.6, \quad p_7 = 75.3$$

There are two resource types with the following availability levels at time t, R_{jt}:

$$R_{1,0} = 10, \quad R_{2,0} = 15$$

Units of *resource type* 1 required by the activities are specified as r_{1j}:

$$r_{11} = 3, \quad r_{12} = 5, \quad r_{13} = 4, \quad r_{14} = 2, \quad r_{15} = 4, \quad r_{16} = 2, \quad r_{17} = 6$$

Units of *resource type* 2 required by the activities are specified as r_{2j}:

$$r_{21} = 0, \quad r_{22} = 4, \quad r_{23} = 1, \quad r_{24} = 0, \quad r_{25} = 3, \quad r_{26} = 7, \quad r_{27} = 2$$

The precedence relationships between the activities are represented as $P(.) = \{$set of predecessors$\}$:

$$P(A) = \emptyset, \quad P(B) = \emptyset, \quad P(C) = \emptyset, \quad P(D) = \{A\}, \quad P(E) = \{C\},$$
$$P(F) = \{A\}, \quad P(G) = \{B, D, E\}$$

The successive knapsack formulations and solutions for the problem are presented as follows:

$$t = 0$$

$$S_0 = \{A, B, C\}$$

Maximize: $z_0 = 55.4x_A + 100x_B + 72.6x_C$

$$3x_A + 5x_B + 4x_C \leq 10$$

Subject to: $\quad 0x_A + 4x_B + 1x_C \leq 15$

$$x_A, x_B, x_C = 0 \text{ or } 1$$

whose solution yields $x_A = 0$, $x_B = 1$, $x_C = 1$. Thus, activity B is scheduled at $t = 0$, and it will finish at $t = 6$; C is scheduled at $t = 0$ and it will finish at $t = 3.83$. The next scheduling time is $t = \text{Min}\{3.83, 6\} = 3.83$.

$$t = 3.83$$

$$S_{3.83} = \{A, E\}$$

Maximize: $z_{3.83} = 55.4x_A + 88x_E$

$$3x_A + 4x_E \leq 5$$

Subject to: $\quad 0x_A + 3x_E + 1x_C \leq 11$

$$x_A, x_E = 0 \text{ or } 1$$

whose solution yields $x_A = 0$, $x_E = 1$. Thus, activity E is scheduled at $t = 3.83$, and it will finish at $t = 9$. The next scheduling time is $t = \text{Min}\{6, 9\} = 6$.

$$t = 6$$

$$S_6 = \{A\}$$

Maximize: $z_6 = 55.4x_A$

Subject to: $\quad \begin{aligned} 3x_A &\leq 6 \\ x_A &= 0 \text{ or } 1 \end{aligned}$

whose solution yields $x_A = 1$. Thus, activity A is scheduled at $t = 6$ and it will finish at $t = 8.17$. The next scheduling time is $t = \text{Min}\{8.17, 9\} = 8.17$.

$$t = 8.17$$

$$S_{8.17} = \{D, F\}$$

Maximize: $z_{8.17} = 54x_D + 66.6x_F$

$$2x_D + 2x_F \leq 6$$

Subject to: $\quad 0x_D + 7x_F \leq 12$

$$x_D, x_F = 0 \text{ or } 1$$

whose solution yields $x_D = 1$, $x_F = 1$. Thus, activities D and F are scheduled at $t = 8.17$. D will finish at $t = 11$. F will finish at $t = 12.17$. The next scheduling time is $t = \text{Min}\{9, 11, 12.17\} = 9$.

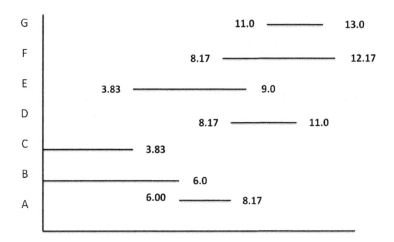

FIGURE 7.24
Gantt chart of knapsack schedule.

$$t = 9$$

$$S_9 = \emptyset$$

No activity is eligible for scheduling at $t = 9$, since G cannot be scheduled until D finishes. The next scheduling time is $t = \text{Min}\{11, 12.17\} = 11$.

$$t = 11$$

$$S_{11} = \{G\}$$

Maximize: $z_{11} = 75.3x_G$

Subject to: $6x_G \leq 10$
             $2x_G \leq 8$
             $x_G = 0 \text{ or } 1$

whose solution yields $x_G = 1$. Thus, activity G is scheduled at $t = 11$, and it will finish at $t = 13$. All activities have been scheduled. The project completion time is $t = \text{Max}\{12.17, 13\} = 13$. Figure 7.24 shows the Gantt chart of the completed schedule, $R_{1,0} = 10$ and $R_{2,0} = 15$.

Transportation Problem for Project Scheduling

The *transportation problem* is a special class of optimization problem dealing with the distribution of items from sources of supply to locations of demand. This type of problem can occur in large or multiple project scheduling environments where supplies must be delivered to various project sites in a coordinated fashion to meet scheduling requirements. A specific algorithm, known as the *transportation method,* is available for solving transportation problems that satisfy the following assumptions:

1. The problem must concern a single product type to be transported.
2. There must be several *sources* from which the product is to be transported.
3. The amount of *supply* available at each source must be known.
4. There must be several *destinations* to which the product is to be transported.
5. The *demand* at each destination must be known.
6. The per-unit cost of transporting the product from any source to any destination must be known.

The general format of the transportation problem, referred to as the *transportation tableau*, is presented in Figure 7.25.

The objective of the transportation method is to determine the minimum cost plan to transport units from origins to destinations while satisfying all supply limitations and demand requirements. The following notation is used:

Src_i is the source i, $i = 1, 2, ..., m$

Des_j is the destination j, $j = 1, 2, ..., n$

S_i is the available supply from source i, $i = 1, 2, ..., m$

D_j is the total demand for destination j, $j = 1, 2, ..., n$

c_{ij} is the cost of transporting one unit of the product from source i to destination j

x_{ij} is the number of units of the product transported from source i to destination j

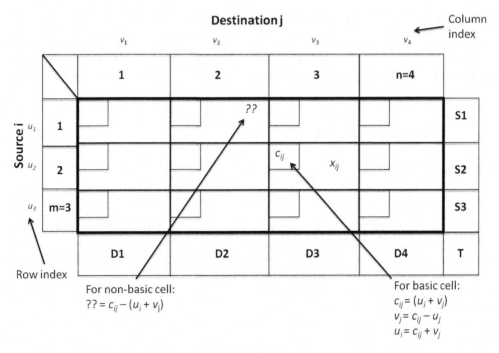

FIGURE 7.25
General layout of the transportation problem.

An LP formulation of the transportation problem is presented next:

$$\text{Minimize: } z = \sum_{i=1}^{m}\sum_{j=1}^{m} c_{ij}x_{ij}$$

Subject to:
$$\sum_{j=1}^{n} x_{ij} \le S_i, \quad i = 1,2,\ldots,m$$

$$\sum_{i=1}^{n} x_{ij} \le D_j, \quad j = 1,2,\ldots,n$$

$$x_{ij} \ge 0$$

The first set of constraints indicates that the sum of the shipments from a source cannot exceed its supply. The second set of constraints indicates that the sum of shipments to a destination must satisfy its demand. The aforementioned LP model implies that the total supply must at least equal the total demand. Although it is possible to model and solve a transportation problem as an LP problem, the special structure of the problem makes it possible to use the *transportation method* as a more efficient solution method. Like the LP simplex method, the transportation method is an iterative solution procedure, moving from one trial solution to another less costly trial solution. The iterative solution continues until the optimal solution is reached. Unlike the simplex method, however, the transportation method is limited to only certain types of problems.

Balanced versus Unbalanced Transportation Problems

A transportation problem is said to be *balance* if the total available supply is equal to the total demand. For a balanced problem, the inequality constraints in the model will be modified to equality constraints. That is,

$$\sum_{j=1}^{n} x_{ij} = S_i, \quad i = 1,2,\ldots,m$$

$$\sum_{j=1}^{m} x_{ij} = D_i, \quad i = 1,2,\ldots,m$$

A transportation problem is *unbalanced* when total supply is not equal to total demand. Unbalanced transportation problems are more realistic in practical situations. A modification of the initial transportation tableau is required for unbalanced transportation problems. If total supply is greater than total demand, we will add one extra *dummy destination* to the initial tableau. The dummy destination is assigned a demand, D_{n+1}, that is equal to the total available supply minus the total demand: That is,

$$D_{n+1} = \sum_{i=1}^{m} S_i - \sum_{j=1}^{n} D_j$$

The dummy destination serves to balance the problem. Allocations to cells associated with the dummy destination are simply not shipped from sources. These unshipped units represent excess supply. The magnitude of the excess supply can be used for making capacity adjustment decisions at the sources if no "real" destinations can be found to absorb the excess units.

If total supply is less than total demand, then some of the demand cannot be fulfilled. In such a case, a *dummy source* is included in the formulation. Other options are to reduce demand levels or increase supply capacities. The supply from the dummy source may be viewed as subcontracted supply. If supply and demand levels are fixed, then the supply to be subcontracted out, S_{m+1}, will be

$$S_{m+1} = \sum_{j=1}^{n} D_j - \sum_{i=1}^{m} S_i$$

The solution to the transportation problem will always be an integer. Hence, no integer restrictions need to be specified.

Initial Solution to the Transportation Problem

To use the transportation method, we must first identify an *initial feasible solution*. The *northwest-corner technique* is a way to obtain an initial solution to a transportation problem. The technique makes the first allocation to the northwest-corner cell in the transportation tableau. As much allocation as possible is made to that northwest-corner cell. This technique is a simple method to determine a *feasible solution* to the transportation problem. The feasible solution satisfies all supply and demand restrictions. The initial solution may not be optimal, but it serves as a starting point for procedures that are designed to find the optimal solution. A summary of the northwest-corner technique is presented as follows.

Northwest-Corner Technique

1. *Step 1*: Set up tableau and enter the S_i, D_j, and c_{ij} values.
   a. Allocate $u_{11} = \text{Min}\{S_i, D_j\}$.
   b. Update supply and demand: $S_1 = S_1 - u_{11}$ and $D_1 = D_1 - u_{11}$. Set $i = 1, j = 1$.

2. *Step 2*: Move to a new cell.
   a. If current $S_i = 0$, then move down one cell.
   b. If current $D_j = 0$, move one cell to the right. If both S_i and D_j are zero, perform step 2a and 2b only.

3. *Step 3*: Make allocation to the new cell *ij*.
   a. Allocate $x_{ij} = \text{Min}\{S_i, D_j\}$.
   b. Update supply and demand: $S_i = S_i - x_{ij}$ and $D_j = D_j - x_{ij}$.

4. *Step 4*: Repeat steps 2 and 3 until all available supply and demand have been allocated to cells within the initial tableau, that is, until $S_i = 0$ and $D_j = 0$ ($i = 1, 2, …, m$; $j = 1, 2, …, n$).

The northwest-corner method, although very simple and fast, usually does not yield a good initial solution because it totally ignored costs. Better initial approximation methods are available. These other methods include the *least-cost method, Vogel's approximation method,* and *Russell's approximation method.*

Least-cost method. The least-cost method uses the cell costs as the basis for making allocations to the cells in the transportation tableau.

1. *Step 1*: Assign as much as possible to the variable with the smallest unit cost in the tableau. Break ties arbitrarily. Cross out the satisfied row or column.

2. *Step 2*: If both rows and columns are satisfied simultaneously, only one may be crossed out. Adjust the supply and demand for all uncrossed rows and columns.

3. *Step 3*: Repeat the process by assigning as much as possible to the cell with the smallest uncrossed unit cost.

4. *Step 4*: The procedure is complete when all the rows and columns are crossed out.

Vogel's approximation. Vogel's approximation is better than the northwest-corner method. In fact, the method usually yields an optimal or close to optimal starting solution to the transportation problem. The steps of the method are as follows:

1. *Step 1*: Set up the transportation tableau and evaluate the row or column penalty.
    a. For each row, subtract the smallest cost element from the next smallest cost element in the same row.
    b. For each column, subtract the smallest cost element from the next smallest cost element in the same column.

2. *Step 2*: Identify the row or column with the largest penalty (break ties arbitrarily).
    a. Allocate as much as possible to the cell with the least cost in the selected row or column. That is, assign the smaller of row supply and column demand.
    b. If there is only one cell remaining in a row or column, choose the cell and allocate as many units as possible to it.
    c. Adjust the supply and demand and cross out the satisfied row or column.

3. *Step 3*: Determine satisfied rows or columns.
    a. If a row supply becomes zero, cross out the row and calculate the new column penalties. The other row penalties are not affected.
    b. If a column demand becomes zero, cross out the column and calculate new row penalties.
    c. If both a row supply and a column demand become zero at the same time, cross out only one of them. The one remaining will have a supply or demand of zero, which means an assignment of zero units in a subsequent step.

4. *Step 4*: Repeat steps 2 and 3 until an initial basic feasible solution has been obtained.

Russell's approximation method. Russell's approximation method is another approach to obtaining an initial solution for the transportation problem. The steps of the method are as follows:

1. *Step 1*: For each source row i remaining under consideration, determine u_i as $u_i = \text{Max}\{c_{ij}\}$ still remaining in that row. For each destination column j remaining under consideration, determine v_j as $v_j = \text{Max}\{c_{ij}\}$.

2. *Step 2*: For each cell not previously selected in the rows and columns, calculate $(c_{ij} - u_i - v_j)$.

3. *Step 3*: Select the cell having the largest negative value of $(c_{ij} - u_i - v_j)$. Break ties arbitrarily. Assign as many units as possible to the cell.

4. *Step 4*: Repeat steps 1–3 until an initial solution is obtained.

Transportation Algorithm

The transportation algorithm generates an optimal solution to the transportation problem. The algorithm gets to the optimal solution by improving on an initial feasible solution. The steps of the algorithm are summarized as follows:

1. *Step 1*: Find an initial basic feasible solution.

2. *Step 2*: Define a row index, u_i, and a column index, v_j. This is called the MODI (modified distribution) or the *uv* method. There are m of the u_i ($i = 1, 2, \ldots, m$) and n of the v_j ($j = 1, 2, \ldots, n$). The indices are composed by solving the equations $c_{ij} = u_i + v_j$ for the basic cells. A basic cell is the cell where x_{ij} values were allocated. A basic cell may have $x_{ij} = 0$ in the case of degeneracy. This happens when an allocation satisfies the supply and demand simultaneously. There must be $m + n - 1$ basic cells. Thus, there are $m + n - 1$ equations in $m + n$ unknowns. This means that we can arbitrarily assign any value to one of the row or column indices (e.g., $u_i = 0$), and solve the equations simultaneously for u_i and v_j values, as explained in step 3.

3. *Step 3*: After setting $u_1 = 0$, solve for the remaining indices by using the relationship $u_i + v_j = c_{ij}$ iteratively for each basic cell. Each time one index is computed, there will be another one that is immediately defined. Place the value of each index in the appropriate cell in the tableau.

4. *Step 4*: Calculate the reduced cost of each nonbasic cell as follows:

$$c_{ij} - z_{ij} = c_{ij} - \left(u_i + v_j\right) = c_{ij} - u_i - v_j$$

Circle the $c_{ij} - (u_i + v_j)$ values in the nonbasic cells. The $c_{ij} - (u_i + v_j)$ value represents the reduced cost for the nonbasic cell. Because this is a cost minimization problem, the current basis is optimal if

$$c_{ij} - z_{ij} = c_{ij} - u_i - v_j \geq 0$$

for all nonbasic variables. Stop if all of which indicates an optimal solution.

5. *Step 5*: Select the nonbasic variable with the most negative reduced cost to enter the basis. That is, select x_{ij} to enter the basis so that

$$c_{ij} - z_{ij} = \text{Min}_{pq}\left\{c_{pq} - z_{pq}\right\}$$

6. *Step 6*: Use the *stepping-stone method* to determine the loop formed by the nonbasic cell (*ij*) and the basic cells. The stepping-stone method steps through cells in a closed loop path in the tableau based on the following rules:

   a. Only one nonbasic variable (i.e., the entering variable) is included in the stepping-stone path.

   b. The path is unique.

c. Start by placing a plus sign in the cell of the entering variable. The plus sign indicates that this variable will increase.

d. Place a minus sign in the cell of a basic variable so that the column or row of the incoming variable is balanced with one plus sign and by one unit for every unit increase in the entering variable.

e. Continue the assignment of plus and minus signs in the basic cells in such a way that each corner is a basic cell except one corner, and each vertical step is followed by a horizontal step with the signs alternated until the loop is close.

f. The stepping-stone path ends when both the row and the column of the entering variable are balanced. Basic variables that are not in the path contain no sign and will not change as the entering variable increases.

7. *Step 7*: Determine the exiting variable. This is the basic variable containing a minus sign and having the smallest x_{ij} value. Set $K = \text{Min}\{x_{ij} | ij\text{th cell sign is minus}\}$.

8. *Step 8*: Update the transportation tableau as follows:

a. Add K units to each cell that has a plus sign in the closed path.

b. Subtract K units from each that has a minus sign in the closed path.

9. *Step 9*: Return to step 2.

Example of Transportation Problem

Suppose four project sites are to be supplied with units of a certain product from three different sources. It is desired to minimize the total cost of supplying the sites with the required units. The supply and demand data and the shipment costs are summarized as follows:

$S_1 = 30$	$S_2 = 50$	$S_3 = 40$	
$D_1 = 30$	$D_2 = 20$	$D_3 = 40$	$D_4 = 30$
$c_{11} = \$5$	$c_{12} = \$8$	$c_{13} = \$3$	$c_{14} = \$6$
$c_{21} = \$4$	$c_{22} = \$5$	$c_{23} = \$7$	$c_{24} = \$4$
$c_{31} = \$6$	$c_{32} = \$2$	$c_{33} = \$4$	$c_{34} = \$5$

Figure 7.26 shows the layout of the problem. This is a balanced transportation problem. The initial feasible solution obtained through the northwest-corner method is shown in Figure 7.27.

There are $6 = m + n - 1$ basic cells, one of which $x_{12} = 0$. This is because when 30 units were allocated to x_{11}, it satisfied both S_1 and D_1. Crossing row 1, the northwest-corner became cell (2, 1); hence, 0 units were allocated to x_{12} and column 1 crossed out. The procedure continues with the next northwest-corner cell (2, 2). Let $u_2 = 0$. Determine the remaining u_i and v_j values for the basic cells using the relationship $c_{ij} = (u_i + v_j)$. Using the u_i and v_j already available, compute $c_{ij} - (u_i + v_j)$ for each of the nonbasic cells. The resulting tableau is shown in Figure 7.28.

If all the $c_{ij} - (u_i + v_j)$ values for the nonbasic cells were positive, the solution would be optimal. Cell (1, 3) has the most negative value, so it enters the solution. This is shown in

Destinations

	1	2	3	4	
1	5	8	3	6	30
2	4	5	7	4	50
3	6	2	4	5	40
	30	20	40	30	120

FIGURE 7.26
Cost matrix for transportation problem example.

Destinations

	1	2	3	4	
1	5 30	8	3	6	30
2	4	5 20	7 30	4	50
3	6	2	4 10	5 30	40
	30	20	40	30	120

FIGURE 7.27
Tableau 1 for transportation example.

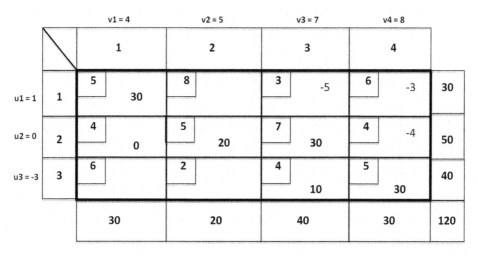

FIGURE 7.28
Tableau 2 for transportation example.

Figure 7.29. Note that cell (2, 1) is treated as a basic cell since an *allocation* of 0 has been made to it. Allocate as much as possible to the entering cell, cell (1, 3), without violating the supply and demand constraints. The new solution is shown in Figure 7.30. The tableau generated by a return to step 2 of the algorithm is shown in Figure 7.31.

The optimal solution has not been reached. Therefore, continue with the algorithm steps. Figures 7.32–7.35 show the subsequent tableaus of the solution. Figure 7.36 shows the final tableau and the optimal solution.

The solution indicates that we should ship 30 units from source 1 to destination 3; 30 units from source 2 to destination 1; 20 units from source 2 to destination 4; 20 units from source 3 to destination 2; 10 units from source 3 to destination 3; and 10 units from source 3 to destination 4. The minimum cost is

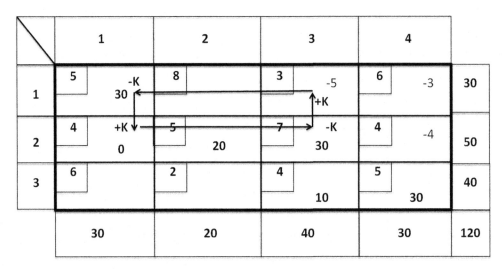

FIGURE 7.29
Tableau 3 for transportation example.

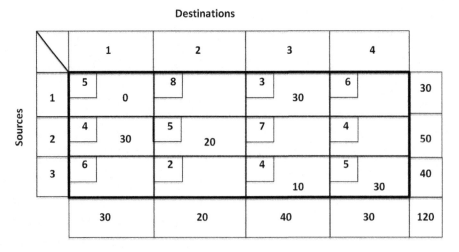

FIGURE 7.30
Solution from Tableau 3 for transportation example.

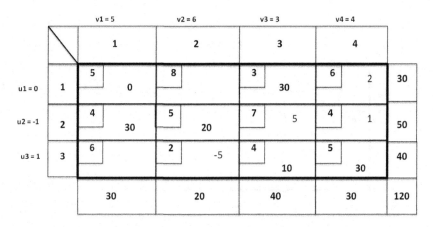

FIGURE 7.31
Tableau 4 for transportation example.

	1	2	3	4	
1	5 -K 30	8	3 -5 +K	6 -3	30
2	4 +K 0	5 -K 20	7 30	4 -4	50
3	6	2 +K	4 -K 10	5 30	40
	30	20	40	30	120

FIGURE 7.32
Tableau 5 for transportation example.

	1	2	3	4	
1	5	8	3 30	6	30
2	4 30	5 20	7	4	50
3	6	2 0	4 10	5 30	40
	30	20	40	30	120

FIGURE 7.33
Tableau 6 for transportation example.

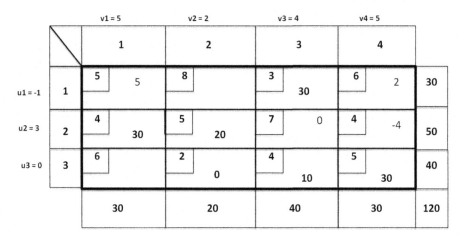

FIGURE 7.34
Tableau 7 for transportation example.

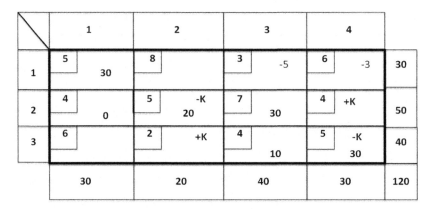

FIGURE 7.35
Tableau 8 for transportation example.

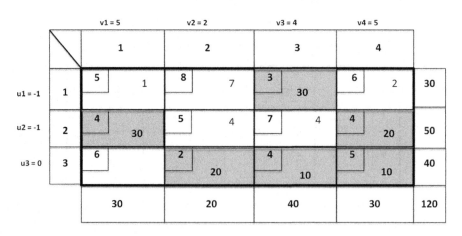

FIGURE 7.36
Final tableau for transportation example.

$$z = 3(30) + 4(30) + 4(20) + 2(20) + 4(10) + 5(10) = 420$$

Transshipment Formulation

The transshipment problem is a general model of the transportation problem. In this model, there can be *pure sources, pure destinations,* and *transshipment points* that can serve as both sources and destinations. It is possible for any source or destination to ship to any other source or destination. Thus, there may be many different ways of shipping from point *i* to point *j* in addition to the direct route. In the transportation problem, the way in which units are distributed from source *i* to destination *j* must be known in advance so that the corresponding cost per unit, c_{ij}, can be determined ahead of time. In the transshipment problem, units may go through intermediate points that offer lower total shipment cost. For example, instead of shipping units directly from source 2 to destination 3, it may be cheaper to include the units going to destination 3 with the units going to destination 4 and then ship those units from destination 4, which now serves as a source, to destination 3. A mathematical formulation of the transshipment problem is as follows. Let

x_{ij} be the amount shipped from point *i* to point *j*, $i, j = 1, 2,..., n; i \neq j$

c_{ij} be the cost of shipping from point *i* to point *j*, $c_{ij} \geq 0$

r_i be the net requirement at point *i* (negative for demand point, positive for supply point)

Each point must satisfy a balance equation stating that the amount shipped minus the amount received equals the net requirement at the point. It is also required that total demand equals total supply. Thus, we have

$$\text{Minimize: } z = \sum_{i=1}^{n} \sum_{j=1}^{n} c_{ij} x_{ij}, \quad i \neq j$$

Subject to:

$$\sum_{j=1, j \neq i}^{n} x_{ij} - \sum_{j=1, j \neq i}^{n} x_{ij} = r_i, \quad i = 1, 2,..., n$$

$$\sum_{i=1}^{n} r_i = 0$$

$$x_{ij} \geq 0, \quad j = 1, 2,..., n; i \neq j$$

Assignment Problem in Project Optimization

Suppose there are *n* tasks which must be performed by *n* workers. The cost of worker *i* performing task *j* is c_{ij}. It is desired to assign workers to the tasks in a fashion that minimized the cost of completing the tasks. This problem scenario is referred to as the *assignment problem.* The technique for finding the optimal solution to the problem is called the *assignment method.* Like the transportation method, the assignment method is an iterative procedure that arrives at the optimal solution by improving on a trial solution at each stage of the procedure.

CPM and PERT can be used in controlling projects to ensure that the project will be completed on time. As mentioned in Chapters 2 and 3, these two techniques do not consider the assignment of resources to the tasks that make up a project. The *assignment method* can be used to achieve an optimal assignment of resources to specific tasks in a project. Although the assignment method is cost-based, task duration can be incorporated into the modeling in terms of time–cost relationships. Of course, task precedence requirements and other scheduling restrictions will have to be accounted for in the final scheduling of the tasks. The objective is to minimize the total cost. Thus, the formulation of the assignment problem is as follows. Let

$x_{ij} = 1$ if worker i is assigned to task j, $i, j = 1, 2,..., n$

$x_{ij} = 0$ if worker i is not assigned to task j

$c_{ij} = $ cost of worker i performing task j

$$\text{Minimize: } z = \sum_{i=1}^{n}\sum_{j=1}^{n} c_{ij}x_{ij}$$

$$\sum_{j=1}^{n} x_{ij} = 1, \quad i = 1, 2,..., n$$

Subject to: $$\sum_{i=1}^{n} x_{ij} = 1, \quad j = 1, 2,..., n$$

$$x_{ij} \geq 0, \quad j = 1, 2,..., n$$

It can be seen that the previous formulation is a transportation problem with $m = n$ and all supplies and demands equal to 1. Note that we have used the nonnegativity constraint, $x_{ij} \geq 0$, instead of the integer constraint, $x_{ij} = 0$ or 1. However, the solution of the model will still be integer valued. Hence, the assignment problem is a special case of the transportation problem with $m = n$, $S_i = 1$, and $D_i = 1$ for all i. Conversely, the transportation problem can also be viewed as a special case of the assignment problem. A transportation problem can be modeled as an assignment problem and vice versa. The basic requirements of an assignment problem are as follows:

1. There must be two or more tasks completed.
2. There must be two or more resources that can be assigned to the tasks.
3. The cost of using any of the resources to perform any of the tasks must be known.
4. Each resource is to be assigned to one and only one task.

If the number of tasks to be performed is greater than the number of workers available, we will need to add *dummy* workers to balance the problem. Similarly, if the number of workers is greater than the number of tasks, we will need to add *dummy tasks* to balance the problem. Thus, the assignment problem can be extended to consider partial allocation of resource units to multiple tasks.

Although the assignment problem can be formulated for and solved by the simplex method or the transportation method, a more efficient algorithm has been developed specifically for the assignment problem. The method, known as the *Hungarian method*, is a simple iterative technique. The method is based on the assignment theory developed by

E. Egervary, a Hungarian mathematician. It is based on properties of matrices and relationships between primal and dual problems. The steps of the assignment method are summarized as follows:

1. *Step 1*: Develop the $n \times n$ cost matrix, in which rows represent workers and columns represent tasks.
2. *Step 2*: For each row of the cost matrix, subtract the smallest number in the row from each and every number in the row. For each column of the resulting matrix, subtract the smallest number from each number in the column. The resulting matrix is called the *matrix of reduced costs*.
3. *Step 3*: Find the minimum number of lines through the rows and columns of the reduced-cost matrix such that all zeros have a line through them. If the number of lines is n, the optimal solution has been reached; stop. Otherwise, go to step 4.
4. *Step 4*: Define a new reduced-cost matrix as follows: Determine the smallest number in the matrix which does not have a line through it. Subtract this number from all the unlined numbers and add it to all the numbers that have two lines through them (i.e., numbers located at the intersection of two lines in the matrix). Go to step 3.

Example of Assignment Problem

Suppose five workers are to be assigned to five tasks on the basis of the cost matrix presented in Figure 7.37. Let us use the algorithm steps presented earlier to solve this assignment problem.

For convenience, we can divide each cost element by 100. When the smallest number in each row of the resulting simplified matrix is subtracted from all the elements in the row, we obtain Figure 7.38.

When the smallest number in each column of the figure is subtracted from all the elements in the column, we obtain Figure 7.39. The minimum number of lines required to cover all the zeros is 3. This is shown in Figure 7.40. The smallest uncovered number in the figure is 1. When this 1 is subtracted from the other uncovered numbers and added to the numbers that are covered by 2 lines, we obtain the new reduced-cost matrix shown in Figure 7.41.

The minimum number of lines needed to cover all the zeros in the figure is 4. The smallest uncovered number is 1. After repeating the appropriate steps of the algorithm, we obtain Figure 7.42. Since $n = 5$ lines are need to cover all the zeros, the optimal solution has been found.

		Tasks				
		1	2	3	4	5
	1	$200	$400	$500	$100	$400
	2	$400	$700	$800	$1,100	$700
Worker	3	$300	$900	$800	$1,000	$500
	4	$100	$300	$500	$100	$400
	5	$700	$100	$200	$100	$200

FIGURE 7.37
Cost matrix for assignment problem.

Tasks

Worker	1	2	3	4	5
1	1	3	4	0	3
2	0	3	4	7	3
3	0	6	5	7	2
4	0	2	4	0	3
5	6	0	1	0	1

FIGURE 7.38
Tableau 1 for assignment problem.

Tasks

Worker	1	2	3	4	5
1	1	3	3	0	2
2	0	3	3	7	2
3	0	6	4	7	1
4	0	2	3	0	2
5	6	0	0	0	0

FIGURE 7.39
Tableau 2 for assignment problem.

Tasks

Worker	1	2	3	4	5
1	1	3	3	0	2
2	0	3	3	7	2
3	0	6	4	7	1
4	0	2	3	0	2
5	0	0	0	0	0

FIGURE 7.40
Tableau 3 for assignment problem.

Tasks

Worker	1	2	3	4	5
1	1	2	2	0	1
2	0	2	2	7	1
3	0	5	3	7	0
4	0	1	2	0	1
5	7	0	0	1	0

FIGURE 7.41
Tableau 4 for assignment problem.

		Tasks				
		1	2	3	4	5
Worker	1	1	1	1	0	1
	2	0	1	1	7	1
	3	0	4	2	7	0
	4	0	0	1	0	1
	5	8	0	0	2	1

FIGURE 7.42
Final tableau for assignment problem.

To determine the optimal assignment, first make assignments to the rows and columns with only one zero. In Figure 7.42, if we start with the rows, we will make assignments in cell (1, 4) and cell (2, 1). Then, considering the columns, we will make assignments in cell (5, 3) and cell (3, 5). The only remaining assignment is for cell (4, 2). These optimal assignments are shown shaded in the figure. The minimum cost is $1,500.

Traveling Resource Formulation

The *traveling salesman problem* (TSP) is a special case of the assignment problem. The problem is stated as follows: Given a set of n cities, numbered 1 to n, a salesman must determine a *minimum-distance* route, called a *tour,* which begins in city 1, goes through each of the other $(n-1)$ cities exactly once, and then returns to city 1. There are $(n-1)!/2$ possible tours. Thus, complete enumeration of the tours to find the minimum distance is impractical for most application scenarios.

The TSP may be viewed as a *traveling resource* problem in project scheduling whereby a single resource unit is expected to perform tasks at several sites. Consider the following model where

$x_{ij} = 1$, if route includes going from city i to city j, $i, j = 1, 2,\ldots, n$; $i \neq j$; 0, otherwise

c_{ij} = distance between city i and j

$$\text{Minimize: } z = \sum_{i=1}^{n} \sum_{\substack{j=1 \\ i \neq j}}^{n} c_{ij} x_{ij}$$

$$\text{Subject to:} \quad \begin{aligned} &\sum_{j=1}^{n} x_{ij} = 1, \quad i = 1,2,\ldots,n \\ &\sum_{i=1}^{n} x_{ij} = 1, \quad j = 1,2,\ldots,n \\ &x_{ij} = 0,1, \quad j = 1,2,\ldots,n; i \neq j \end{aligned}$$

The first constraint represents the salesman's departures, while the second constraint represents arrivals. Although the previous formulation can be solved as an assignment problem, a feasible assignment solution does not necessarily represent a *valid tour.* Specifically, a solution of the assignment problem may include subtours, which are disconnected tour cycles. One approach to formulating the salesman problem is to add additional constraints to the assignment problem formulation to eliminate the possibility of subtours.

Hence, we need to add a constraint to aforementioned model, indicating that a feasible assignment must define a complete tour with precisely one cycle. The additional constraint destroys the simplicity of the previous model. In fact, there is no method available which will determine the optimal tour for a TSP problem in polynomial time. However, numerous heuristics exist that will find good, but not necessarily optimal, solutions in a reasonable amount of time. The techniques can be divided into two classes: tour construction techniques and tour improvement techniques. Tour construction techniques spend a considerable amount of time constructing a tour to make sure that the final tour is a good tour. Tour improvement heuristics start with an arbitrary tour and switch nodes of the tour if the switch results in a reduction in the total tour length.

The simplest and most efficient tour construction heuristic is the nearest neighbor insertion technique. The simplest and most efficient tour improvement heuristic is the 2-opt technique. Another approach to solving a TSP problem involves the use of a simulated annealing technique. We will demonstrate the nearest neighbor technique and the 2-opt technique next. The simulated annealing technique is explained in a later section.

Nearest Neighbor Algorithm

1. Choose any node as the first node in the tour.
2. Locate an unchosen node that is closed to the recently chosen node.
3. Enter this node as the next node to visit in the tour. The node becomes the recently chosen node.
4. Repeat steps 2 and 3 until all the nodes are chosen.
5. Close the tour by connecting the first node and the last node.

Example 7.1

Suppose a project involves a crane picking up material from five different locations. Table 7.6 shows the distance between locations i and j. Determine the sequence in which the crane should visit the five locations and return to the original position in a way that minimizes the total distance traveled.

Starting with location 1 and choosing the nearest location each time, one will get the tour 1–3–2–4–5–1 with the total distance of 49 units. Since the nearest neighbor algorithm is a heuristic, it does not guarantee optimality but provides good tours.

The 2-opt improvement technique searches for tour reversals that will lead to a reduction in the tour length. In 2-opt switches, two tour connections are broken and nodes

TABLE 7.6

Distance Matrix for Traveling Resource Problem

Location	1	2	3	4	5
1	-	20	10	30	15
2	20	-	5	10	8
3	10	5	-	12	15
4	30	10	12	-	9
5	15	8	15	9	-

FIGURE 7.43
The 2-opt example for traveling resource problem.

are connected in a way that a complete tour is defined as shown in Figure 7.43, where connections 3–2 and 4–5 are replaced by new connections 3–4 and 2–5.

The tour length of 1–3–4–2–5–1 is 55. Since this 2-opt switch did not improve the tour length, the new tour will be discarded.

The 2-Opt Technique

The 2-opt technique can be summarized as follows:

1. Pick any two connections. Perform the 2-opt switch.
2. Retain the new tour if the tour length is improved. Return to step 1. Otherwise, return to step 1 with the tour before the switch.
3. Stop when no 2-opt switch improves the tour length.

The TSP has many applications in real life and often used to solve storage, retrieval, and distribution problems.

Shortest-Path Problem

The *shortest-path* problem involves finding the shortest path from a specified origin to a specified destination in an acyclic network. A *network* consists of a set of *nodes* (or vertices) and a set of *arcs* (or edges or links) which connect the nodes. Arc (i, j) represents the link from node i to node j. It is often assumed that the arcs are *directed*, that is, each arc has a specific orientation. A *path* is a collection of arcs from one node to another without regard to their directions. A *directed chain* from one node to another is a set of arcs that are all oriented in a specific direction. A directed chain beginning and ending at the same node is a *directed cycle*. A network that has no directed cycles is said to be *acyclic*. In acyclic networks, $i < j$ for all arcs (i, j).

Practical network problems involve assigning costs, profits, distances, capacity restrictions, or other performance measures to arcs. Supply and demand levels can also be associated with nodes in the network. Let us consider a directed network with a distance d_{ij} associated with each arc (i, j). It is assumed that the distances between arcs are nonnegative. When no arc exists from node i to node j, the length of the *dummy arc* (i, j) is set to M, where $M \ggg 0$. The shortest-path problem consists of finding the shortest directed chain from a specified origin to a specified destination.

In project planning, transportation and equipment routing among project sites are suitable problems for the application of the shortest-path algorithm. Possible extensions of the basic shortest-path problem may include the following:

1. *Smartest path between two project sites*: In some applications, the shortest path may not be the best path in terms of ease of transportation, resource requirement, and other performance considerations. The objective here is to quantify *smartness* and incorporate the measure into the shortest-path procedure.

2. *Safest path between two project sites*: In this case, the objective is to find the path that offers the highest measure of safety. We will need to quantify *safety* so that the measure can be incorporated into the shortest-path procedure.

3. *Cheapest path between project sites*: For some problems, the shortest path might turn out to be the most expensive. Commuters using toll roads are familiar with this dilemma. The objective in the cheapest-path problem is to find the least-cost path between two points. Distance–cost relationships can be used as the basis for obtaining quantitative measures to incorporate into the conventional shortest-path procedure so as to minimize cost rather than distance. Some examples of path-to-path problems of interest are

    a. Shortest path between the originating node and the terminal node
    b. Shortest path between the origin and all nodes in the network
    c. Shortest paths between all nodes in the network
    d. The *n* shortest paths from the origin to the terminal node

Dijkstra's algorithm can be used to solve the shortest-path problem for networks with non-negative d_{ij} values. The algorithm can be summarized as follows:

1. *Step 1*: Assign to the origin (node 1), the *permanent* label $y_1 = 0$, and assign to every other node the *temporary* label $y_j = M$. Set $i = 1$.

2. *Step 2*: From node i, recompute the temporary labels $y_i = \text{Min}\{y_j, y_i + d_{ij}\}$, where node j is temporarily labeled and $d_{ij} < M$.

3. *Step 3*: Find the smallest of the temporary labels, say y_j. Label node I permanently with value y_j. Label node i permanently with value y_j.

4. *Step 4*: If all nodes are permanently labeled, stop. Otherwise, go to step 2.

Example 7.2

Figure 7.44 gives the information flow network for the ABC company. The numbers on each arc indicate the probability of successful communication between the two end nodes calculated, based on past performances.

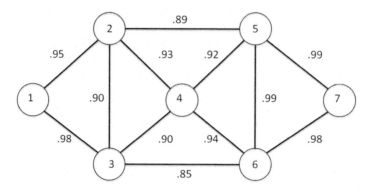

FIGURE 7.44
Information flow network.

Select the most reliable path from information source 1 to destination source 7. Let r_i be the reliability of arc i. Then, reliability of path P is defined as $R_p = \prod\limits_{i \in P} r_i$.

The most reliable path is the path that has the maximum product of the reliabilities of the arcs on the path. Taking the logarithm of r_i, one gets

$$R'_p = \log R_p = \sum_{i \in P} \log r_i = \sum_{i \in P} r'_i$$

The problem then reduces to finding the longest path in the network with negative arc distances. Dijkstra's algorithm will determine the path when the minimum operator is changed to the maximum operator in step 2. It is important to note that the arc distances are required to be nonpositive. Otherwise, positive cycles may exist and the algorithm may never terminate. Figure 7.45 indicates r'_i for each arc.

The algorithm starts by assigning a permanent label 0 to node 1, and temporary labels M to all the other nodes. Nodes 2 and 3 are directly connected to node 1. Hence, $y_2 = y_1 + d_{12} = -0.0513$ and $y_3 = y_1 + d_{13} = -0.0202$. Since $-0.0202 > -0.0513$, node 3 receives the permanent label of -0.0202. Nodes 2, 4, and 6 are adjacent to node 3. The new labels are $y_2 = \max(y_2, y_3 + d_{32}) = \max(-0.0513, -0.0202 \pm 0.1054) = -0.0513$, $y_4 = \max(M, -0.0202 \pm 0.1054) = -0.1256$, $y_6 = \max(M, -0.0202 \pm 0.1625) = -0.1827$.

Since $y_2 > y_4 > y_6$, y_2 receives the permanent label -0.0513. The rest of the steps are summarized in Table 7.7. The permanent labels are indicated by shading. The longest path from node 1 to node 7 is found as path 1–2–5–7 with a value of -0.1779. Taking the antilogarithm, one gets 0.84 as the reliability of the path.

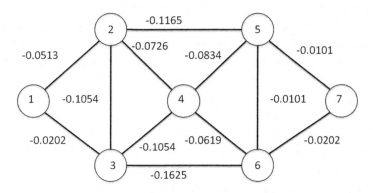

FIGURE 7.45
The network with r-prime values.

TABLE 7.7

Summary of the Dijkstra's Algorithm for the Example Problem

	Node						
Step	1	2	3	4	5	6	7
1	0	M	M	M	M	M	M
2	0	−0.0513	−0.0202	M	M	M	M
3	0	−0.0513	−0.0202	−0.1256	M	−0.1827	M
4	0	−0.0513	−0.0202	−0.1238	−0.1678	M	M
5	0	−0.0513	−0.0202	−0.1238	−0.1678	−0.1827	M
6	0	−0.0513	−0.0202	−0.1238	−0.1678	−0.1779	−0.1779
7	0	−0.0513	−0.0202	−0.1238	−0.1678	−0.1779	−0.1779

Goal Programming

One major shortcoming of LP is that only one objective can be considered at a time. In many real-world situations, the decision maker is faced with problems involving more than one objective. *Goal programming* is an extension of LP that permits decision makers to set and prioritize multiple goals. It is applicable to situations where not all goals can be satisfied equally and the aim is to minimize the overall dissatisfaction. An assignment of relative priorities to the goals being considered is required. The multiple goals do not have to be on the same measurement scale. For example, cost minimization, revenue maximization, and minimization of project duration may all coexist as goals in a goal programming problem. Goal programming solution techniques choose the values of the decision variables in such a way that deviations from the goals are minimized. If all the goals cannot be satisfied, goal programming will attempt to satisfy them in order of priority. The formulation of goal programming problems is similar to that of LP problems, except for the following requirements:

1. Explicit consideration of multiple goals
2. A measure of deviation from desired goals
3. Relative prioritization of the multiple goals

The objective of goal programming is to minimize the deviations from the desired goals. The priorities assigned to the goals are considered to be *preemptive*. Lower priority goals are satisfied only after higher priority goals have been satisfied. The general formulation of the goal programming model with preemptive weights is presented as follows. Let

n be the number of goals to be considered

x_i be the value of the ith decision variable in the problem

d_i^+ be the amount by which goal i is exceeded

d_i^- be the amount by which goal i is underachieved

P_i be the priority factor for the goal having the ith priority (The goal with the highest priority has a priority factor of P_1)

$P_i \ggg P_{i+1}$ so that there is no number $k > 0$ such that $nP_{i+1} \geq P_i$. That is, P_i is infinitely larger than P_{i+1}

The priority factors are included in the objective function with the appropriate deviational variables.

z is the objective function.

$x_1, x_2, ..., x_n$ are the n *decision variables*.

$c_1, c_2, ..., c_n$ are coefficients of decision variables in the *objective function*.

$a_{i1}, a_{i2}, ..., a_{in}$ are *coefficients* of decision variables in the ith constraint.

b_i are the right-hand side constants of the ith constraint ($i = 1, 2, ..., m$).

$$\text{Minimize: } z = \sum_{k}\sum_{i} P_k w_{i,k}^+ d_i^+ + \sum_{s}\sum_{i} P_s w_{i,s}^- d_i^-$$

$$\sum_{j=1}^{n} m_{ij}x_j - d_i^+ + d_i^- = g_i, \quad i = 1,2,\ldots,p$$

$$\text{Subject to: } \sum_{j=1}^{n} a_{ij}x_j \le b_i, i = p+1,\ldots,p+m$$

$$x_j, d_i^+, d_i^- \ge 0, \quad j = 1,\ldots,n; i = 1,\ldots,p$$

where
p represents the number of goals
m represents the number of nongoal constraints
n represents the number of decision variables
d_i^+, d_i^- represent the deviations from ith goal
P_k, P_s represent the priority factors
$w_{i,k}^+$ represents the relative weight of d_i^+ in the kth ranking
$w_{i,s}^-$ represents the relative weight of d_i^- in the sth ranking

The formulation of a goal programming model requires a very careful analysis. The formulation can be better seen by illustrative examples.

Consider the time–cost trade-off problem discussed in Integer Programming Approach for Resource Scheduling. Suppose that rather than finding a schedule with minimum crashing cost, the management is interested in finding a schedule that satisfies a set of goals as much as possible. The goals have been identified as follows:

1. The crashing cost should not exceed a certain budget.
2. The project should be crashed to T_p days.
3. Certain set of nodes should be realized by certain dates.

For example, for the problem given earlier in Figure 7.5, suppose the goals in the order of decreasing priority are as follows:

1. The crashing cost should not exceed $225.
2. The project should be completed by day 9.
3. Node 4 should be realized exactly on day 8.

The goal programming model for the time–cost trade-off problem can then be formulated as

$$\text{Minimize: } P_1 d_1^+ + P_2 d_2^+ + P_3\left(d_3^+ + d_3^-\right)$$

$$\sum_{(i,j)\in A} (u_{ij} - y_{ij}) \cdot a_{ij} - d_1^+ + d_1^- = 3000$$

$$t_5 - d_2^+ + d_2^- = 9$$

Subject to: $$t_4 - d_3^+ + d_3^- = 8$$

$$t_i + y_{ij} - t_j \le 0 \qquad \forall (i,j) \in A$$

$$d_{ij} \le y_{ij} \le u_{ij} \qquad \forall (i,j) \in A$$

It is suggested that the reader should compare this model with the model defined by Equations 7.5–7.8. More specifically, the model can be written as follows:

$$\text{Minimize: } P_1 d_i^+ + P_2 d_2^+ + P_3 \left(d_3^+ + d_3^- \right)$$

$$60(6 - y_6) + 250(4 - y_c) + 60(3 - y_2) +$$

$$100(5 - y_e) + 35(4 - y_f) + 25(2 - y_g) -$$

$$d_i^+ + d_1 = 225$$

$$t_5 - d_2^+ + d_2^- = 9$$

$$t_4 - d_3^+ + d_3^- = 8$$

$$t_1 + y_a - t_2 \le 0$$

$$t_1 + y_c - t_4 \le 0$$

$$t_1 + y_b - t_3 \le 0$$

$$t_2 + y_d - t_4 \le 0$$

Subject to: $$t_3 + y_e - t_4 \le 0$$

$$t_2 + y_f - t_5 \le 0$$

$$t_4 + y_g - t_5 \le 0$$

$$2 \le y_a \le 2$$

$$4 \le y_b \le 6$$

$$3 \le y_c \le 4$$

$$2 \le y_d \le 3$$

$$3 \le y_e \le 5$$

$$3 \le y_f \le 4$$

$$1 \le y_g \le 2,$$

TABLE 7.8

LINDO Solution for Goal Programming Example

Variable	Value	Reduced Cost
d_1^+	0.0000	100.0000
d_2^+	0.0000	10.0000
d_3^+	0.0000	1.0000
d_3^-	0.0000	0.0000
y_b	5.0000	0.0000
y_c	3.0000	0.0000
y_d	3.0000	0.0000
y_e	3.0000	0.0000
y_f	4.0000	0.0000
y_g	1.0000	0.0000
d_1^-	310.0000	0.0000
t_5	9.0000	0.0000
d_2^-	0.0000	0.0000
t_4	8.0000	0.0000
t_1	0.0000	0.0000
y_a	2.0000	0.0000
t_2	5.0000	0.0000
t_3	5.0000	0.0000

Objective function value = 0.0000.

where P_1, P_2, and P_3 are the priority weights of three goals, respectively. Assuming that $P_1 = 100$, $P_2 = 10$, and $P_3 = 1$, the LINDO solution for the previous problem is given in Table 7.8. Note that a value of zero for the objective function indicates that all the goals are satisfied. This is an infrequent occurrence in practice.

Resource Allocation Using Simulated Annealing

In Chapter 5, we discussed several priority orders and performed resource allocation based on the order. Simulated annealing is an iterative process that starts with an arbitrary order and performs a switch in the order and reallocates the resource with respect to the new order. If the project completion time is reduced, then the new order is kept and another switch is performed. Otherwise, the new order is kept with some probability, which decreases as the iteration number increases. The process stops when the order has not changed for a given number of iterations. The annealing step prevents the process from converging to a local minimum quickly and enables the process to get out of a local minimum point in the search for a global minimum. The simulated annealing procedure has been successfully applied to several combinatorial optimization problems such as the TSP, sequencing, scheduling, and facility layout problems. We will show its application to the resource allocation problem. The discussions will assume one resource problem to simplify the illustration. However, it can easily be extended to cover the multiple-resource allocation problem.

Simulated annealing method has its roots in thermodynamics and is specifically related to the way liquid freezes and crystallizes, and metals cool and anneal. The slow cooling process allows ample time for atoms to move around and achieve perfect crystallization. Similarly, in discrete minimization problems, the objective is to prevent rapid descent and allow the procedure to escape the local minimum and continue the descent process. This is only possible if the method allows ascent, that is, accepts solutions which increase the objective function value. The procedure controls the ascent process using a probability function defined as

$$P(z_i) = e^{-(z_i - z_{i-1})/kT}$$

where
  z_i is the ith solution value
  k is a constant
  T is the temperature

When $z_i < z_{i-1}$, the probability is greater than one which is then set equal to one. When $z_i < z_{i-1}$ indicating an ascent, the solution z_i is only accepted with probability $P(z_i)$. The temperature T is decreased (cooling) as the number of iterations increase, which in turn decreases the probability of ascent at later stages. This enables the process to converge to a good local minimum. The initial value of T and the annealing schedule differ with the problem and are generally defined after experimentation.

The initial parameters needed for simulated annealing are as follows:

1. $T(0)$, initial temperature value and a temperature function. We will assume that $T(0) = 65$ and decrease it by $k = 0.9$ after every iteration.
2. Stopping criterion, which can be a time limit or a test on some variable. We will stop when the sequence is not updated for five iterations.

The algorithm as applied to the resource allocation problem can be outlined as follows:

1. *Step 1*: Select an initial order S_0. S_0 must satisfy precedence relationship. Set $i = 0$. Select an initial temperature T. Set $n = 0$.
2. *Step 2*: Allocate the resource with respect to the priority order S_0. Basically, the first feasible job in the order is scheduled if the available resource is sufficient for day k. When no feasible allocation exists for day k, day $k + 1$ is considered. Let z_0 be the project completion time under S_0.
3. *Step 3*: Generate S_j. S_j must differ from S_i in only two positions, and also must satisfy the precedence relationship. S_j is referred to as the neighbor of S_i. If $z_j \leq z_i$, then set $i = j$. Set $n = 0$, $T = 0.9T$. Return to step 3. Otherwise, go to step 4.
4. *Step 4*: Let $r = $ random $(0, 1)$. If $r < e^{-(z_j - z_i)/T}$, then set $i = j$, $n = 0$, $T = 0.9T$. Return to step 3. Otherwise, set $n = n + 1$, $T = 0.9T$. If $n \geq 5$, stop. Otherwise, return to step 3.

Consider the resource allocation example problem discussed earlier. The maximum daily resource available is 10 units. The set of jobs to be performed, the predecessors, durations, and resource requirements are given in Table 7.9.

Suppose the procedure starts with the arbitrary feasible initial order ABCDEFG. When the resource is allocated according to this order, $z_0 = 13.17$ days. Possible switches with A are B and C. Job B can be switched with jobs, A, C, D, and F. Job C can be switched with jobs F and D. Similarly, job D with jobs C, F, and E, job E with jobs D and F, job F with jobs C, D, E, and G, and job G with job F. Suppose S_1 is defined by switching jobs A and C.

TABLE 7.9

Problem Data for Simulated Annealing Example

Jobs	Immediate Predecessors	Time	Resource Requirement
A	-	2.17	3
B	-	6.00	5
C	-	3.83	4
D	A	2.83	2
E	C	5.17	4
F	A	4.00	2
G	B,D,E	2.00	6

Therefore, the new order is CBADEFG. Resource allocation results in $z_1 = 13.17$. The new order is accepted with probability 1 since $z_1 = z_0$.

The feasible switched on S_2 to define a neighbor order S_3 are

1. Job C with job A or B
2. Job B with jobs C, A, or E
3. Job A with job C
4. Job D with jobs E or F
5. Job E with job D or F
6. Job F with jobs D, E, or G
7. Job G with job F

Note that job B can be switched with job E since the successor of B (which is job G) comes after job E in the sequence. Suppose job D is switched with job F. The new order is CBAFEDG, which results in $z_3 = 13.17$. Hence, the order is accepted. Suppose job F is switched with job E next, which leads to CBEAFDG and $z_4 = 13.17$. Switching job A with job E, one gets the order CBEAFDG and $z_5 = 13.00$. The procedure continues in this fashion until no new order is accepted in five consecutive iterations. For this problem $z = 13$ is the optimal solution. Figure 7.46 is the schedule chart for the order CBEAFDG.

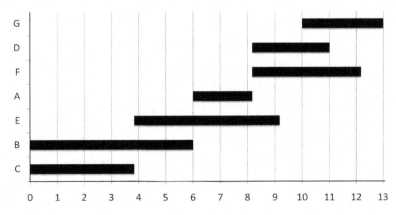

FIGURE 7.46
Schedule chart for CBEAFDG.

Exercises

7.1 Repeat the activity scheduling example presented in this chapter with the assumption that a 15-min break is required between the completion of one activity and the start of the next activity.

7.2 A delivery person makes deliveries to two project sites, site A and site B. He receives $1,000 per delivery to site A and $200 per delivery to site B. He is required to make at least five deliveries per week to site B. Each delivery to either site A or site B takes him 3h to complete. There is an upper limit of 35h of delivery time per week. The objective of the delivery person is to increase his delivery income as much as possible. Formulate his problem as an LP model.

7.3 Solve the LP model of the example presented in this chapter on the number of workers to assign to a project workstation. Discuss how you would go about implementing the LP solution in a practical setting.

7.4 Develop and solve the knapsack formulation to schedule the project presented in Table 7.10, the resource for which is presented in Table 7.11.

   Draw the Gantt chart for the schedule. Use the following notation in your formulation:

   C_t is the set of activities eligible for scheduling at time t

   CAF_i is the relative priority weight for activity i

   R_{jt} is the units of resource type j available at time t

TABLE 7.10

Activity Data for Exercise 7.4

Activity	Predecessor	Duration (Days)	Resource Requirement (Type1, Type 2)	Priority Weight
A	-	6	2, 5	50.1
B	-	5	1, 3	57.1
C	-	2	3, 1	56.8
D	A, B	4	2, 0	100
E	B	5	1, 4	53.2
F	C	6	3, 1	51.8
G	B,F	1	4, 3	44.9
H	D, E	7	2, 3	54.9

TABLE 7.11

Resource Data for Exercise 7.4

Time Period	Units Available (Type1, Type 2)
0 to 5	3, 5
5+ to 10	2, 3
10+ to 14	3,4
14+ to 20	4, 4
20+ to 22	5, 3
22+ to 30	4, 5
30+ and up	6, 6

$\overline{R_{jt}}$ is the units of resource type j in use at time t

r_{ij} is the units of resource type j required by activity i

x_{it} is the indicator variable for scheduling activity i at time t

$x_{it} = 1$ if activity i is scheduled at time t

$x_{it} = 0$ if activity i is not scheduled at time t

There are two resource types. Resource availability varies from day to day based on the schedule as follows:

7.5 Solve the resource assignment problem with the following cost matrix:

$$\begin{bmatrix} 3 & 7 & 2 & 4 \\ 3 & 6 & 5 & 7 \\ 6 & 4 & 5 & 6 \\ 4 & 3 & 4 & 5 \end{bmatrix}$$

7.6 Starting with Vogel's approximation as the starting solution, solve the transportation problem example presented in this chapter. The example was previously solved with the northwest-corner starting solution.

7.7 Golden Eagles, Inc. produces three styles of portable stadium seats: high-rider style, low-rider style, and line-hugger style. The seats are made of high-impact composite material and are produced in two steps, assembly and painting. The schedule for labor and material inputs for each seat style are as follows:

Assembly labor
   1 h per high-rider seat
   3 h per low-rider seat
   1.5 h per line-hugger seat

Painting labor
   1.5 h per high-rider seat
   1.5 h per low-rider seat
   4 h per line-hugger seat

Material requirement
   4 pounds per high-rider seat
   3 pounds per low-rider seat
   6 pounds per line-hugger seat
   2,000 h of assembly labor hours are available per month
   1,000 h of painting labor hours are available per month

The contributions to profit for the styles are $35, $40, and $65, respectively. The company has two equally desirable goals: minimization of idle time in painting, and making a monthly profit of $20,000. Set up this problem as a goal programming problem and show the initial tableau.

7.8 A project manager has four crews that can be hired on a temporary basis during peak construction periods. The manager currently has three projects more than can be handled with his regular crews. It is desired to determine the crew assignments that will minimize the total project cost. The costs of using certain crews for certain projects are presented as follows:

Crew 1
   Project A: $4,000

Project B: $3,000
Project C: $9,000

Crew 2

Project A: $7,000
Project B: $1,000
Project C: $8,000

Crew 3

Project A: $2,000
Project B: $6,000
Project C: $4,000

Crew 4

Project A: $9,000
Project B: $5,000
Project C: $5,000

7.9   Find the shortest path between the originating node and the terminal node for the following acyclic network.

7.10 Develop a computer program for finding the shortest path between all pairs of nodes in a directed acyclic network. Use the program to verify your solution in Exercise 7.9.

7.11 Develop a general LP formulation for scheduling the project in Table 7.12. Resource requirements and precedence relationships must be satisfied in the model ($R_1 = 5$, $R_2 = 3$).

7.12 Use any software tool available to you to solve the LP model in Exercise 7.11.

7.13 White Water Fun Park is planning to add another ride that consists of three main parts A, B, and C. The manager anticipates that it will take 14 days to receive the material. It should then take 7 days to assemble the components of part A which is then to be tested separately, which will take 3 days. It should take 6 and 4 days to complete the assembly of the components of parts B and C, respectively. Then, parts B and C have to be assembled (2 days) and attached to part A (5 days). The completed ride has to be tested for 21 days before it can be opened to the public. Draw the project network (A) and determine the critical path. Suppose on day 18 the project progress is evaluated, the following report has been generated. Assembly of part A is on time, part B will be ready 2 days early, part C will be ready a day late. Will the critical path remain the same? If not, find the new critical path.

7.14 Develop an integer programming model for the resource allocation problem in Figure 7.47, assuming that daily resource availability is 5 units. Use LINDO or any other integer programming software to determine an optimal solution to your model. Discuss the results. Draw the project schedule chart (Gantt chart).

7.15 Consider the project network given in Figure 7.48.

TABLE 7.12

Activity and Resource Data for Exercise 7.11

Activity	Predecessor	Duration (Days)	Resource Requirement (Type1, Type 2)
A	-	3	0, 3
B	A	4	2, 1
C	A	2	1, 1
D	B, C	7	5, 2

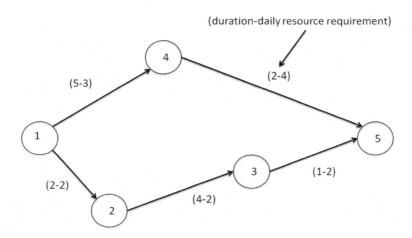

FIGURE 7.47
Resource allocation problem for Exercise 7.14.

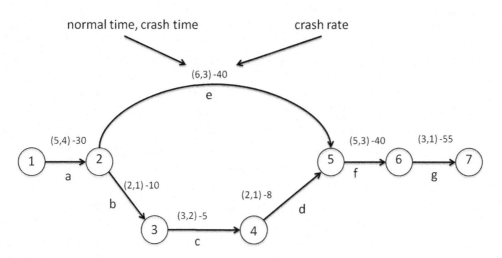

FIGURE 7.48
Project network for Exercise 7.15.

a. Determine the critical path for the network when each activity is realized at its normal time.

b. Suppose the project must be completed in 15 days. Formulate an LP model that will minimize the cost of meeting the deadline.

c. Solve the model in (b) using an LP software.

d. Generate the time–cost trade-off curve using the network flow approach.

7.16 A fashion clothing warehouse is located in Oklahoma City and serves five stores located in Norman, Tulsa, Oklahoma City, Muskogee, and Stillwater. Each week a truck delivers clothes from the warehouse to the stores and returns to the warehouse. The distances between each pair of locations (in miles) is provided in Table 7.13. Determine the order in which the truck should visit the five stores so that the total distance traveled is minimized.

TABLE 7.13

Data for Exercise 7.13

	Warehouse	Norman	Oklahoma City	Tulsa	Muskogee	Stillwater
Warehouse	0	45	10	105	157	54
Norman	45	0	35	130	140	80
Oklahoma City	10	35	0	115	90	64
Tulsa	105	130	115	0	52	64
Muskogee	157	140	90	52	0	85
Stillwater	54	80	64	64	85	0

7.17 Solve Exercise 7.14 (integer programming) using the simulated annealing procedure. Use $T = 60$, $k = 0.8$ and stop when the order does not change in three iterations.

References

Badiru, A.B. (1993). Activity resource assignments using critical resource diagramming, *Project Management Journal*, 14(3): 15–21.

Badiru, A.B. and Pulat, P.S. (1995). *Comprehensive Project Management: Integrating Optimization Models, Management Principles, and Computers*, Prentice Hall, Upper Saddle River, NJ, pp. 162–209.

8

Project Cost Systems

Economic Analysis Process

Like a system of differential equations, project cost elements consist of several parameters and are subject to differing factors in time-varying profiles. All these elements and factors must be understood and evaluated systematically over the project life cycle. Capital, in the form of money, is one of the factors that sustain organizational projects. It is necessary to structurally consider the implications of committing capital to a project over a period of time. The approach of project cost systems analysis and control helps us achieve that goal.

The time value of money is an important factor in economic consideration of projects. This is particularly crucial for long-term projects that are subject to changes in several cost parameters. Both the timing and quantity of cash flow are important for project management. The evaluation of a project alternative requires consideration of the initial investment, depreciation, taxes, inflation, economic life of the project, salvage value, and cash flow. These fundamental concepts play a central role in the contents of this chapter.

The process of economic analysis has the following basic components. The specific components will differ depending on the nature and prevailing circumstances of a project (Badiru and Omitaomu, 2007). This list is, however, representative of what a project cost analyst might expect to encounter:

1. Problem identification
2. Problem definition
3. Development of metrics and parameters
4. Search for alternate solutions
5. Selection of the preferred solutions
6. Implementation of the selected solution
7. Monitoring and sustaining the project

Simple and Compound Interest Rates

Interest rates are used to quantify the time value of money, which may be defined as the value of capital committed to a project or business over a period of time. Interest *paid* is the cost on borrowed money, and interest *earned* is the benefit on saved or invested money.

Interest rates can be calculated as simple rates or compound rates. *Simple interest* is interest paid only on the principal, whereas *compound interest* is interest paid on both the principal and the accrued interest.

Let

$$F_n = \text{Future value after } n \text{ periods;}$$

$$P = \text{Initial investment amount (the principal);}$$

$$I = \text{Interest rate per interest period;}$$

$$n = \text{number of investment or loan periods;}$$

$$r = \text{nominal interest rate per year;}$$

$$m = \text{number of compounding periods per year;}$$

$$i_a = \text{effective interest rate per compounding period;}$$

$$i = \text{effective interest rate per year.}$$

The expressions for computing interest amounts based on simple interest and compound interest are presented next:

$$\text{Simple interest: } I_n = P(i)(n)$$

$$\text{Compound interest: } I_n = iF_{(n-1)}$$

$F_{(n-1)}$ is the future value at period $(n - 1)$, which is the period immediately preceding the one for which the interest amount is being computed. The future values at time n are computed as

$$\text{Simple interest: } F_n = P(1 + ni)$$

$$\text{Compound interest: } F_n = P(^1$$

F_n is the accumulated value after n periods. At $n = 0$ and 1, the future values for both simple and compound interest are equal as shown in Table 8.1. In other words, simple and compound interest calculations yield the same results for F_n only when $n = 0$ or $n = 1$. This is computationally verified as shown in the following:

Set the simple interest formula equal to the compound interest formula. That is,

$$F_n = P(1 + in) \equiv F_n = P(^1$$

$$\Rightarrow P(1 + in) = P(^1$$

$$\Rightarrow (1 + in) = (^1$$

$$\Rightarrow (^1 - (1 + in) = 0$$

$$\Rightarrow n = 0 \quad \text{or} \quad n = 1$$

TABLE 8.1

Comparison of Simple Interest and Compound Interest Computations

Periods, N	Cash Flow	Simple Interest	Compound Interest
0		$F_0 = P + I_0$	$F_0 = P + I_0$
		$= P + P(i)(0)$	$= P + P(i)(0)$
		$= P$	$= P$
1		$F_1 = P + I_1$	$F_1 = F_0 + I_1$
		$= P + P(i)(1)$	$= F_0 + iF_0$
		$= P(1 + i)$	$= F_0(1 + i)$
			$= P(1 + i)$
2		$F_2 = P + I_2$	$F_2 = F_1 + I_2$
		$= P + P(i)(2)$	$= F_1 + iF_1$
		$= P(1 + 2i)$	$= F_1(1 + i)$
			$= P(1 + i)(1 + i)$
			$= P(1 + i)^2$
⋮	⋮	⋮	⋮
N		$F_n = P(1 + ni)$	$F_n = P(1 + i)^n$

It can be seen that compound interest calculations represent a compound sum of a series of one-period simple interest calculations. Simple interest is not widely used in economic analysis, primarily because no lending institution will want to charge simple interest. However, an understanding of the fact that simple interest is equivalent to compound interest when $n = 1$, sometimes leads to the trickery of achieving compound interest by executing a series of one-period simple interest calculations.

Investment Life for Multiple Returns

A topic that is often of intense interest in many investment scenarios is how long it will take a given amount to reach a certain multiple of its initial level. The "Rule of 72" is one simple approach to calculating how long it will take an investment to double in value at a given interest rate per period. The Rule of 72 gives the following formula for estimating the doubling period:

$$n = \frac{72}{i}$$

where i is the interest rate expressed in percentage. Referring to the single payment compound amount factor, we can set the future amount equal to twice the present amount and then solve for n, the number of periods. That is, $F = 2P$. Thus,

$$2P = P(1+i)^n$$

Solving for n in the previous equation yields an expression for calculating the exact number of periods required to double P:

$$n = \frac{\ln(2)}{\ln(1+i)}$$

where i is the interest rate expressed in decimals. When exact computation is desired, the length of time it would take to accumulate m multiple of P is expressed in its general form as

$$n = \frac{\ln(m)}{\ln(1+i)}$$

where m is the desired multiple. For example, at an interest rate of 5% per year, the time it would take an amount P to double in value ($m = 2$) is 14.21 years. This, of course, assumes that the interest rate will remain constant throughout the planning horizon. Table 8.2 presents a tabulation of the values calculated from both approaches. Figure 8.1

TABLE 8.2

Evaluation of the Rule of 72

$i\%$	n (Rule of 72)	n (Exact Value)
0.25	288.00	277.61
0.50	144.00	138.98
1.00	72.00	69.66
2.00	36.00	35.00
5.00	14.20	17.67
8.00	9.00	9.01
10.00	7.20	7.27
12.00	6.00	6.12
15.00	4.80	4.96
18.00	4.00	4.19
20.00	3.60	3.80
25.00	2.88	3.12
30.00	2.40	2.64

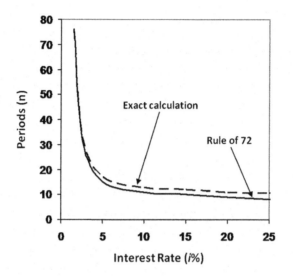

FIGURE 8.1

Evaluation of investment life for double return.

shows a graphical comparison of the results from use of the Rule of 72 to use of the exact calculation.

Nominal and Effective Interest Rates

The compound interest rate, which we will refer to as simply "interest rate," is used in economic analysis to account for the time value of money. Interest rates are usually expressed as a percentage, and the interest period (the time unit of the rate) is usually a year. However, interest rates can also be computed more than once a year. Compound interest rates can be quoted as *nominal interest rates* or as *effective interest rates*.

A *nominal interest rate* is the interest rate as quoted without considering the effect of any compounding. It is not the real interest rate used for economic analysis; however, it is usually the quoted interest rate because it is numerically smaller than the effective interest rate. It is equivalent to the annual percentage rate, which is usually quoted for loan and credit card purposes. The expression for calculating the nominal interest rate follows:

$$r = (\text{Interest rate per period}) \times (\text{Number of periods})$$

The format for expressing r is as follows:

$$r\% \text{ per time period } t$$

The effective interest rate can be expressed either per year or per compounding period. It is the effective interest rate per year that is used in engineering economic analysis calculations. It is the annual interest rate taking into consideration the effect of any compounding during the year. It accounts for both the nominal rate and the compounding frequency. Effective interest rate *per year* is given by

$$i = (^1 - 1$$

$$= \left(1 + \frac{r}{m}\right)^m - 1$$

$$= \left(\frac{F}{P}, \frac{r}{m\%}, m\right) - 1$$

Effective interest rate *per compounding period* is given by

$$i_a = (1 + i)^{1/m} - 1 = \frac{r}{m}$$

When compounding occurs more frequently, the compounding period becomes shorter; hence, we have the phenomenon of continuous compounding. This situation can be seen in the stock markets. The effective interest rate for *continuous compounding* is given by

$$i = e^r - 1$$

Note that the time period for *i* and *r* must be the same in using the earlier equations.

Example 8.1

The nominal annual interest rate of an investment is 9%. What is the effective annual interest rate if the interest is

　　i. Payable, or compounded, quarterly?
　　ii. Payable, or compounded, continuously?

Solution

　　i. Using the effective interest rate formula, the effective annual interest rate compounded quarterly is $(1 + (0.09/4))^4 - 1 = 9.31\%$.
　　ii. Using the equation for continuous rate, the effective annual interest rate compounded continuously is $e^{0.09} - 1 = 9.42\%$.

The slight difference between each of these values and the nominal interest rate of 9% becomes a big concern if the period of computation is in double digits. The effective interest rate must always be used in all computations. Therefore, a correct identification of the nominal and effective interest rates is very important. See the following example.

Example 8.2

Identify the following interest rate statements as either nominal or effective:

　　i. 14% per year
　　ii. 1% per month, compounded weekly

iii. Effective 15% per year, compounded monthly
iv. 1.5% per month, compounded monthly
v. 20% per year, compounded semiannually

Solution

i. This is an *effective interest rate*. This may also be written as 14% per year, compounded yearly.
ii. This is a *nominal interest rate* since the rate of compounding is not equal to the rate of interest period.
iii. This is an effective interest for yearly rate.
iv. This is an *effective interest for monthly rate*. A new rate should be computed for yearly computations. This may also be written as 1.5% per month.
v. This is a *nominal interest rate* because the rate of compounding and the rate of interest period are not the same.

Cash-Flow Patterns and Equivalence

The basic reason for performing economic analysis is to provide information that helps in making choices between mutually exclusive projects competing for limited resources. The cost performance of each project will depend on the timing and levels of its expenditures. By using various techniques of computing cash-flow equivalence, we can reduce competing project cash flows to a common basis for comparison. The common basis depends, however, on the prevailing interest rate. Two cash flows that are equivalent at a given interest rate are not equivalent at a different interest rate. The basic techniques for converting cash flows from an interest rate at one point in time to the interest rate at another are presented in this section.

A cash-flow diagram (CFD) is a graphical representation of revenues (cash inflows) and expenses (cash outflows). If several cash flows occur during the same time period, a net CFD is used to represent the differences in cash flows. CFDs are based on several assumptions:

1. Interest rate is computed once in a time period.
2. All cash flows occur at the end of the time period.
3. All periods are of the same length.
4. The interest rate and the number of periods are of the same length.
5. Negative cash flows are drawn downward from the time line.
6. Positive cash flows are drawn upward from the time line.

Cash-flow conversion involves the transfer of project funds from one point in time to another. There are several factors used in the conversion of cash flows.
Let

P be the cash flow value at the present time period
F be the cash flow value at some time in the future

A be the series of equal, consecutive, and end-of-period cash flow (i.e., annuity)

G be the uniform arithmetic gradient increase in period-by-period cash flow

t be the time period (years, months, days, etc.)

n be the total number of time periods

i be the rate per time period (expressed in percentage)

In many cases, the interest rate used in performing economic analysis is set equal to the minimum attractive rate of return (MARR) of the decision maker. The MARR is also sometimes referred to as the *hurdle rate,* the *required internal rate of return (IRR)*, the *return on investment (ROI)*, or the *discount rate.* The value of MARR is chosen with the objective of maximizing the economic performance of a project.

Compound Amount Factor

The procedure for the single payment compound amount factor finds a future sum of money, F, that is equivalent to a present sum of money, P, at a specified interest rate, i, after n periods. This is calculated as

$$F = P(1+i)^n$$

A graphical representation of the relationship between P and F is shown in Figure 8.2.

Example 8.3

A sum of $5,000 is deposited in a project account and is left there to earn interest for 15 years. If the interest rate per year is 12%, the compound amount after 15 years can be calculated as follows:

$$F = \$5{,}000(1+0.12)^{15}$$

$$= \$27{,}367.85$$

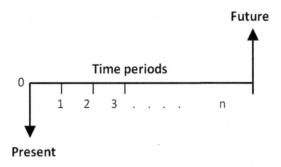

FIGURE 8.2
Single payment compound amount cash flow.

Present Worth Factor

The present worth (PW) factor computes P when F is given. The PW factor is obtained by solving for P in the equation for the compound amount factor. That is,

$$P = F(1+i)^{-n}$$

Suppose it is estimated that $15,000 would be needed to complete the implementation of a project 5 years in the future, how much should be deposited in a special project fund now so that the fund would accrue to the required $15,000 exactly in 5 years? If the special project fund pays interest at 9.2% per year, the required deposit would be

$$P = \$15,000(1+0.092)^{-5}$$

$$= \$9,660.03$$

Uniform Series PW Factor

The uniform series PW factor is used to calculate the PW equivalent, P, of a series of equal end-of-period amounts, A. Figure 8.3 shows the uniform series cash flow. The derivation of the formula uses the finite sum of the PW of the individual amounts in the uniform series cash flow, as shown in the following.

Some formulae for series and summation operations are presented in Appendix A.

$$P = \sum_{t=1}^{n} A(1+i)^{-t}$$

$$= A\left[\frac{(1+i)^{n}-1}{i(1+i)^{n}}\right]$$

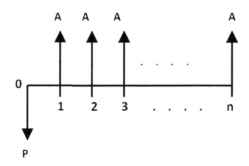

FIGURE 8.3
Uniform series cash flow.

Example 8.4

Suppose that the sum of $12,000 must be withdrawn from an account to meet the annual operating expenses of a multiyear project. The project account pays interest at 7.5% per year compounded on an annual basis. If the project is expected to last 10 years, how much must be deposited in the project account now so that the operating expenses of $12,000 can be withdrawn at the end of every year for 10 years? The project fund is expected to be depleted to zero by the end of the last year of the project. The first withdrawal will be made 1 year after the project account is opened, and no additional deposits will be made in the account during the project life cycle. The required deposit is calculated to be

$$P = \$12,000 \left[\frac{(1+0.075)^{10} - 1}{0.075(1+0.075)^{10}} \right]$$

$$= \$82,368.92$$

Uniform Series Capital Recovery Factor

The capital recovery formula is used to calculate the uniform series of equal end-of-period payments, A, that are equivalent to a given present amount, P. This is the converse of the uniform series present amount factor. The equation for the uniform series capital recovery factor is obtained by solving for A in the uniform series present amount factor. That is,

$$A = P \left[\frac{i(1+i)^n}{(1+i)^n - 1} \right]$$

Example 8.5

Suppose a piece of equipment needed to launch a project must be purchased at a cost of $50,000. The entire cost is to be financed at 13.5% per year and repaid on a monthly installment schedule over 4 years. It is desired to calculate what the monthly loan payments will be. It is assumed that the first loan payment will be made exactly 1 month after the equipment is financed. If the interest rate of 13.5% per year is compounded monthly, then the interest rate per month will be 13.5%/12 = 1.125% per month. The number of interest periods over which the loan will be repaid is 4(12) = 48 months. Consequently, the monthly loan payments are calculated to be

$$A = \$50,000 \left[\frac{0.01125(1+0.01123)^{48}}{(1+0.01125)^{48} - 1} \right]$$

$$= \$1,353.82$$

Uniform Series Compound Amount Factor

The series compound amount factor is used to calculate a single future amount that is equivalent to a uniform series of equal end-of-period payments. The cash flow is shown in

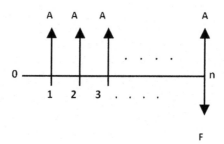

FIGURE 8.4
Uniform series compound amount cash flow.

Figure 8.4. Note that the future amount occurs at the same point in time as the last amount in the uniform series of payments.

The factor is derived as shown in the following formula:

$$F = \sum_{t=1}^{n} A(^1$$

$$= A\left[\frac{(1+i)^n - 1}{i}\right]$$

Example 8.6

If equal end-of-year deposits of $5,000 are made to a project fund paying 8% per year for 10 years, how much can be expected to be available for withdrawal from the account for capital expenditure immediately after the last deposit is made?

$$F = \$5,000\left[\frac{(1+0.08)^{10} - 1}{0.08}\right]$$

$$= \$72,432.50$$

Uniform Series Sinking Fund Factor

The sinking fund factor is used to calculate the uniform series of equal end-of-period amounts, A, that are equivalent to a single future amount, F. This is the reverse of the uniform series compound amount factor. The formula for the sinking fund is obtained by solving for A in the formula for the uniform series compound amount factor. That is,

$$A = F\left[\frac{i}{(1+i)^n - 1}\right]$$

Example 8.7

How large are the end-of-year equal amounts that must be deposited into a project account so that a balance of $75,000 will be available for withdrawal immediately after the 12th annual deposit is made? The initial balance in the account is zero at the beginning of the first year. The account pays 10% interest per year. Using the formula for the sinking fund factor, the required annual deposits are

$$A = \$75,000 \left[\frac{0.10}{(1+0.10)^{12} - 1} \right]$$

$$= \$3,507.25$$

Capitalized Cost Formula

Capitalized cost refers to the present value of a single amount that is equivalent to a perpetual series of equal end-of-period payments. This is an extension of the series PW factor with an infinitely large number of periods. This is shown graphically in Figure 8.5.

Using the limit theorem from calculus as n approaches infinity, the series PW factor reduces to the following formula for the capitalized cost:

$$P = \lim_{n \to \infty} A \left[\frac{(1+i)^n - 1}{i(1+i)^n} \right]$$

$$= A \left\{ \lim_{n \to \infty} \left[\frac{(1+i)^n - 1}{i(1+i)^n} \right] \right\}$$

$$= A \left(\frac{1}{i} \right)$$

There are several real-world investments that can be computed using this idea of capitalized cost formula. These include scholarship funds, maintenance of public buildings, and maintenance of roads and bridges, among others.

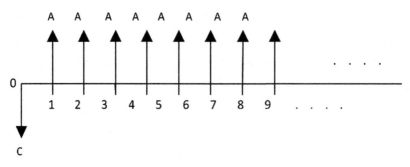

FIGURE 8.5
Capitalized cost cash flow.

Example 8.8

How much should be deposited in a general fund to service a recurring public service project to the tune of $6,500 per year forever if the fund yields an annual interest rate of 11%? Using the capitalized cost formula, the required onetime deposit to the general fund is as follows:

$$P = \frac{\$6,500}{0.11}$$

$$= \$59,090.91$$

Example 8.9

A football stadium is expected to have an annual maintenance expense of $75,000. What amount must be deposited today in an account that pays a fixed interest rate of 12% per year to provide for this annual maintenance expense forever?

The amount of money to be deposited today is

$$\frac{75,000}{0.12} = \$625,000$$

That is, if $625,000 is deposited today into this account, it will pay $75,000 annually forever. This is the power of compounded interest rate.

Permanent Investments Formula

This measure is the reverse of capitalized cost. It is the net annual value (NAV) of an alternative that has infinitely long period. Public projects such as bridges, dams, irrigation systems, and railroads fall into this category. In addition, permanent and charitable organization endowments are evaluated using this approach. The NAV in case of permanent investments is given by

$$A = Pi$$

Example 8.10

If we deposit $25,000 in an account that pays a fixed interest rate of 10% today, what amount can be withdrawn each year to sponsor college scholarships forever?

Solution

Using the permanent investments formula, the required annual college scholarship worth is

$$A = \$25,000 \times 0.10 = \$2,500$$

The formulas presented previously represent the basic cash flow conversion factors. The factors are tabulated in engineering economy books. Variations in the cash-flow profiles include situations where payments are made at the beginning of each period rather than at the end and situations where a series of payments contains unequal amounts. Conversion formulas can be derived mathematically for those special cases using the basic factors presented earlier. Conversion factors for some complicated cash-flow profiles are discussed in the following text.

Arithmetic Gradient Series

The gradient series cash flow involves an increase of a fixed amount in the cash flow at the end of each period. Thus, the amount at a given point in time is greater than the amount during the preceding period by a constant amount. This constant amount is denoted by G. Figure 8.6 shows the basic gradient series, in which the base amount at the end of the first period is zero.

The size of the cash flow in the gradient series at the end of period t is calculated as follows:

$$A_t = (t-1)G, \quad t = 1,2,\ldots,n$$

The total present value of the gradient series is calculated using the present amount factor to convert each individual amount from time t to 0 at an interest rate of $i\%$ per period and then by summing up the resulting present values. The finite summation reduces to a closed form, as is shown in the following:

$$P = \sum_{t=1}^{n} A_t (1+i)^{-t}$$

$$= \sum_{t=1}^{n} (t-1)G(1+i)^{-t}$$

$$= G\sum_{t=1}^{n} (t-1)(1+i)^{-t}$$

$$= G\left[\frac{(1+i)^n - (1+ni)}{i^2 (1+i)^n}\right]$$

Example 8.11

The cost of supplies for a 10-year project increases by \$1,500 every year, starting at the end of the second year. There is no cost of supplies at the end of the first year. If the interest rate is 8% per year, determine the present amount that must be set aside at time

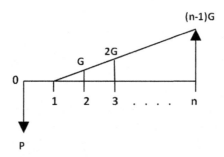

FIGURE 8.6
Arithmetic gradient cash flow with zero base amount.

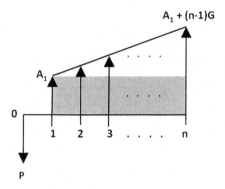

FIGURE 8.7
Arithmetic gradient cash flow with nonzero base amount.

zero to take care of all the future expenditure of supplies. We have $G = 1,500$, $i = 0.08$, and $n = 10$. Using the arithmetic gradient formula, we obtain the following:

$$P = 1,500 \left[\frac{1 - (1 + 10(0.08))(1 + 0.08)^{-10}}{(0.08)^2} \right]$$

$$= \$1,500(25.9768)$$

$$= \$38,965.20$$

In many cases, an arithmetic gradient starts with some base amount at the end of the first period and then increases by a constant amount thereafter. The nonzero base amount is denoted as A_1. Figure 8.7 shows this type of cash flow.

The calculation of the present amount for such cash flows requires breaking the cash flow into a uniform series cash flow of amount A_1 and an arithmetic gradient cash flow with zero base amount. The uniform series PW formula is used to calculate the PW of the uniform series portion, and the basic gradient series formula is used to calculate the gradient portion. The overall PW is then calculated as follows:

$$P = P_{\text{uniformseries}} + P_{\text{gradientseries}}$$

$$= A_1 \left[\frac{(1+i)^n - 1}{i(1+i)^n} \right] + G \left[\frac{(1+i)^n - (1+ni)}{i^2(1+i)^n} \right]$$

Increasing Geometric Series Cash Flow

In an increasing geometric series cash flow, the amounts in the cash flow increase by a constant percentage from period to period. There is a positive base amount, A_1, at the end of period one. Figure 8.8 shows an increasing geometric series.

The amount at time t is denoted as

$$A_t = A_{t-1}(1 + j), t = 2, 3, \ldots, n$$

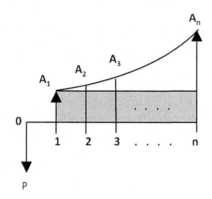

FIGURE 8.8
Increasing geometric series cash flow.

where j is the percentage increase in the cash flow from period to period. By doing a series of back substitutions, we can represent A_t in terms of A_1 instead of in terms of A_{t-1}, as shown next:

$$A_2 = A_1(1+j)$$

$$A_3 = A_2(1+j) = A_1(1+j)(1+j)$$

$$\vdots$$

$$A_t = A_1(^1, \quad t = 1,2,3,\dots,n$$

The formula for calculating the PW of the increasing geometric series cash flow is derived by summing the present values of the individual cash flow amounts. That is,

$$P = \sum_{t=1}^{n} A_t(^1$$

$$= \sum_{t=1}^{n}\left[A_1(1+j)^{t-1}\right](1+i)^{-t}$$

$$= \frac{A_1}{(1+j)}\sum_{t=1}^{n}\left(\frac{1+j}{1+i}\right)^{t}$$

$$= A_1\left[\frac{1-(1+j)^n(1+i)-n}{i-j}\right], \quad i \neq j$$

If $i = j$, the previous formula reduces to the limit as $i \to j$, shown as follows:

$$P = \frac{nA_1}{1+i}, \quad i = j$$

Example 8.12

Suppose funding for a 5-year project is to increase by 6% every year, with an initial funding of \$20,000 at the end of the first year. Determine how much must be deposited into a budget account at time zero in order to cover the anticipated funding levels if the budget account pays 10% interest per year. We have $j = 6\%$, $i = 10\%$, $n = 5$, and $A_1 =$ \$20,000. Therefore,

$$P = 20,000 \left[\frac{1 - (1 + 0.06)^5 (1 + 0.10)^5}{0.10 - 0.06} \right]$$

$$= \$20,000 (4.2267)$$

$$= \$84,533.60$$

Decreasing Geometric Series Cash Flow

In a decreasing geometric series cash flow, the amounts in the cash flow decrease by a constant percentage from period to period. The cash flow starts at some positive base amount, A_1, at the end of period one. Figure 8.9 shows a decreasing geometric series. The amount at time t is denoted as follows:

$$A_t = A_{t-1}(1 - j), \quad t = 2, 3, \ldots, n$$

where j is the percentage decrease in the cash flow from period to period.

As in the case of the increasing geometric series, we can represent A_t in terms of A_1.

$$A_2 = A_1(1 - j)$$

$$A_3 = A_2(1 - j) = A_1(1 - j)(1 - j)$$

$$\vdots$$

$$A_t = A_1(^1, \quad t = 1, 2, 3, \ldots, n$$

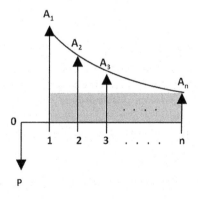

FIGURE 8.9
Decreasing geometric series cash flow.

The formula for calculating the PW of the decreasing geometric series cash flow is derived by finite summation, as in the case of the increasing geometric series. The final formula is

$$P = A_1 \left[\frac{1 - (1-j)^n (1+i)^{-n}}{i+j} \right]$$

Example 8.13

A contract amount for a 3-year project is expected to decrease by 10% every year with an initial contract of $100,000 at the end of the first year. Determine how much must be available in a contract reservoir fund at time zero in order to cover the contract amounts. The fund pays 10% interest per year. Since $j = 10\%$, $i = 10\%$, $n = 3$, and $A_1 = \$100,000$, we should have

$$P = 100,000 \left[\frac{1 + (1+0.10)^3 (1+0.10)^{-3}}{0.10 + 0.10} \right]$$

$$= \$100,000(2.2615)$$

$$= \$226,150$$

Internal Rate of Return

The IRR for a cash flow is defined as an interest rate that equates the future worth (FW) at time n or PW at time 0 of the cash flow to zero. If we let i^* denote IRR, we then have the following:

$$FW_{t=n} = \sum_{t=0}^{n} (\pm A_t)(1 + i^*)^{n-t} = 0$$

$$PW_{t=0} = \sum_{t=0}^{n} (\pm A_t)(1 + i^*)^{-t} = 0$$

where
   "+" is used in the summation for positive cash-flow amounts or receipts
   "−" is used for negative cash-flow amounts or disbursements
   A_t denotes the cash-flow amount at time t, which may be a receipt (+) or a disbursement (−)
   The value of i^* is referred to as *discounted cash-flow rate of return, IRR,* or *true rate of return*

The aforementioned procedure essentially calculates the net FW (NFW) or the net PW (NPW) of the cash flow. That is,

Net future worth = future worth of receipts − future worth of disbursements

$$NFW = FW_{(receipts)} - FW_{(disbursements)}$$

Net present worth = present worth of receipts – present worth of disbursements

$$NPW = PW_{(receipts)} - PW_{(disbursements)}$$

Setting the NPW or NFW equal to zero and solving for the unknown variable i determines the IRR of the cash flow.

Benefit–Cost Ratio

The computational methods described earlier are mostly used for private projects, since the objective of most private projects is to maximize profits. Public projects, on the other hand, are executed to provide services to the citizenry at no profit; therefore, they require a special method of analysis. Benefit–cost (B/C) ratio analysis is normally used for evaluating public projects. It has its origins in the Flood Act of 1936, which requires that for a federally financed project to be justified, its benefits must, at minimum, equal its costs. B/C ratio is the systematic method of calculating the ratio of project benefits to project costs at a discounted rate. For over 60 years, the B/C ratio method has been the accepted procedure for making "go" or "no-go" decisions on independent and mutually exclusive projects in the public sector.

The B/C ratio of a cash flow is the ratio of the PW of benefits to the PW of costs. This is defined as

$$\frac{B}{C} = \frac{\sum_{t=0}^{n} B_t (1+i)^{-t}}{\sum_{t=0}^{n} C_t (1+i)^{-t}}$$

$$= \frac{PW_{benefits}}{PW_{costs}} = \frac{AW_{benefits}}{AW_{costs}}$$

where
B_t is the benefit (receipt) at time t
C_t is the cost (disbursement) at time t

If the B/C ratio is greater than one, then the investment is acceptable. If the ratio is less than one, the investment is not acceptable. A ratio of one indicates a break-even situation for the project.

Example 8.14

Consider the following investment opportunity by Knox County.

Initial cost = $600,000

Benefit per year at the end of years 1 and 2 = 30,000

Benefit per year at the end of years 3 to 30 = 50,000

If Knox County sets the interest at 7% per year, would this be an economically feasible investment for the county?

Solution

For this problem, we will use both the PW and the annual worth (AW) approaches to show that both equations would give the same results.

Using PW

$$PW_{benefits} = 30,000\left(\frac{P}{A},7\%,2\right) + 50,000\left(\frac{P}{A},7\%,28\right)\left(\frac{P}{F},7\%,2\right)$$

$$= \$584,267.16$$

$$PW_{costs} = \$600,000$$

Using AW

$$AW_{benefits} = \left[30,000\left(\frac{P}{A},7\%,2\right) + 50,000\left(\frac{P}{A},7\%,28\right)\left(\frac{P}{F},7\%,2\right)\right]\left(\frac{A}{P},7\%,30\right)$$

$$= \$47,086.09$$

$$AW_{costs} = \$600,000\left(\frac{A}{P},7\%,30\right)$$

$$= \$48,354.00$$

$$\frac{B}{C} = \frac{584,267.16}{600,000.00} = \frac{47,086.09}{48,354.00} = 0.97$$

The *B/C* ratio indicates that the investment is not economically feasible since the *B/C* < 1.0. However, one can see that both PW and AW methods produced the same result.

Simple Payback Period

The term "payback period" refers to the length of time it will take to recover an initial investment. The approach does not consider the impact of the time value of money. Consequently, it is not an accurate method of evaluating the worth of an investment. However, it is a simple technique that is used widely to perform a "quick-and-dirty" or superficial assessment of investment performance. The technique considers only the initial cost. Other costs that may occur after time zero are not included in the calculation. The payback period is defined as the smallest value of $n(n_{min})$ that satisfies the following expression:

$$\sum_{t=1}^{n_{min}} R_t \geq C_0$$

where

R_t is the revenue at time t

C_0 is the initial investment

The procedure calls for a simple addition of the revenues period by period until enough total has been accumulated to offset the initial investment.

Example 8.15

An organization is considering of installing a new computer system that will generate significant savings in material and labor requirements for order processing. The system has an initial cost of $50,000. It is expected to save the organization $20,000 a year. The system has an anticipated useful life of 5 years with a salvage value of $5,000. Determine how long it would take for the system to pay for itself from the savings it is expected to generate. Since the annual savings are uniform, we can calculate the payback period by simply dividing the initial cost by the annual savings. That is,

$$n_{min} = \frac{\$50,000}{\$20,000}$$

$$= 2.5 \text{ years}$$

Note that the salvage value of $5,000 is not included in the preceding calculation since the amount is not realized until the end of the useful life of the asset (i.e., after 5 years). In some cases, it may be desired to consider the salvage value. In that case, the amount to be offset by the annual savings will be the net cost of the asset, represented here as

$$n_{min} = \frac{\$50,000 - \$5,000}{\$20,000}$$

$$= 2.25 \text{ years}$$

If there are tax liabilities associated with the annual savings, those liabilities must be deducted from the savings before calculating the payback period. The simple payback period (SPP) does not take the time value of money into consideration; however, it is a concept readily understood by people unfamiliar with economic analysis.

Discounted Payback Period

The discounted payback period (DPP) is a payback analysis approach in which the revenues are reinvested at a certain interest rate. The payback period is determined when enough money has been accumulated at the given interest rate to offset the initial cost as well as other interim costs. In this case, the calculation is done with the aid of the following expression:

$$\sum_{t=1}^{n_{min}} R_t (1+i)^{n_{min}-1} \geq \sum_{t=0}^{n_{min}} C_t$$

Example 8.16

A new solar cell unit is to be installed in an office complex at an initial cost of $150,000. It is expected that the system will generate annually a cost savings of $22,500 on the electricity bill. The solar cell unit will need to be overhauled every 5 years at a cost of $5,000 per overhaul. If the annual interest rate is 10%, find the DPP for the solar cell unit considering the time value of money. The costs of overhaul are to be considered in calculating the DPP.

Solution

Using the single payment compound amount factor for one period iteratively, the following solution is obtained:
Time: cumulative savings

1: $22,500
2: $22,500 + $22,500(1.10)^1 = $47,250
3: $22,500 + $47,250(1.10)^1 = $74,475
4: $22,500 + $74,475(1.10)^1 = $104,422.50
5: $22,500 + $104,422.50(1.10)^1 − $5,000 = $132,364.75
6: $22,500 + $132,364.75(1.10)^1 = $168,101.23

The initial investment is $150,000. By the end of period 6, we have accumulated $168,101.23, more than the initial cost. Interpolating between periods 5 and 6, we obtain

$$n_{min} = 5 + \frac{150,000 - 132,364.75}{168,101.25 - 132,364.75}(6-5)$$
$$= 5.49$$

That is, it will take 5.49 years, or 5 years and 6 months, to recover the initial investment.

Example 8.17

Consider a cash flow with the following data:

$n = 6$ in years
$A = \$30,000$ (series of end-of-period revenues)
$P = \$100,000$ cost at time 0
$X = \$20,000$ expenditure occurring at the end of year 6
MARR = 15% annual interest rate
a. Calculate SPP.
b. Calculate DPP.

Solution

To solve this problem, we can make a payback table to cumulatively compute the sum of revenues until we recover the initial cost of $100,000. For part (a), the cumulative sum is calculated by present value (PV) at $i = 0\% = \sum_{k=0}^{n} a_k$, which yields a payback period occurring at $n = 4$. For part (b), the cumulative sum is calculated by PV at $i = 15\% = \sum_{k=0}^{n} b_k$, which yields a payback period occurring at $n = 5$. The SPP of 4 years is when the cumulative PV becomes a positive value. The DPP of 5 years is when the discounted cumulative PV becomes a positive value. It must be noted that, in computing these values, the cash flows after SPP or DPP are not taken into consideration; therefore, both SPP and DPP techniques are usually used for initial screening of potential investment alternatives. They must never be used for final selection without considering other techniques such as net present value (NPV) and/or IRRs techniques.

Fixed and Variable Interest Rates

An interest rate may be fixed or may vary from period to period over the useful life of an investment, and companies, when evaluating investment alternatives, need also to consider the variable interest rates involved, especially in long-term investments. Some of the factors responsible for varying interest rates include changes in nominal and international economies, effects of inflation, and changes in market share. Loan rates, such as mortgage loan rates, may be adjusted from year to year based on the inflation index of the consumer price index. If the variations in interest rate from period to period are not large, cash-flow calculations usually ignore the effects of varying interest rates. However, the results of the computation will vary considerably if the variations in interest rates are large. In such cases, the varying interest rates should be considered in economic analysis, even though such consideration may become computationally involved.

Example 8.18

Find the PW (present value) of the CFD in Figure 8.10 if for $n < 5$, $i = 0.5\%$ and for $n > 4$, $i = 0.25\%$.

Solution

This is typical of several real-world cash flows. The computation must be carefully done to avoid the common errors. The present value is given as follows:

$$PV = A_0 + A_{1-4}\left(P/A, 0.5\%, 4\right) + \left[A_5\left(P/A, 0.25\%, 6\right) + G\left(P/G, 0.25\%, 6\right)\right]\left(P/F, 0.5\%, 4\right)$$

$$= 100 + 100\left(3.950\right) + \left[150(5.948) + 50(14.826)\right](0.9802)$$

$$= 100 + 395 + \left[892.2 + 741.3\right](0.9802)$$

$$= \$2096.16.$$

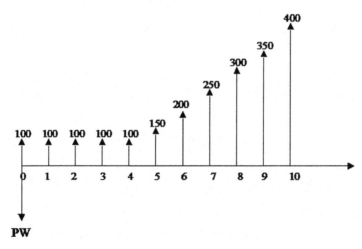

FIGURE 8.10
Cash flow diagram for Example 8.18.

Therefore, the present value for these cash flows with two different interest rates is $2,096.16. The interest rate for $n > 4$ affects only cash flows in periods 5–10, whereas the interest rate for $n < 5$ affects cash flows in periods 1–10.

Amortization of Capitals

Many capital investment projects are financed with external funds repaid according to an amortization schedule. A careful analysis must be conducted to ensure that the company involved can financially handle the amortization schedule. Several software programs are available for analyzing amortization schedules. Such a program can analyze installment payments, unpaid balance, principal amount paid per period, total installment payment, and current cumulative equity. It can also calculate the "equity break-even point" for the debt being analyzed. The equity break-even point indicates the time when the unpaid balance on a loan is equal to the cumulative equity on the loan. With this information, the basic cost of servicing the project debt can be evaluated quickly. The computational procedure for analyzing project debt follows the steps outlined in the following:

1. Given a principal amount, P, a periodic interest rate, i (in decimals), and a discrete time span of n periods, the uniform series of equal end-of-period payments needed to amortize P is computed as

$$A = \frac{P\left[i(1+i)^n\right]}{(1+i)-1}$$

   It is assumed that the loan is to be repaid in equal monthly payments. Thus, $A(t) = A$, for each period t throughout the life of the loan.

2. The unpaid balance after making t installment payments is given by

$$U(t) = \frac{A\left[1-(1+i)^{(t-n)}\right]}{i}$$

3. The amount of equity or principal amount paid with installment payment number t is given by

$$E(t) = A(1+i)^{t-n-1}$$

4. The amount of interest charge contained in installment payment number t is derived to be

$$I(t) = A\left[1-(1+i)^{t-n-1}\right]$$

where $A = E(t) + I(t)$.

5. The cumulative total payment made after t periods is denoted by

$$C(t) = \sum_{k=1}^{t} A(k)$$

$$= \sum_{k=1}^{t} A = t \cdot A$$

6. The cumulative interest payment after t periods is given by

$$Q(t) = \sum_{x=1}^{t} I(x)$$

7. The cumulative principal payment after t periods is computed as

$$S(t) = \sum_{k=1}^{t} E(k)$$

$$= A \sum_{k=1}^{t} (1+i)^{-(n-k+1)} = A \left[\frac{(1+i)^{t} - 1}{i(1+i)^{n}} \right]$$

where

$$\sum_{n=1}^{t} x^{n} = \frac{x^{x+1} - x}{x - 1}$$

8. The percentage of interest charge contained in installment payment number t is

$$f(t) = \frac{I(t)}{A}(100\%)$$

9. The percentage of cumulative interest charge contained in the cumulative total payment up to and including payment number t is

$$F(t) = \frac{Q(t)}{C(t)}(100\%)$$

10. The percentage of cumulative principal payment contained in the cumulative total payment up to and including payment number t is

$$H(t) = \frac{S(t)}{C(t)} = \frac{C(t) - Q(t)}{C(t)} = 1 - \frac{Q(t)}{C(t)}$$

$$= 1 - F(t)$$

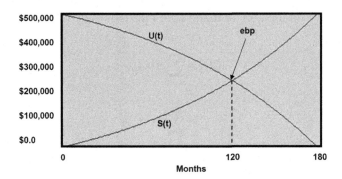

FIGURE 8.11
Plot of unpaid balance and cumulative equity.

Example 8.19

Suppose a manufacturing productivity improvement project is to be financed by borrowing $500,000 from an industrial development bank. The annual nominal interest rate for the loan is 10%. The loan is to be repaid in equal monthly installments over a period of 15 years. The first payment on the loan is to be made exactly 1 month after financing is approved. A detailed analysis of the loan schedule is desired.

The tabulated result shows a monthly payment of $5,373.03 on the loan. If time $t = 10$ months, one can see the following results:

$U(10) = \$487,473.83$ (unpaid balance)
$A(10) = \$5,373.03$ (monthly payment)
$E(10) = \$1,299.91$ (equity portion of the 10th payment)
$I(10) = \$4,073.11$ (interest charge contained in the 10th payment)
$C(10) = \$53,730.26$ (total payment to date)
$S(10) = \$12,526.17$ (total equity to date)
$f(10) = 75.81\%$ (percentage of the 10th payment going into interest charge)
$F(10) = 76.69\%$ (percentage of the total payment going into interest charge)

Thus, over 76% of the sum of the first 10 installment payments goes into interest charges. The analysis shows that by time $t = 180$, the unpaid balance has been reduced to zero. That is, $U(180) = 0.0$. The total payment made on the loan is $967,144.61 and the total interest charge is $967,144.61 − $500,000 = $467,144.61. So, 48.30% of the total payment goes into interest charges. The information about interest charges might be very useful for tax purposes. A tabulation of the calculations will show that equity builds up slowly while unpaid balance decreases slowly. Note that very little equity is accumulated during the first 3 years of this loan schedule. Figure 8.11 shows a plot of the unpaid balance and cumulative equity for this example. Note when the equity break-even point (ebp) occurs in the plot (Badiru, 2016).

Equity Break-Even Point

The point at which the curves intersect is referred to as the *equity break-even point*. It indicates when the unpaid balance is exactly equal to the accumulated equity or the cumulative principal payment. For example, the equity break-even point is approximately 120 months

(10 years). The importance of the equity break-even point is that any equity accumulated after that point represents the amount of ownership or equity that the debtor is entitled to after the unpaid balance on the loan is settled with project collateral. The implication of this is very important, particularly in the case of mortgage loans. "Mortgage" is a word of French origin, meaning "death pledge," which, perhaps, is an ironic reference to the burden of mortgage loans. The equity break-even point can be calculated directly from the formula derived as follows:

Let the equity break-even point, x, be defined as the point where $u(x) = S(x)$. That is,

$$A\left[\frac{1-(1+i)^{-(n-x)}}{i}\right] = A\left[\frac{(1+i)^x - 1}{i(1+i)^n}\right]$$

Multiplying both the numerator and denominator of the left-hand side of the preceding expression by $(1 + i)^n$ and then simplifying yields

$$\frac{(1+i)^n - (1+i)^x}{i(1+i)^n}$$

on the left-hand side. Consequently, we have

$$(1+i)^n - (1+i)^x = (1+i)^x - 1$$

$$(1+i)^x = \frac{(1+i)^n + 1}{2}$$

which yields the equity break-even expression

$$x = \frac{\ln\left[0.5(1+i)^n + 0.5\right]}{\ln(1+i)}$$

where

ln is the natural log function
n is the number of periods in the life of the loan
i is the interest rate per period

Figure 8.12 presents a plot of the total loan payment and the cumulative equity with respect to time. The total payment starts from $0.0 at time 0 and goes up to $967,144.61 by the end of the last month of the installment payments. Since only $500,000 was borrowed, the total interest payment on the loan is $967,144.61 − $500,000 = $467,144.61. The cumulative principal payment starts at $0.0 at time 0 and slowly builds up to $500,000.00, which is the original loan amount.

Figure 8.13 presents a plot of the percentage of interest charge in the monthly payments and the percentage of interest charge in the total payment. The percentage of interest charge in the monthly payments starts at 77.55% for the first month and decreases to 0.83% for the last month. By comparison, the percentage of interest in the total payment starts also at 77.55% for the first month and slowly decreases to 48.30% by the time the last payment is made at time 180. Figure 8.13 show that an increasing proportion of the monthly payment goes into the principal payment as time goes on. If the interest charges are tax

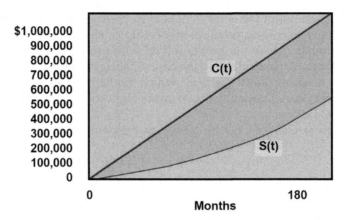

FIGURE 8.12
Plot of total loan payment and total equity.

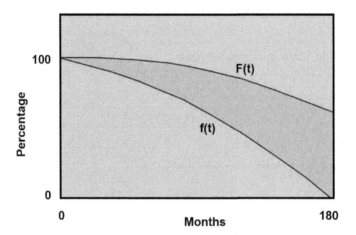

FIGURE 8.13
Plot of percentage of interest charge.

deductible, the decreasing values of $f(t)$ mean that there would be decreasing tax benefits from the interest charges in the later months of the loan.

Analysis of Tent Cash Flows

Analysis of arithmetic gradient series (AGS) cash flows is one of the more convoluted problems in engineering economic analysis. This section presents useful graphical representations of AGS cash flows. The approach presents AGS cash flows in the familiar shapes of tents. Several designs as well as a closed-form analysis of AGS cash-flow profiles are presented for a better understanding of this and other cash-flow profiles and their analysis. A general tent equation (GTE) used to solve various AGS cash flows is presented. The general equation can be used for various tent structures with appropriate manipulations. The GTE eliminates the problem of using the wrong number of periods and is amenable to software implementation.

Special Application of AGS

AGS cash flows feature prominently in many contract payments, but misinterpretation of them can seriously distort the financial reality of a situation. Good examples can be found in the contracts of sports professionals. The pervasiveness and popular press publicity associated with such contracts make the analysis of AGS both appealing and economically necessary. A good example is the 1984 contract of Steve Young, a quarterback for the Los Angeles (LA) express team in the former USFL (United States Football League). The contract was widely reported as being worth $40 million at that time. The cash-flow profile of the contract revealed an intricate use of various segments of AGS cash flows. When everything was taken into account, the $40 million touted in the press amounted only to a PW of the contract of about $5 million at that time. The trick was that the club included some deferred payments stretching over 37 years (1990–2027) at a 1984 present cost of only $2.9 million. The deferred payments were reported as being worth $34 million, which was the raw sum of the amounts in the deferred cash-flow profile. Thus, it turns out that clever manipulation of an AGS cash flow can create unfounded perceptions of the worth of a professional sports contract. This may explain why some sports professionals end up almost bankrupt even after receiving what they assume to be multimillion-dollar contracts. Similar examples have been found in reviewing the contracts of other sports professionals.

Other real-life examples of gradient cash-flow constructions can be found in general investment cash flows and maintenance operations. In a particular economic analysis of maintenance operations, an escalation factor was applied to the annual cost of maintaining a major piece of equipment. The escalation is justified because of inflation, increased labor costs, and other forecasted requirements.

Design and Analysis of Tent Cash-Flow Profiles

AGS cash flows usually start with some base amount at the end of the first period and then increase or decrease by a constant amount thereafter. The nonzero base amount is denoted as A_T starting at period T. The analysis of the PW for such cash flows requires breaking the cash flow into a uniform series cash flow of amount A_T starting at period T and an AGS cash flow with a zero base amount. The uniform series PW formula is used to calculate the PW of the uniform series portion while the basic AGS formula is used to calculate the AGS part of the cash-flow profile. The overall PW is then calculated as follows:

$$P = P_{\text{uniform series}} \pm P_{\text{arithmetic gradientseries}}$$

Figure 8.14 presents a conventional AGS cash flow. Each cash-flow amount at time t is defined as $A = (t - 1)$

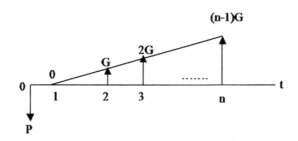

FIGURE 8.14
Conventional AGS cash flow.

The standard formula for this basic AGS profile is derived as follows:

$$P = \sum_{t=1}^{n} A_t (1+i)^{-t}$$

$$= \sum_{t=1}^{n} (t-1)\, G\, (1+i)^{-t}$$

$$= G \sum_{t=1}^{n} (t-1)\, (1+i)^{-t}$$

$$= \dots\dots\dots\dots\dots$$

$$= G \left[\frac{(1+i)^n - (1+ni)}{i^2 (1+i)^n} \right]$$

$$= G(P/G,\ i,\ n), \text{ in tabulated form.}$$

The computational process of deriving and using the P/G formula is where novice practitioners often stumble. Recognizing the tent-like structure of the cash flow and the fact that several applications of AGS are beyond the conventional AGS leads to the idea of tent cash flows. Figure 8.15 presents the basic tent (BT) cash-flow profile. It is composed of an upslope gradient and a downslope portion, both on a uniform series base of $A = A_1$.

Computational formula analysis of BT cash flow is shown as follows. For the first half of the cash flow ($t = 1$ to $t = T$), we have the following:

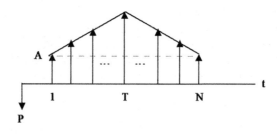

FIGURE 8.15
BT cash-flow profile.

$$P_a = \sum_{t=1}^{T} A_1 (1+i)^{-t} + \sum_{t=1}^{T} A_t (1+i)^{-t}, \ where \ A_t = (t-1)G$$

$$= \sum_{t=1}^{T} A_1 (1+i)^{-t} + G \sum_{t=1}^{T} (t-1)(1+i)^{-t}$$

$$= A_1 \left[\frac{(1+i)^T - 1}{i(1+i)^T} \right] + G \left[\frac{(1+i)^T - (1+Ti)}{i^2 (1+i)^T} \right]$$

For the second half of the cash flow ($t = T + 1$ to $t = N$), we have the following:

$$P_b = \sum_{t=T+1}^{N} A_{T+1} (1+i)^{-t} - \sum_{t=T+1}^{N} A_t (1+i)^{-t}, \ where \ A_t = (t-1)G$$

$$= A_{T+1} \sum_{t=T+1}^{N} (1+i)^{-t} - \sum_{t=T+1}^{N} (t-1)G(1+i)^{-t}$$

$$\vdots$$

$$\vdots$$

$$= A_{T+1} \left[\frac{(1+i)^{N-T} - 1}{i(1+i)^{N-T}} \right] - G \left[\frac{(1+i)^{N-T} - [1+(N-T)i]}{i^2 (1+i)^{N-T}} \right]$$

Now, let $x = (N - T)$. We end up with the following expression:

$$P_b = A_{T+1} \left[\frac{(1+i)^x - 1}{i(1+i)^x} \right] - G \left[\frac{(1+i)^x - (1+xi)}{i^2 (1+i)^x} \right]$$

Consequently, we have

$$P = P_a + P_b (1+i)^{-T}$$

The basic approach to solving this type of cash-flow profile is to partition it into simpler forms. By partitioning, BT can be solved directly using standard cash-flow conversion factors. That is, the solution can be obtained as the sum of a uniform series cash flow (base amount), an increasing AGS, and a decreasing AGS. That is,

$$P = G \left(\frac{P}{A}, i, N \right) + G \left(\frac{P}{G}, i, T \right) + \left[G(T-2) \left(\frac{P}{A}, i, N-T \right) - G \left(\frac{P}{G}, i, N-T \right) \right] \left(\frac{P}{F}, i, T \right)$$

The previous two approaches should yield the same result. Obviously, the partitioning approach using existing standard factors is a more ingenious method. One common error when using the existing AGS factor is the use of incorrect number of periods, n. It should be recognized that the standard AGS factor was derived for a situation where P is located one period before the "nose" of the increasing series. Students often tend to locate P right at the same point on the time line as the "nose" of the series, which means that n will be off by one interest period. One way to avoid this error is to redraw the time line and renumber

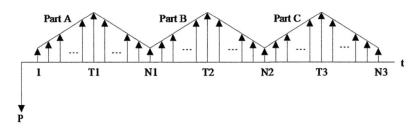

FIGURE 8.16
ET cash-flow profile.

it from a reference point of zero. That is, relocate time zero ($t = 0$) to one period before AGS begins. Figure 8.16 shows a profile of the executive tent (ET) cash-flow profile. It has a constant amount of increasing and decreasing AGS. The magnitudes of the cash-flow amounts at times T_j ($j = 1, 2, \ldots$) are equal.

For part A, we have

$$P_A = P_1 + P_{T_1+1}\left(1+i\right)^{-T_1}$$

where

$$P_1 = A_1\left[\frac{(1+i)^{T_1} - 1}{i(1+i)^{T_1}}\right] + G\left[\frac{(1+i)^{T_1} - (1+T_1 i)}{i^2(1+i)^{T_1}}\right]$$

and

$$P_{T_1+1} = A_{T_1+1}\left[\frac{(1+i)^{x_1} - 1}{i(1+i)^{x_1}}\right] - G\left[\frac{(1+i)^{x_1} - (1+x_1 i)}{i^2(1+i)^{x_1}}\right]$$

where $x_1 = (N_1 - T_1)$.

For part B, we have

$$P_B = P_{N_{i+1}}\left(1+i\right)^{-N_1} + P_{T_2+1}\left(1+i\right)^{-T_2}$$

where

$$P_{N_1+1} = A_{N_1+1}\left[\frac{(1+i)^{T_2} - 1}{i(1+i)^{T_2}}\right] + G\left[\frac{(1+i)^{T_2} - (1+T_2 i)}{i^2(1+i)^{T_2}}\right]$$

and

$$P_{T_2+1} = A_{T_2+1}\left[\frac{(1+i)^{x_2} - 1}{i(1+i)^{x_2}}\right] - G\left[\frac{(1+i)^{x_2} - (1+x_2 i)}{i^2(1+i)^{x_2}}\right]$$

where $x_2 = (N_2 - T_2)$.

For part C, we have

$$P_C = P_{N_2+1}\left(1+i\right)^{-N_2} + P_{T_3+1}\left(1+i\right)^{-T_3}$$

where

$$P_{N_2+1} = A_{N_2+1}\left[\frac{(1+i)^{T_3}-1}{i(1+i)^{T_3}}\right]+G\left[\frac{(1+i)^{T_3}-(1+T_3 i)}{i^2(1+i)^{T_3}}\right]$$

and

$$P_{T_3+1} = A_{T_3+1}\left[\frac{(1+i)^{x_3}-1}{i(1+i)^{x_3}}\right]-G\left[\frac{(1+i)^{x_3}-(1+x_3 i)}{i^2(1+i)^{x_3}}\right]$$

where $x_3 = (N_3 - T_3)$.

Consequently,

$$P = P_A + P_B + P_C$$

Figure 8.17 presents a saw-tooth tent (STT) cash-flow profile.

The present value analysis of the cash flow is computed as follows:

$$P = P_1 + P_{T_1+1}(1+i)^{-T_1} + P_{T_2+1}(1+i)^{-T_2}$$

where

$$P_1 = A_1\left[\frac{(1+i)^{T_1}-1}{i(1+i)^{T_1}}\right]+G\left[\frac{(1+i)^{T_1}-(1+T_1 i)}{i^2(1+i)^{T_1}}\right]$$

$$P_{T_1+1} = A_{T_1+1}\left[\frac{(1+i)^{T_2}-1}{i(1+i)^{T_2}}\right]+G\left[\frac{(1+i)^{T_2}-(1+T_2 i)}{i^2(1+i)^{T_2}}\right]$$

$$P_{T_2+1} = A_{T_2+1}\left[\frac{(1+i)^{N}-1}{i(1+i)^{N}}\right]+G\left[\frac{(1+i)^{N}-(1+N i)}{i^2(1+i)^{N}}\right]$$

In Figure 8.18, a reversed STT (R-STT) is constructed.

Its present value computation is handled as shown in the following:

$$P = P_1 + P_{T_1}(1+i)^{-T_1} + P_{T_2}(1+i)^{-T_2}$$

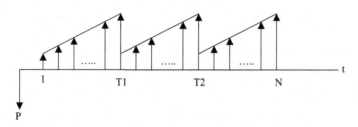

FIGURE 8.17
STT cash-flow profile.

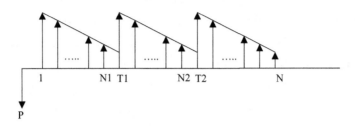

FIGURE 8.18
R-STT cash-flow profile.

where

$$P_1 = A_1 \left[\frac{(1+i)^{N_1} - 1}{i(1+i)^{N_1}} \right] - G \left[\frac{(1+i)^{N_1} - (1+N_1 i)}{i^2 (1+i)^{N_1}} \right]$$

$$P_{T_1} = A_{T_1} \left[\frac{(1+i)^{x_1} - 1}{i(1+i)^{x_1}} \right] - G \left[\frac{(1+i)^{x_1} - (1+x_1 i)}{i^2 (1+i)^{x_1}} \right]$$

where $x_1 = (N_2 - N_1)$.

$$P_{T_2} = A_{T_2} \left[\frac{(1+i)^{x_2} - 1}{i(1+i)^{x_2}} \right] - G \left[\frac{(1+i)^{x_2} - (1+x_2 i)}{i^2 (1+i)^{x_2}} \right]$$

where $x_2 = (N - N_2)$.

Increasingly complicated profiles can be designed depending on the level of complexity desired to test different levels of student understanding. Note that ET, STT, and R-STT cash flows can also be solved by the partitioning approach shown earlier for the BT cash flow.

Derivation of GTE

To facilitate a less convoluted use of the tent cash-flow computations, this section presents a GTE for AGS cash flows. This is suitable for use of project economic analysis. We combine all the tent cash-flow equations into a general tent cash-flow equation that is amenable to software implementation. The general equation is as follows:

$$P_0 = A_0 + A_1 \left[\frac{(1+i)^T - 1}{i(1+i)^T} \right] + G_1 \left[\frac{(1+i)^T - (1+Ti)}{i^2 (1+i)^T} \right]$$

$$P_T = A_{T_1} \left[\frac{(1+i)^x - 1}{i(1+i)^x} \right] + G_2 \left[\frac{(1+i)^x - (1+xi)}{i^2 (1+i)^x} \right]$$

When $G_1 \geq 0$ or $G_2 \leq 0$, we have

$$\text{TPV}_0 = P_0 + P_T (1+i)^{-T}$$

When $G_1 < 0$ or $G_2 > 0$, we have

$$\text{TPV}_0 = P_0 - P_T \left(1 + i\right)^{-T}$$

where
  TPV_0 = Total Present Values at time $t = 0$
  P_0 = Present Value for the first half of the tent at time $t = 0$
  P_T = Present Value for the first half of the tent at time $t = T$
  i = interest rate in fraction
  N = number of periods
  T = the center time value of the tent
  $x = (N - T)$
  A_0 = amount at time $t = 0$
  A_1 = amount at time $t = 1$

  $\qquad A_{T1}$ = amount at time $t = T + 1$

  $\qquad G_1$ = gradient series of the first half of the tent

  $\qquad\qquad \Rightarrow$ increasing G_1 (up-slope) is a positive value

  $\qquad\qquad$ and decreasing G_1 (down-slope) is a negative value

  $\qquad G_2$ = gradient series of the second half of the tent

  $\qquad\qquad \Rightarrow$ increasing G_2 (up-slope) is a positive value

  $\qquad\qquad$ and decreasing G_2 (down-slope) is a negative value

This general tent cash-flow equation is based on the BT but can be used for either one-sided or two-sided tent cash flows. It can also be used for ETs with more than two cycles. The process involved in this case is to divide the tent into smaller sections of two cycles, such as part A, part B, and part C, and then use the equation to solve for the present value of each part at time $t = 0$, $t = N_1$, and $t = N_2$. The present value of the future values at $t = N_1$ and $t = N=$ are added to the present value at $t = 0$ to determine the overall present value of the ET at $t = 0$. This GTE eliminates the problem associated with having an incorrect number of periods and makes AGS analysis interesting. This equation can also be used for situations where the uniform series base of A_1 is zero by finding the present value at time $t = 1$ and taking it back one period to determine the total present value at time $t = 0$. Other scenarios can also be considered by appropriate manipulations of the general equation.

Example 8.20

Using the general tent cash-flow equation, find the present value of the multiple BTs (Figure 8.19) at time $t = 0$ where $i = 10\%$ per period.
  Solving this tent cash flow usually poses a lot of problems for students; however, the use of GTE reduces such problems considerably. Since this tent consists of two BTs, we will use the general equation to determine the present value at time $t = 0$ for the first BT and the present value at time $t = 11$ and then find the present value at time $t = 0$ for the second tent and add it to the PV for the first tent and the present value at time $t = -1$.
  Using the GTE for the first BT, we obtain $P_0 = \$101.17$ for the first half of the first BT and $P_5 = \$82.24$ for the second half of the tent: $PV_0 = \$152.23$.

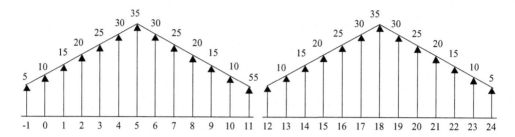

FIGURE 8.19
Dual basic tent cash-flow computation.

To solve the second BT, we renumbered the tent so that $t = 11$ becomes $t = 0$, $t = 12$ becomes $t = 1$, and so on. Therefore, $t = T$ is at $t = 7$, and we obtain $P_{11} = \$88.16$ for the first half of the second BT and $P_{18} = \$82.24$ for the second half of the tent: $PV_{11} = \$130.36$. The total present value for the dual BT at time $t = 0$ is

$$\text{TNPV}_0 = P_{-1}\left(\frac{F}{P}, i\%, N\right) + P_0 + P_{11}\left(\frac{P}{F}, i\%, N\right)$$

$$\text{TNPV}_0 = 5\left(\frac{F}{P}, 10\%, 1\right) + 152.23 + 130.36\left(\frac{P}{F}, 10\%, 11\right)$$

$$\text{TNPV}_0 = \$203.42$$

Therefore, the overall PV of this dual BT at time $t = 0$ is \$203.42.

Multiattribute Project Selection

Project selection is an important aspect of project planning. The right project must be undertaken at the right time to satisfy the constraints of time and resources. A combination of criteria can be used for project selection, including technical merit, management desire, schedule efficiency, cost–benefit ratio, resource availability, criticality of need, availability of sponsors, and user acceptance. This section presents techniques for assessing and comparing project investments in order to improve the selection process. The techniques presented include utility models, polar plots, benchmarking techniques, and the analytic hierarchy process (AHP).

Many aspects of project selection cannot be expressed in quantitative terms. For this reason, project analysis and selection must be addressed by techniques that permit the incorporation of both quantitative and qualitative factors. These techniques in this section supplement the traditional techniques of NPV, profit ratio, and equity break-even point.

Utility Models

The term "utility" refers to the rational behavior of a decision maker faced with making a choice in an uncertain situation. The overall utility of an investment can be measured in terms of both quantitative and qualitative factors. This section presents an approach to investment assessment based on utility models that have been developed within an extensive body of literature. The approach fits an empirical utility function to each factor that is to be included in a multiattribute selection model. The specific utility values (weights) that are obtained from the utility functions are used as the basis for selecting an investment. Utility theory is a branch of decision analysis that involves the building of mathematical models to describe the behavior of a decision maker faced with making a choice among alternatives in the presence of risk. The utility of a composite set of outcomes of n decision factors is expressed in the general form as follows:

$$u(x) = u(x_1, x_2, \ldots, x_n)$$

where
  x_i is the specific outcome of attribute X_i, $i = 1, 2, \ldots, n$
  $u(x)$ is the utility of the set of outcomes to the decision maker

The basic assumption of utility theory is that people make decisions with the objective of maximizing the *expected utility* from those decisions. Drawing on an example presented by Park and Sharp-Bette (1990), we may consider a decision maker whose utility function with respect to investment selection is represented by the following expression:

$$u(x) = 1 - e^{-0.0001x}$$

where x represents a measure of the benefit derived from an investment. Benefit, in this sense, may be a combination of several factors (e.g., quality improvement, cost reduction, or productivity improvement) that can be represented in dollar terms. Suppose this decision maker is faced with a choice between two investment alternatives, each of which has benefits as outlined in the following:

1. *Investment option I*: Probabilistic levels of investment benefits are provided in Table 8.3.
2. *Investment option II*: A definite benefit of $5,000.

Assuming an initial benefit of zero and identical levels of required investment, the decision maker must choose between the two investments. For investment I, the expected utility is computed as shown in Table 8.4, using the following computational formula:

$$E[u(x)] = \sum u(x)\{P(x)\}$$

For investment option I, the computation yields $E[u(x)_I] = 0.1456$. For investment II, we have $u(x)_{II} = u(\$5,000) = 0.3935$. Consequently, the investment providing the certain amount of $5,000 is preferred to the riskier investment I even though investment I has a higher

TABLE 8.3

Benefits and Probabilities for Investment Option I

Benefits, x	−$10,000	$0	$10,000	$20,000	$30,000
Probability, $P(x)$	0.2	0.2	0.2	0.2	0.2

TABLE 8.4

Utility Computations for Investment Option I

Benefit, x	Utility, $u(x)$	$P(x)$	$E[U(x)] = u(x) \cdot P(x)$
−$10,000	−1.7183	0.2	−0.3437
$0	0	0.2	0
$10,000	0.6321	0.2	0.1264
$20,000	0.8647	0.2	0.1729
$30,000	0.9502	0.2	0.1900
		Sum	0.1456

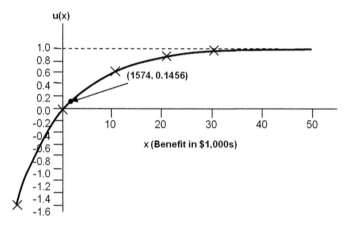

FIGURE 8.20
Utility function and certainty equivalent.

expected benefit of $\sum xP(x) = \$10,000$. A plot of the utility function used in the earlier example is presented in Figure 8.20.

If the expected utility of 0.1456 is set equal to the decision maker's utility function, we obtain the following:

$$0.1456 = 1 - e^{-0.0001x^*}$$

which yields $x^* = \$1574$, referred to as the *certainty equivalent* (CE) of investment I (CE$_1$ = 1,574). The CE of an alternative with variable outcomes is a *certain amount* (CA), which a decision maker will consider to be desirable to the same degree as the variable outcomes of the alternative. In general, if CA represents the CA of benefit that can be obtained from investment II, then the criteria for making a choice between the two investments can be summarized as follows:

1. If CA < $1,574, select investment I
2. If CA = $1,574, select either investment
3. If CA > $1,574, select investment II

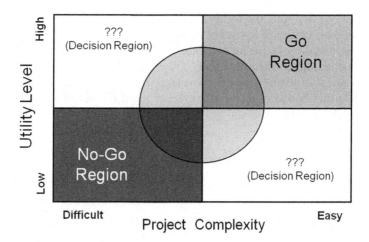

FIGURE 8.21
"GO-NO-GO" chart for project selection analysis.

The key in using utility theory for project selection is choosing the proper utility model. Figure 8.21 presents a "GO-NO-GO" chart for assessing project selection on the basis of utility levels (or any other cost metrics) and the level of project complexity. If utility level is high and project complexity is low, then it is a "GO" project. On the other hand, a complex project that does not exhibit sufficiently high level of utility will fall in the "NO-GO" region. The gray areas of decision making consist of the region of high utility and high complexity and the region of low complexity and low utility. Other factors often come into play in project selection decisions. These may include social necessity, political correctness, legislative requirement, and sustainment of market share.

Additive Utility Model

The additive utility of a combination of outcomes of n factors $(X_1, X_2, ..., X_n)$ is expressed as follows:

$$u(x) = \sum_{i=1}^{n} u\left(x_i, \overline{x}_i^0\right)$$

$$= \sum_{i=1}^{n} k_i u(x_i)$$

where
  x_i is the measured or observed outcome of attribute i
  n is the number of factors to be compared
  x is the combination of outcomes of n factors

$u(x_i)$ is the utility of the outcome for attribute i, x_i

$u(x)$ is the combined utility of the set of outcomes, x

k_i is the weight or scaling factor for attribute $i(0 < k_i < 1)$

x_i^0 is the worst outcome of attribute i

$\bar{x}_i^0$ is the set of worst outcomes for the complement of x_i

$u\left(x_i, \bar{x}_i^0\right)$ is the utility of the outcome of attribute i and the set of worst outcomes for the complement of attribute i

$k_i = u\left(x_i^*, \bar{x}_i^0\right)$

$$\sum_{i=1}^{n} k_i = 1.0 \text{ (required for the additive model)}$$

Additional notations

X_i is the variable notation for attribute i

x_i^* is the best outcome of attribute i

Example 8.21

Let **A** be a collection of four investment attributes defined as **A** = {profit, flexibility, quality, productivity}. Now define **X** = {profit, flexibility} as a subset of A. Then, **X**– is the complement of **X** defined as **X**– = {quality, productivity}. An example of the comparison of two investments under the additive utility model is summarized in Table 8.5 and yields the following results:

$$u\left(_x= \sum_{i=1}^{n} k_i u_i(x_i) = 0.4(0.95) + 0.2(0.45) + 0.3(0.35) + 0.1(0.75) = 0.650\right.$$

$$u\left(_x= \sum_{i=1}^{n} k_i u_i(x_i) = 0.4(0.90) + 0.2(0.98) + 0.3(0.20) + 0.1(0.10) = 0.626\right.$$

since $u(x)_A > u(x)_B$, investment A is selected.

TABLE 8.5

Example of Additive Utility Model

Attribute (*i*)	k_i	Investment A $U_i(x_i)$	Investment B $U_i(x_i)$
Profitability	0.4	0.95	0.90
Flexibility	0.2	0.45	0.98
Quality	0.3	0.35	0.20
Throughput	0.1	0.75	0.10
	1.00		

Multiplicative Utility Model

Under the multiplicative utility model, the utility of a combination of outcomes of n factors $(X_1, X_2, ..., X_{n1})$ is expressed as

$$u(x) = \frac{1}{C}\left[\prod_{i=1}^{n}(Ck_i u_i(x_i)+1)-1\right]$$

where C and k_i are scaling constants satisfying the following conditions:

$$\prod_{i=1}^{n}(1+Ck_i)-C = 1.0$$

$$-1.0 < C < 0.0$$

$$0 < k_i < 1$$

The other variables are as defined previously for the additive model. Using the multiplicative model for the data in Table 8.5 yields $u(x)_A = 0.682$ and $u(x)_B = 0.676$. Thus, investment A is the best option.

Fitting a Utility Function

An approach presented in this section for multiattribute investment selection is to fit an empirical utility function to each factor to be considered in the selection process. The specific utility values (weights) that are obtained from the utility functions may then be used in any of the standard investment justification methodologies. One way to develop empirical utility function for an investment attribute is to plot the "best" and "worst" outcomes expected from the attribute and then to fit a reasonable approximation of the utility function using concave, convex, linear, S-shaped, or any other logical functional form.

Alternately, if an appropriate probability density function (pdf) can be assumed for the outcomes of the attribute, then the associated cumulative distribution function may yield a reasonable approximation of the utility values between 0 and 1 for corresponding outcomes of the attribute. In that case, the cumulative distribution function gives an estimate of the cumulative utility associated with increasing levels of attribute outcome. Simulation experiments, histogram plotting, and goodness-of-fit tests may be used to determine the most appropriate density function for the outcomes of a given attribute. For example, the following five attributes are used to illustrate how utility values may be developed for a set of investment attributes. The attributes are ROI, productivity improvement, quality improvement, idle time reduction, and safety improvement.

Example 8.22

Suppose we have historical data on the ROI for investing in a particular investment. Assume that the recorded ROI values range from 0% to 40%. Thus, the worst outcome is 0% and the best outcome is 40%. A frequency distribution of the observed ROI values is developed and an appropriate pdf is fitted to the data. For our example, suppose the ROI is found to be exceptionally distributed with a mean of 12.1%. That is

$$f(x) = \begin{cases} \dfrac{1}{\beta} e^{-x/\beta}, & \text{if } x \geq 0 \\ 0, & \text{otherwise} \end{cases}$$

$$F(x) = \begin{cases} 1 - e^{-x/\beta}, & \text{if } x \geq 0 \\ 0, & \text{otherwise} \end{cases}$$

$$\approx u(x)$$

where $\beta = 12.1$. $F(x)$ approximates $u(x)$. The pdf and cumulative distribution function are shown graphically in Figure 8.22. The utility of any observed ROI within the applicable range may be read directly from the cumulative distribution function.

For the productivity improvement attribute, suppose it is found (based on historical data analysis) that the level of improvement is normally distributed with a mean of 10% and a standard deviation of 5%. That is,

$$f(x) = \frac{1}{\sqrt{2\pi}\sigma} e^{-1/2((x-\mu)/\sigma)^2}, \quad -\infty < x < \infty$$

where $\pi = 10$ and $\sigma = 5$. Since the normal distribution does not have a closed-form expression for $F(x)$, $u(x)$ is estimated by plotting representative values based on the standard normal table. Figure 8.23 shows $f(x)$ and the estimated utility function for productivity improvement. The utility of productivity improvement may also be evaluated on the basis of cost reduction.

Suppose quality improvement is subjectively assumed to follow a beta distribution with shape parameters $\alpha = 1.0$ and $\beta = 2.9$. That is,

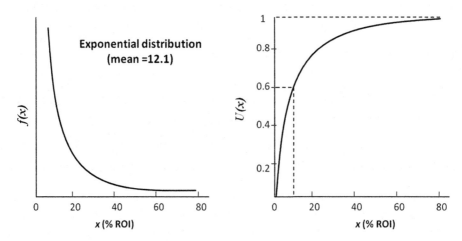

FIGURE 8.22
Estimated utility function for investment ROI.

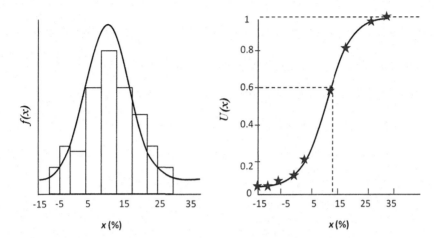

FIGURE 8.23
Utility function for productivity improvement.

$$f(x) = \frac{\Gamma(\alpha+\beta)}{\Gamma(\alpha)\Gamma(\beta)} \cdot \frac{1}{(b-a)^{\alpha+\beta-1}} \cdot (x-a)^{\alpha-1} (b-x)^{\beta-1}, \quad \text{for } a \leq x \leq b \text{ and } \alpha > 0, \beta > 0$$

where
   a is the lower limit for the distribution
   b is the upper limit for the distribution
   α, β are the shape parameters for the distribution

As with the normal distribution, there is no closed-form expression for $F(x)$ for the beta distribution. However, if either of the shape parameters is a positive integer, then a binomial expansion can be used to obtain $F(x)$. Figure 8.24 shows a plot of $f(x)$ and the estimated $u(x)$ for quality improvement due to the proposed investment.

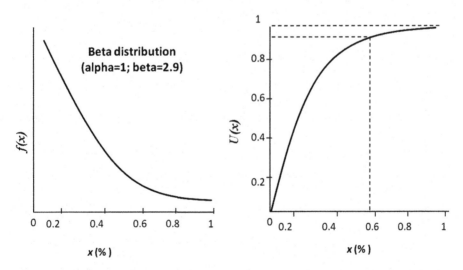

FIGURE 8.24
Utility function for quality improvement.

Based on work analysis observations, idle time reduction is found to be best described by a log-normal distribution with a mean of 10% and a standard deviation of 5%. This is represented as shown in the following

$$f(x) = \frac{1}{x\sqrt{2\pi\sigma^2}} e^{\left[-(\ln(x-\mu))^2/2\sigma^2\right]}, \quad x > 0$$

There is no closed-form expression for $F(x)$. Figure 8.25 shows $f(x)$ and the estimated $u(x)$ for idle time reduction due to the investment.

For example, suppose safety improvement is assumed to have a previously known utility function, defined as follows:

$$u_p(x) = 30 - \sqrt{400 - x^2}$$

where x represents percent improvement in safety. For the expression, the unscaled utility values range from 10 (for 0% improvement) to 30 (for 20% improvement). To express any particular outcome of an attribute i, x_i, on a scale of 0.0–1.0, it is expressed as a proportion of a range of best-to-worst outcomes shown as follows:

$$X = \frac{x_i - x_i^0}{x_i^* - x_i^0}$$

where

X is the outcome expressed on a scale of 0.0–1.0
x_i is the measured or observed raw outcome of attribute i
x_i^0 is the worst raw outcome of attribute i
x_i^* is the best raw outcome of attribute i

The utility of the outcome may then be represented as $u(X)$ and read off of the empirical utility curve. Using the preceding approach, the utility function for safety improvement is scaled from 0.0 to 1.0. This is shown in Figure 8.26. The numbers within parentheses represent the scaled values.

The respective utility values for the five attributes may be viewed as relative weights for comparing investment alternatives. The utility obtained from the modeled functions

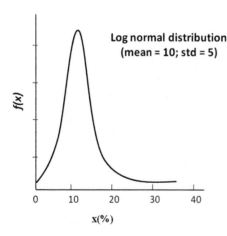
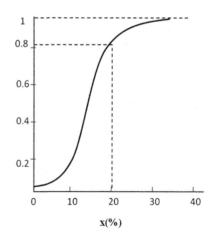

FIGURE 8.25
Utility function for idle time reduction.

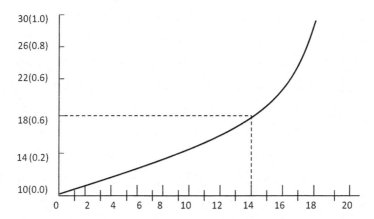

FIGURE 8.26
Utility function for safety improvement.

TABLE 8.6

Composite Utility for a Proposed Investment

Attribute (*i*)	k_i	Value	$U_i(x_i)$
ROI	0.30	12.1%	0.61
Productivity improvement	0.20	10.0%	0.49
Quality improvement	0.25	60.0%	0.93
Idle-time reduction	0.15	15.0%	0.86
Safety improvement	0.10	15.0%	0.40
	1.00		

can be used in the additive and multiplicative utility models discussed earlier. For example, Table 8.6 shows a composite utility profile for a proposed investment.

Using the additive utility model, the composite utility (CU) of the investment, based on the five attributes, is given by

$$u(X) = \sum_{i=1}^{n} k_i u_i (x_i)$$

$$= 0.30(0.61) + 0.20(0.49) + 0.25(0.93) + 0.15(0.86) + 0.10(0.40) = 0.6825$$

This CU value may then be compared with the utilities of other investments. On the other hand, a single investment may be evaluated independently on the basis of some minimum acceptable level of utility (MALU) desired by the decision maker. The criteria for evaluating an investment based on MALU may be expressed by the following rule:

Investment j is acceptable if its composite utility, $u(X)_j$, is greater than MALU.
Investment j is not acceptable if its composite utility, $u(X)_j$, is less than MALU.

The utility of an investment may be evaluated on the basis of its economic, operational, or strategic importance to an organization. Utility functions can be incorporated into existing justification methodologies. For example, in the AHP, utility functions can be used to generate values that are, in turn, used to evaluate the relative preference levels of attributes and alternatives. Utility functions can be used to derive component

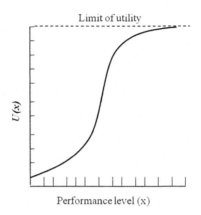

Limit of utility

$U(x)$

Performance level (x)

FIGURE 8.27
S-curve model for investment utility.

weights when the overall effectiveness of investments is being compared. Utility functions can generate descriptive levels of investment performance as well as indicating
the limits of investment effectiveness, as shown by the S-curve in Figure 8.27.

Polar Plots

Polar plots provide a means of visually comparing project investment alternatives. In a
conventional polar plot, as shown in Figure 8.28, the vectors drawn from the center of the
circle are on individual scales based on the outcome ranges for each attribute. For example,
the vector for NPV is on a scale of \$0–\$500,000 while the scale for quality is from 0 to 10. It
should be noted that the overall priority weights for the alternatives are not proportional
to the areas of their respective polyhedrons.

A modification of the basic polar plot is presented in this section. The modification
involves a procedure that normalizes the areas of the polyhedrons with respect to the total
area of the base circle. With this modification, the normalized areas of the polyhedrons
are proportional to the respective priority weights of the alternatives, so, the alternatives
can be ranked on the basis of the areas of polyhedrons. The steps involved in the modified
approach are presented as follows:

1. Let n be the number of attributes involved in the comparison of alternatives, such
   that $n \geq 4$. Number the attributes in a preferred order (1, 2, 3, ..., n).
2. If the attributes are considered to be equally important (i.e., equally weighted),
   compute the sector angle associated with each attribute as

$$\theta = \frac{360°}{n}$$

3. Draw a circle with a large enough radius. A radius of 2 in. is usually adequate.
4. Convert the outcome range for each attribute to a standardized scale of 0–10 using
   appropriate transformation relationships.

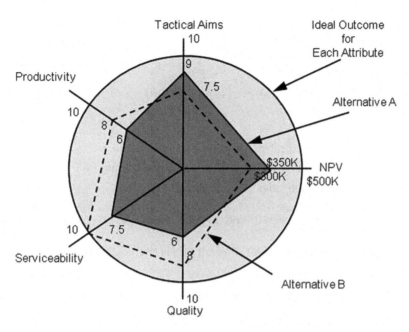

FIGURE 8.28
Basic polar plot.

5. For Attribute 1, draw a vertical vector up from the center of the circle to the edge of the circle.

6. Measure θ clockwise and draw a vector for Attribute 2. Repeat this step for all attributes in the numbered order.

7. For each alternative, mark its standardized relative outcome with respect to each attribute along the attribute's vector. If a 2 in. radius is used for the base circle, then we have the following linear transformation relationship:

0.0 in. = Rating score of 0.0

2.0 in. = Rating score of 10.0

8. Connect the points marked for each alternative to form a polyhedron. Repeat this step for all alternatives.

9. Compute the area of the base circle as follows:

$$\Omega = \pi r^2$$

$$= 4\pi \text{ squared inches}$$

$$= 100\pi \text{ squared rating units}$$

10. Compute the area of the polyhedron corresponding to each alternative. This can be done by partitioning each polyhedron into a set of triangles and then calculating the areas of the triangles. To calculate the area of each triangle, note that we will know the lengths of two sides of the triangle and the angle subtended by the two sides. With these three known values, the area of each triangle can be calculated through basic trigonometric formulas.

For example, the area of each polyhedron may be represented as λ_I ($I = 1, 2, ..., m$), where m is the number of alternatives. The area of each triangle in the polyhedron for a given alternative is then calculated as

$$\Delta_t = \frac{1}{2}(L_j)(L_{j+1})(\sin\theta)$$

where

L_j is the standardized rating with respect to attribute j
L_{j+1} is the standardized rating with respect to attribute $j + 1$
L_j and L_{j+1} are the two sides that subtend θ

Since $n \geq 4$, θ will be between $0°$ and $90°$, and $\sin(\theta)$ will be strictly increasing over that interval.

The area of the polyhedron for alternative i is then calculated as

$$\lambda_i = \sum_{t(i)=1}^{n} \Delta_{t(i)}$$

Note that θ is constant for a given number of attributes and the area of the polyhedron will be a function of the adjacent ratings (L_j and L_{j+1}) only.

11. Compute the standardized area corresponding to each alternative as

$$w_i = \frac{\lambda_i}{\Omega}(100\%)$$

12. Rank the alternatives in decreasing order of λ_i. Select the highest ranked alternative as the preferred alternative.

Example 8.23

The problem presented here is used to illustrate how modified polar plots can be used to compare investment alternatives. Table 8.7 presents the ranges of possible evaluation ratings within which an alternative can be rated with respect to each of the five attributes. The evaluation rating of an alternative with respect to attribute j must be between the given range a_j to b_j. Table 8.8 presents the data for raw evaluation ratings of three alternatives with respect to the five attributes specified in Table 8.7.

TABLE 8.7

Ranges of Raw Evaluation Ratings for Polar Plots

Attribute (j)	Description	Rank (k_j)	Evaluation Range	
			Lower Limit (a_j)	Upper Limit (b_j)
I	Quality	1	0.5	9
II	Profit (×$1,000)	2	0	100
III	Productivity	3	1	10
IV	Flexibility	4	0	12
V	Satisfaction	5	0	10

TABLE 8.8

Raw Evaluation Ratings for Modified Polar Plots

Alternatives	Attributes				
	I ($j = 1$)	II ($j = 2$)	III ($j = 3$)	IV ($j = 4$)	V ($j = 5$)
A ($i = 1$)	5	50	3	6	10
B ($i = 2$)	1	20	1.5	9	2
C ($i = 3$)	8	75	4	11	1

The attributes of quality (I), profit (II), and productivity (III) are quantitative measures that can be objectively determined. The attributes of flexibility (IV) and customer satisfaction (V) are subjective measures that can be intuitively rated by an experienced investment analyst. The steps in the solution are presented as follows:

1. *Step 1*: It is given that $n = 5$. The attributes are numbered in the following preferred order:
   Quality: Attribute I
   Profit: Attribute II
   Productivity: Attribute III
   Flexibility: Attribute IV
   Satisfaction: Attribute V
2. *Step 2*: The sector angle is computed as

$$\theta = \frac{360°}{n}$$
$$= 72°$$

3. *Step 3*: This step is shown in Figure 8.29.
4. *Step 4*: Let Y_{ij} be the raw evaluation rating of alternative i with respect to attribute j. Let Z_{ij} be the standardized evaluation rating.
   The standardized evaluation ratings (between 0.0 and 10.0) shown in Table 8.9 were obtained using the following linear transformation relationship:

$$Z_{ij} = 10\left[\frac{(Y_{ij} - a_j)}{b_j - a_j}\right]$$

5. *Steps 5–8*: These are shown in Figure 8.9.
6. *Step 9*: The area of the base circle is $\Omega = 100\pi$ squared rating units.
   Note that it is computationally more efficient to calculate the areas in terms of rating units rather than inches.
7. *Step 10*: Using the expressions presented in Step 10, the areas of the triangles making up each of the polyhedrons are computed and summed up. The respective areas are
   $\lambda_A = 72.04$ squared units
   $\lambda_B = 10.98$ squared units
   $\lambda_C = 66.14$ squared units
8. *Step 11*: The standardized areas for the three alternatives are as follows:
   $w_A = 22.93\%$
   $w_B = 3.50\%$
   $w_C = 21.05\%$

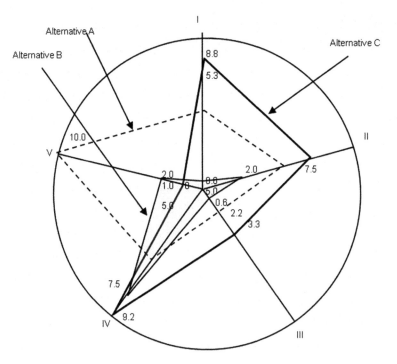

FIGURE 8.29
Modified polar plot.

TABLE 8.9

Standardized Evaluation Ratings for Modified Polar Plots

	Attributes				
Alternatives	**I ($j = 1$)**	**II ($j = 2$)**	**III ($j = 3$)**	**IV ($j = 4$)**	**V ($j = 5$)**
A ($i = 1$)	5.3	5.0	2.2	5	10.0
B ($i = 2$)	0.6	2.0	0.6	7.5	2.0
C ($i = 3$)	8.8	7.5	3.3	9.2	1.0

9. *Step 12*: On the basis of the standardized areas in Step 11, alternative A is found to be the best choice.

As an extension to the modification presented earlier, the sector angle may be a variable indicating relative attribute weights while the radius represents the evaluation rating of the alternatives with respect to the weighted attribute. That is, if the attributes are not equally weighted, the sector angles will not all be equal. In that case, the sector angle for each attribute is computed as

$$\theta_j = p_j\left(360°\right)$$

where p_j is the relative numeric weight of each of n attributes

$$\sum_{j=1}^{n} p_j = 1.0$$

TABLE 8.10

Relative Weighting of Attributes for Polar Plots

Attribute (i)	Weight (p_j)	Angle (θ_j)
I	0.333	119.88
II	0.267	96.12
III	0.200	72.00
IV	0.133	47.88
V	0.067	24.12
	1.000	360.00

Suppose the attributes in the preceding example are considered to have unequal weights, as shown in Table 8.10.

The resulting polar plots for weighted sector angles are shown in Figure 8.30.

The respective weighted areas for the alternatives are

$\lambda_A = 51.56$ squared units

$\lambda_B = 9.07$ squared units

$\lambda_C = 60.56$ squared units

The standardized areas for the alternatives are as follows:

$w_A = 16.41\%$

$w_B = 2.89\%$

$w_C = 19.28\%$

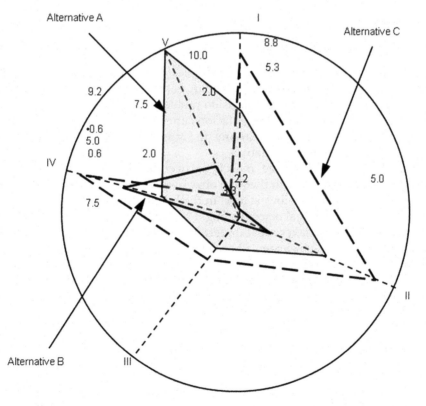

FIGURE 8.30
Polar plot with weighted sector angles.

Thus, if the given attributes are weighted as shown in Table 8.10, alternative C will turn out to be the best choice. However, it should be noted that the relative weights of the attributes are too skewed, resulting in some sector angles being greater than 90°. It is preferable to have the attribute weights assigned in such a way that all sector angles are less than 90°. This leads to more consistent evaluation since $\sin(\theta)$ is strictly increasing between 0° and 90°.

It should also be noted that the weighted areas for the alternatives are sensitive to the order in which the attributes are drawn in the polar plot. Thus, a preferred order of the attributes must be defined before starting the analysis. The preferred order may be based on the desired sequence in which alternatives must satisfy management goals. For example, it may be desirable to attend to product quality issues before addressing throughput issues. The surface area of the base circle may be interpreted as a measure of the global organizational goal with respect to performance indicators such as available capital, market share, capacity utilization, and so on. Thus, the weighted area of the polyhedron associated with an alternative may be viewed as the degree to which that alternative satisfies organizational goals.

Some of the attributes involved in a selection problem might constitute a combination of quantitative and/or qualitative factors or a combination of objective and/or subjective considerations. The prioritizing of the factors and considerations are typically based on the experience, intuition, and subjective preferences of the decision maker. Goal programming is another technique that can be used to evaluate multiple objectives or criteria in decision problems.

Analytic Hierarchy Process

The AHP is a practical approach to solving complex decision problems involving the pairwise comparisons of alternatives. The technique, popularly known as AHP, has been used extensively in practice to solve many decision problems. AHP enables decision makers to represent the hierarchical interaction of factors, attributes, characteristics, or alternatives in a multifactor decision-making environment. Figure 8.31 presents an example of a decision hierarchy for investment alternatives.

In an AHP hierarchy, the top level reflects the overall objective of the decision problem. The factors or attributes on which the final objective is dependent are listed at intermediate levels in the hierarchy. The lowest level in the hierarchy contains the competing alternatives through which the final objective might be achieved. After the hierarchy has been constructed, the decision maker must undertake a subjective prioritization procedure to determine the weight of each element at each level of hierarchy. Pairwise comparisons are performed at each level to determine the relative importance of each element at that level with respect to each element at the next higher level in the hierarchy. In our example, three alternate investments are to be considered. The investments are to be compared on the basis of the following five investment attributes:

1. Management support
2. Resource requirements
3. Technical merit
4. Schedule effectiveness
5. Cost–benefit ratio

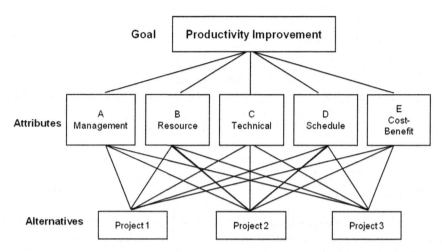

FIGURE 8.31
AHP for investment alternatives.

The first step in the AHP procedure involves the relative weighting of the attributes with respect to the overall goal. The attributes are compared pairwise with respect to their respective importance to the goal. The pairwise comparison is done through subjective and/or quantitative evaluation by the decision maker(s). The following matrix shows the general layout for pairwise comparisons:

$$F = \begin{bmatrix} r_{11} & r_{12} & \cdots & r_{1n} \\ r_{21} & r_{22} & \cdots & r_{2n} \\ \cdot & \cdot & \cdots & \cdot \\ \cdot & \cdot & \cdots & \cdot \\ \cdot & \cdot & \cdots & \cdot \\ r_{n1} & r_{n2} & \cdots & r_{nn} \end{bmatrix}$$

where
F is the matrix of pairwise comparisons
r_{ij} is the relative preference of the decision maker for i to j
$r_{ij} = 1/r_{ji}$
$r_{ij} = 1$

If $r_{ik}/r_{ij} = r_{jk}$ for all i, j, k, then matrix **F** can be said to be perfectly consistent. In other words, the transitivity of the preference orders is preserved. Thus, if factor A is preferred to factor B by a scale of 2 and factor a is preferred to factor C by a scale of 6, then factor B will be preferred to factor C by a factor of 3. That is, $A = 2B$ and $A = 6$. Then, $2B = 6C$ (i.e., $B = 3C$). The weight ratings used by AHP are summarized in Table 8.11.

In practical situations, one cannot expect all pairwise comparison matrices to be perfectly consistent. To address this, a tolerance level for consistency was developed by Saaty (1980). The tolerance level, referred to as consistency ratio, is acceptable if it is less than 0.10 (10%). If a consistency ratio is greater than 10%, the decision maker has the option of going back to reconstruct the comparison matrix or of proceeding with the analysis, with the recognition, however, that he accepts the potential bias that may exist in the final

TABLE 8.11

AHP Weight Scale

Scale	Definition
1	Equal importance
3	Weak importance of one over another
5	Essential importance
7	Demonstrated importance
9	Absolute importance
2, 4, 6, 8	Intermediate weights
Reciprocals	Represented by negative numbers

decision. Once the pairwise comparisons matrix is complete, the relative weights of the factors included in the matrix are obtained from the estimate of the maximum eigenvector of the matrix. This is done by the following expression:

$$FW = \lambda_{max}W$$

where
F is the matrix of pairwise comparisons
λ_{max} is the maximum eigenvector of **F**
W is the vector of relative weights

Example 8.24

For the example in Figure 8.31, Table 8.12 shows the tabulation of the pairwise comparison of the investment attributes.

Each of the attributes listed along the rows of the table is compared against each of the attributes listed in the columns. Each number in the body of the table indicates the degree of preference or importance of one attribute over the other on a scale of 1–9. A typical question that may be used to arrive at the relative rating is the following:

With respect to the goal of improving productivity, do you consider investment resource requirements to be more important than technical merit?

If so, how much more important is it on a scale of 1–9?

Similar questions are asked iteratively until each attribute has been compared with each of the other attributes. For example, in Table 8.12, Attribute B (resource requirements) is considered to be more important than Attribute C (technical merit) with a degree of 6. In general, the numbers indicating the relative importance of the attributes are based on the following weight scales:

TABLE 8.12

Pairwise Comparisons of Investment Attributes

Attributes	Management	Resource	Technical	Schedule	Cost-Benefit
Management	1	1/3	5	6	5
Resource	2	1	6	7	6
Technical	1/5	1/6	1	3	1
Schedule	1/6	1/7	1/3	1	1/4
Cost-benefit	1/5	1/6	1	4	1

1: Equally important
3: Slightly more important
5: Strongly more important
7: Very strongly more important
9: Absolutely more important

Intermediate ratings are used as appropriate to indicate intermediate levels of importance. If the comparison order is reversed (e.g., B versus A rather than A versus B), then the reciprocal of the important rating is entered in the pairwise comparison table. The relative evaluation ratings in the table are converted to the matrix of pairwise comparisons shown in Table 8.13.

The entries in Table 8.13 are normalized to obtain Table 8.14.

The normalization is done by dividing each entry in a column by the sum of all the entries in the column. For example, the first cell in Table 8.12 (i.e., 0.219) is obtained by dividing 1.000 by 4.567. Note that the sum of the normalized values in each attribute column is 1.

The last column in Table 8.14 shows the normalized average rating associated with each attribute. This column represents the estimated maximum eigenvector of the matrix of pairwise comparisons. The first entry in the column (0.288) is obtained by dividing 1.441 by 5, which is the number of attributes. The averages represent the relative weights (between 0.0 and 1.0) of the attributes that are being evaluated. The relative weights show that Attribute B (resource requirements) has the highest importance rating of 0.489. Thus, for this example, resource consideration is seen to be the most important factor in the selection of one of the three alternate investments. It should be emphasized that these attribute weights are valid only for the particular goal specified in the AHP model for the problem. If another goal is specified, the attributes would need to be reevaluated with respect to that new goal.

After the relative weights of the attributes are obtained, the next step is to evaluate the alternatives on the basis of the attributes. In this step, relative evaluation rating

TABLE 8.13

Pairwise Comparisons of Investment Attributes

Attributes	A	B	C	D	E
A	1.000	0.333	5.000	6.000	5.000
B	3.000	1.000	6.000	7.000	6.000
C	0.200	0.167	1.000	3.000	1.000
D	0.167	0.143	0.333	1.000	0.250
E	0.200	0.167	1.000	4.000	1.000
Column sum	4.567	1.810	13.333	21.000	13.250

TABLE 8.14

Normalized Matrix of Pairwise Comparisons

Attributes	A	B	C	D	E	Sum	Average
A: Management support	0.219	0.184	0.375	0.286	0.377	1.441	0.288
B: Resource requirement	0.656	0.551	0.450	0.333	0.454	2.444	0.489
C: Technical merit	0.044	0.094	0.075	0.143	0.075	0.431	0.086
D: Schedule effectiveness	0.037	0.077	0.025	0.048	0.019	0.206	0.041
E: Cost–benefit ratio	0.044	0.094	0.075	0.190	0.075	0.478	0.096
Column sum	1.000	1.000	1.000	1.000	1.000		1.000

TABLE 8.15

Investment Ratings Based on Management Support

Alternatives	Investment 1	Investment 2	Investment 3
Investment 1	1	1/3	1
Investment 2	3	1	2
Investment 3	1	1/2	1

is obtained for each alternative with respect to each attribute. The procedure for the pairwise comparison of the alternatives is similar to the procedure for comparing the attributes. Table 8.15 presents the tabulation of the pairwise comparisons of the three alternatives with respect to Attribute A (management support).

The table shows that investments I and III have the same level of management support. Examples of questions that may be used in obtaining the pairwise ratings of the alternatives are these:

Is investment 1 preferable to investment 2 with respect to management support?

What is the level of preference on a scale of 1–9?

It should be noted that the comparisons shown in Table 8.14 are valid only when management support of the investments is being considered. Separate pairwise comparisons of the investment must be done whenever another attribute is being considered. Consequently, for our example, we would have five separate matrices of pairwise comparisons of the alternatives, one matrix for each attribute. Table 8.15 is the first of the five matrices. The other four are not shown. The normalization of the entries in Table 8.15 yields the following relative weights of the investments with respect to management support: investment I (0.21), investment II (0.55), and investment III (0.24). Table 8.16 shows a summary of the normalized relative ratings of the three investments with respect to each of the five attributes.

The attribute weights shown in Table 8.15 are combined with the weights in Table 8.16 to obtain the overall relative weights of the investments shown in the following:

$$\alpha_j = \sum_i \left(w_i k_{ij} \right)$$

where

α_j represents *overall* weight for investment j

w_i represents relative weight for attribute i

k_{ij} represents rating (local weight) for investment j with respect to attribute i

$w_i k_{ij}$ represents global weight of alternative j with respect to attribute i

Table 8.17 shows the summary of the final AHP analysis for the example. The summary shows that investment II should be selected, since it has the highest overall weight of 0.484. AHP can be used to prioritize multiple investments with respect to several objectives. Table 8.18 shows a generic layout for multiple investment evaluation.

TABLE 8.16

Investment Weights Based on Attributes

	Attributes				
	Management Support	Resource Requirements	Technical Merit	Schedule Effectiveness	C/B Ratio
Investment 1	0.21	0.12	0.50	0.63	0.62
Investment 2	0.55	0.55	0.25	0.30	0.24
Investment 3	0.24	0.33	0.25	0.07	0.14

TABLE 8.17

Summary of AHP Evaluation of Three Investments

	Attributes					
	A $i = 1$	**B** $i = 1$	**C** $i = 1$	**D** $i = 1$	**E** $i = 1$	
$w_i \Rightarrow$	0.288	0.489	0.086	0.041	0.096	
Investment j			k_{ij}			α_j
Investment 1	0.21	0.12	0.50	0.63	0.62	*0.248*
Investment 2	0.55	0.55	0.25	0.30	0.24	*0.484*
Investment 3	0.24	0.33	0.25	0.07	0.14	*0.268*
Column sum	1.000	1.000	1.000	1.000	1.000	1.000

TABLE 8.18

Layout for Multiple Investments Comparison

	Objectives				
	Objective 1	**Objective 2**	**Objective 3**	...	**Objective 4**
Investment 1	K_{11}	K_{12}	K_{13}		K_{14}
Investment 2	K_{21}	K_{22}	K_{23}		K_{24}
Investment 3	K_{31}	K_{32}	K_{33}		K_{34}
$\vdots$					
Investment n	K_{n1}	K_{n2}	K_{n3}	...	K_{nn}

Cost Benchmarking

The techniques presented in the preceding sections can be used for benchmarking investments. For example, to develop a baseline schedule, evidence of successful practices from other investments may be needed. Metrics based on an organization's most critical investment implementation issues should be developed. *Benchmarking* is a process whereby target performance standards are established based on the best examples available. The objective is to equal or surpass the best example. In its simplest term, benchmarking means learning from and emulating a superior example. The premise of benchmarking is that if an organization replicates the best quality examples, it will become one of the best in the industry. A major approach of benchmarking is to identify performance gaps between investments. Figure 8.32 shows an example of such a gap. Benchmarking requires that an attempt be made to close the gap by improving the performance of the subject investment.

Benchmarking requires frequent comparison with the target investment. Updates must be obtained from investments already benchmarked, and new investments to be benchmarked must be selected on a periodic basis. Measurement, analysis, feedback, and modification should be incorporated into the performance improvement program. The benchmark-feedback model presented in Figure 8.33 is useful for establishing a continuous drive toward performance benchmarks.

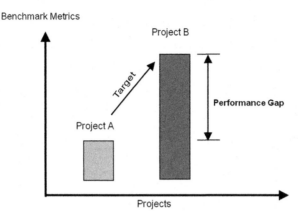

FIGURE 8.32
Identification of benchmark gaps.

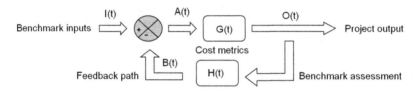

FIGURE 8.33
Investment benchmark-feedback model.

The figure shows the block diagram representation of input–output relationships of the components in a benchmarking environment. In the model, $I(t)$ represents the set of benchmark inputs to the subject investment. The inputs may be in terms of data, information, raw material, technical skill, or other basic resources. The index t denotes a time reference. $A(t)$ represents the feedback loop actuator. The actuator facilitates the flow of inputs to the various segments of the investment. $G(t)$ represents the forward transfer function, which coordinates input information and resources to produce the desired output, $O(t)$. $H(t)$ represents the management control process that monitors the status of improvement and generates the appropriate feedback information, $B(t)$, which is routed to the input transfer junction. The feedback information is necessary to determine what control actions should be taken at the next improvement phase. The primary responsibility of an economic analyst is to ensure proper forward and backward flow of information concerning the performance of an investment on the basis of the benchmarked inputs.

Exercises

8.1   Calculate how long it would take your current personal or family savings to double at the current interest rate you are being offered by your bank. What will it take for you or your family to reduce the calculated period by one-half?

8.2 If the nominal interest rate on a savings account is 0.25% payable, or compounded, quarterly, what is the effective annual interest rate? If $1,000 is deposited into this account quarterly, how much would be available in the account after 10 years?

8.3 A football player signed an $11 million 10-year contract package with a football team he joined recently. Based on this contract, the football player will receive the following benefits; his yearly salaries start at $300,000 and go up yearly to $400,000, $500,000, $600,000, $700,000, $1 million, $1.1 million, $1.2 million, $1.3 million, and finally to $1.4 million in the 10th year. Asides his salary, a $2.5 million bonus is available that will pay him $500,000 immediately and $500,000 each year from the 11th year to the 14th year. Calculate the total PW of the salaries and bonuses if the prevailing interest rate is 8% compounded per year. Did the footballer get the value of his contract?

8.4 How much must be deposited into a project account today if the project costs for each of the first 5 years is $12,000 and this amount increases by 10% per year for the following 10 years if the account pays 2.5% per year, compounded yearly?

8.5 In order to maintain RAB University's football stadium, the athletic department of the university needs annual maintenance costs of $60,000, annual insurance costs of $5,000, and annual utilities costs of $1,500. In addition, the department needs $100,000 worth of donations every 10 years for expansion projects. If the department opens a special account that pays 2.3% per year, how much must be deposited into this account now to pay for these annual costs forever?

8.6 A newly implemented technology in a manufacturing plant pays zero revenue in the first 2 years but $1,000 revenue in the third and fourth years; this amount increases by $500 annually for the following 5 years. If the company uses an MARR of 5.5% per year, what is the present value of these cash flows? What is the annual equivalent of the benefits over a 7-year period?

8.7 Repeat Exercise 8.6 if the MARR for the company is 5.5% in the first 4 years and increases to 6% starting in year 5.

8.8 A company borrowed $10,000 at 12% interest per year compounded yearly. The loan was repaid at $2,000 per year for the first 4 years and $2,200 in the fifth year. How much must be paid in the sixth year to pay off the loan?

8.9 A university alum wants to save $25,000 over 15 years so that he could start a scholarship for students in industrial engineering. To have this amount when it is needed, annual payments will be made into a savings account that earns 8% interest per year. What is the amount of each annual payment?

8.10 A small-scale industry thinks that it will produce 10,000 tons of metal during the coming year. If the processes are controlled properly, then the metal is going to increase 5% per year thereafter for the next 6 years. Profit per ton of metal is $14 for years 1–7. If the industry earns 15% per year on this capital, what is the future equivalent of the industries cash flows at the end of year (EOY) 7?

8.11 My grandmother just purchased a new house for $500,000. She made a down payment of 50% of the negotiated price and then makes a payment of $2,000.00 per month for 36 months. Furthermore, she thinks that she can resell the house for $600,000 due to the increasing real estate bubble at the end of 3 years. Draw a cash flow for this from my grandmother's point of view.

8.12 A manufacturing unit in a facility is thinking of purchasing automatic lathes that could save $67,000 per year on labor and scrap. This lathe has an expected life of 5 years and no market value. If the company tells the manufacturing units that it is expecting to see a 15% ROI per year, how much could be justified now for the purchase of this lathe? Explain it from the manufacturing unit's perspective.

8.13 An ambitious student wants to start a restaurant, so he wants to buy a nice kitchen set consisting of state-of-art grilles with two tables, three stoves, two dishwashers, three refrigerators, two ovens, and three microwaves along with some other stuff. So he plans on spending $112,000 on the equipment alone. He feels that these all would produce him a net income of $25,000 per year. If he does not change his mind and keeps these equipment for 4 years considering he does not lose much in this business, what would be the resale value of all this equipment at the EOY 4 to justify his investment? A 15% annual ROI is desired.

8.14 In the process of saving some money for my kids college, I plan to make six annual deposits of $4,000 into a secret savings account that pays an interest of 4% compounded annually. Two years after making the last deposit, the interest rate increases to 7% compounded annually. Twelve years after the last deposit, the accumulated money is taken out for the first time to pay for his tuition. How much is withdrawn?

8.15 A newly employed engineer wants to find out how much he should invest at 12% nominal interest, compounded monthly to provide an annuity of $25,000 (per year) for 6 years starting 12 years from now.

References

Badiru, A.B. (2016). Equity breakeven point: A graphical and tabulation tool for engineering managers, *Engineering Management Journal*, 28(4), 249–255.

Badiru, A.B. and Omitaomu, O.A. (2007). *Computational Economic Analysis for Engineering and Industry*, CRC Press, Taylor & Francis Group, Boca Raton, FL.

Park, C.S. and Sharp-Bette, G.P. (1990). *Advanced Engineering Economics*, Wiley & Sons, New York.

Saaty, T.L. (1980). *The Analytic Hierarchy Process*, McGraw-Hill, New York.

9

Advanced Forecasting and Inventory Modeling

Forecasting Techniques

Managing complex projects effectively calls for good information, which can be provided by forecasting and inventory control. Forecasting is not just for marketing and production planning purposes. This chapter presents techniques of forecasting and inventory management as a part of overall quantitative techniques for project planning and control. Several analytical tools are essential for analyzing project systems. Forecasting and inventory control are two of such tools. Prior to proceeding to the project management phase, a good understanding of the enterprise system, within which the project resides, is indispensable for getting a successful output.

Forecasting should be an important part of overall project planning and control, particularly in business and industrial projects. Effective prediction provides information needed to make good enterprise-wide decisions. Several techniques are available for forecasting. Regression, time series analysis, computer simulation, and artificial neural networks are common examples of forecasting techniques. There are two basic types of forecasting: *intrinsic forecasting* and *extrinsic forecasting*.

Intrinsic forecasting is based on the assumption that historical data adequately describe the problem scenario to be forecasted. Forecasting models based on historical data require extrapolation to generate estimates for the future. The requirements of intrinsic forecasting are as follows:

1. Collect historical data.
2. Develop quantitative forecasting model based on the data collected.
3. Generate forecasts recursively for the future.
4. Revise the forecasts as new pieces of data become available.

Extrinsic forecasting assumes that the forecasts to be generated are correlated to some other external factors such that the forecasts of the external factors provide reliable forecasts for the current problem. For example, the demand for a new product may be based on forecasts of household incomes. Before any forecasting system is implemented, a complete analysis of the data required must be performed. This is useful for setting activity times and task allocation strategies.

Forecasting Based on Averages

The most common forecasting techniques are based on averages. Sophisticated quantitative forecasting models can be formulated from basic average formulas. The traditional methods of averages are presented as follows.

Simple Average Forecast

In this method, the forecast for the next period is computed as an arithmetic average of the preceding data points. This is often referred to as average to date. That is,

$$f_{n+1} = \frac{\sum_{t=1}^{n} d}{n}$$

where
f_{n+1} is the forecast for period $n + 1$
d is the data element for the period in question
n is the number of preceding periods for which data are available

Period Moving Average Forecast

In this method, the forecast for the next period is based only on the most recent data values. Each time a new value is included, the oldest value is dropped. Thus, the average is always computed from a fixed number of values. This is represented as

$$f_{n+1} = \frac{\sum_{t=n-T+1}^{n} d_t}{T}$$

$$= \frac{d_{n-T+1} + d_{n-T+2} + \cdots + d_{n-1} + d_n}{T}$$

where
f_{n+1} is the forecast for period $n + 1$
d_t is the datum for period t
T is the number of preceding periods included in the moving average calculation
n is the current period at which forecast of f_{n+1} is calculated

The moving average technique is an after-the-fact approach. Since T data points are needed to generate a forecast, we cannot generate forecasts for the first $T-1$ periods. But this shortcoming is quickly overcome as the number of data points available becomes large.

Weighted Average Forecast

The weighted average forecast method is based on the assumption that some data points might be more significant than others in generating future forecasts. For example, the most recent data points may weigh more than very old data points in the calculation of future estimates. This is expressed as

$$f_{n+1} = \frac{\sum\limits_{t=1}^{n} w_i d_t}{\sum\limits_{t=1}^{n} w_t}$$

$$= \frac{w_1 d_1 + w_2 d_2 + \cdots + w_n d_n}{w_1 + w_2 + \cdots + w_n}$$

where

f_{n+1} is the weighted average forecast for period $n + 1$
d_t is the datum for period t
T is the additional notation representing the planning horizon for the forecast problem
n is the current period at which forecast of f_{n+1} is calculated
w_t is the weight of data point t

The w_t's are the respective weights of the data points such that

$$\sum\limits_{t=1}^{n} w_t = 1.0$$

Weighted T-Period Moving Average Forecast

In this technique, the forecast for the next period is computed as the weighted average of past data points over the last T time periods. That is,

$$f_{n+1} = w_1 d_n + w_2 d_{n-1} + \cdots + w_T d_{n-T+1}$$

where w_i's are the respective weights of the data points such that

$$\sum\limits_{i=1}^{n} w_i = 1.0$$

Exponential Smoothing Forecast

This is a special case of weighted moving average forecast. The forecast for the next period is computed as the weighted average of the immediate past data point and the forecast of the previous period. In order words, the previous forecast is adjusted based on the deviation (forecast error) of that forecast from the actual data. That is,

$$f_{n+1} = \alpha d_n + (1 - \alpha) f_n$$

$$= f_n + \alpha(d_n - f_n)$$

where

f_{n+1} is the exponentially weighted average forecast for period $n + 1$
d_n is the datum for period n
f_n is the forecast for period n
α is the smoothing factor (real number between 0 and 1)

A low smoothing factor gives a high degree of smoothing while a high value causes the forecast to closely match actual data.

Regression Analysis

The primary function of regression analysis is to develop a model that expresses the relationship between a dependent variable and one or more independent variables. It is sometimes called line fitting or curve fitting. Regression analysis is an important statistical tool that can be applied to many prediction problems and forecasting problems in the project environment. The utility of a regression model is often tested by analysis of variance (ANOVA), which is a technique for breaking down the variance in a statistical sample into components that can be attributed to each factor affecting that sample. One major purpose of ANOVA is testing of the model. Model testing is important because of the serious consequences of erroneously concluding that a regression model is good when, in fact, it has little or no significance to the data. Model inadequacy often implies an error in the assumed relationships between the variables, poor data, or both. A validated regression model can be used for the following purposes:

1. Prediction/forecasting
2. Description
3. Control

Description of Regression Relationship

Sometimes, the desired result from a regression analysis is an equation describing the best fit to the data under investigation. The "least squares" line drawn through the data is the line of best fit. This line may be linear or curvilinear depending on the dispersion of data. The linear situation exists in those cases where the slope of the regression equation is a constant. The nonconstant slope indicates curvilinear relationships. A plot of the data, called scatter plot, will usually indicate whether a linear or nonlinear model will be appropriate. The major problem with the nonlinear relationship is the necessity of assuming a functional relationship before accurately developing the model. Example of regression models (simple linear, multiple, and nonlinear) are presented as follows:

$$Y = \beta_0 + \beta_1 x + \varepsilon$$

$$Y = \beta_0 + \beta_1 x_1 + \beta_2 x_2 + \varepsilon$$

$$Y = \beta_0 + \beta_1 x_1^{\alpha_1} + \beta_2 x_2^{\alpha_2} + \varepsilon$$

$$Y = \beta_0 + \beta_1 x_1^{\alpha_1} + \beta_2 x_2^{\alpha_2} + \beta_{12} x_1^{\alpha_3} x_2^{\alpha_4} + \varepsilon$$

where
  Y is the dependent variable
  x_i's are independent variables
  β_i's are model parameters
  ε is the error term

The error terms are assumed to be independent and identically distributed normal random variables with mean of zero and variance of σ^2.

Prediction

Another major use of regression analysis is prediction or forecasting. Prediction can be of two basic types: interpolation and extrapolation. Interpolation predicts values of the dependent variable over the range of independent variables. Extrapolation involves predictions outside the range of independent variables. Extrapolation carries a risk in the sense that projections are made over a data range that is not included in the development of the regression model. There is some level of uncertainty about the nature of relationships that may exist outside the study range. Interpolation can also create a problem when the values of independent variables are widely spaced.

Control

Extreme care is needed in using regression for control. The difficulty lies in the assumption of a functional relationship when in fact none exists. Suppose, for example, that regression shows a relationship between chemical content in a product and noise level in the room. Further the real reason for this relationship is that the noise level increases as the machine speed increases and a higher machine speed produces higher chemical content. It would be erroneous to assume a functional relationship between the noise level in the room and the chemical content in the product. If this relationship does exist, then changes in the noise level could control chemical content. In this case, the real functional relationship exists between machine speed and chemical content. It is often difficult to prove functional relationships outside a laboratory environment because many extraneous and intractable factors may have an influence on the dependent variable. A simple example of the use of functional relationship for control can be seen in the following familiar equation of electrical circuits

$$I = \frac{V}{R}$$

where
  V is voltage
  I is electrical current
  R is the resistance

The current can be controlled by changes in either the voltage or the resistance or both. This particular equation, which has been experimentally validated, can be used as a control device.

Procedure for Regression Analysis

Problem definition: Failure to properly define the scope of the problem could result in useless conclusions. Time can be saved throughout all phases of a regression study by knowing, as precisely as possible, the purpose of the required model. A proper definition of the problem will facilitate the selection of appropriate variables to include in the study.

Selection of variables: Two very important factors in the selection of variables are ease of data collection and expense of data collection. Ease of data collection deals with the accessibility and the desired form of data. We must first determine whether the data can be collected and, if so, how difficult the process will be. The economic question is of prime importance. How expensive will the data be to collect and compile into a useable form? If the expense cannot be justified, then the variable under consideration may necessarily be omitted from the selection process.

Test of significance of regression: After the selection and compilation of all possible relevant variables, the next step is a test for the significance of regression. The test should help avoid wasted effort on the use of an invalid model. The test for the significance of regression is a test to see if at least one of the variable coefficient(s) in the regression equation is statistically different from zero. A test indicating that none of the coefficients is significantly different from zero implies that the best approximation of the data is a straight line through the data at the average value of the dependent variable, regardless of the values of independent variables. The significance level of the data is an indication of the probability of erroneously assuming model validity.

Coefficient of Determination

The coefficient of multiple determination, denoted by R^2, is used to judge the effectiveness of regression models containing multiple variables (multiple regression model). It indicates the proportion of the variation in the dependent variable explained by the model. The coefficient of multiple determination is defined as

$$R^2 = \frac{\text{SSR}}{\text{SST}}$$

$$= 1 - \frac{\text{SSE}}{\text{SST}}$$

where
   SSR represents the sum of squares due to the regression model
   SST represents the sum of squares total
   SSE represents the sum of squares due to error

R^2 measures the proportionate reduction of total variation in the dependent variable accounted for by a specific set of independent variables. The coefficient of multiple determination, R^2, reduces to the *coefficient of simple determination*, r^2, when there is only one independent variable in the regression model. R^2 is equal to 0 when all the coefficients, b_k, in the model are zero. That is, no regression fit at all. R^2 is equal to one when all data points fall directly on the fitted response surface. Thus, we have

$$0.0 \le R^2 \le 1.0$$

The following points should be noted about regression modeling:

1. A large R^2 does not necessarily imply that the fitted model is a useful one. For example, observations may have been taken at only a few levels of independent variables. In such a case, the fitted model may not be useful because most predictions would require extrapolation outside the region of observations. For example, for only two data points, the regression line passes perfectly through the two points and the R^2 value will be one. In that case, despite the high R^2, there will be no useful prediction capability.

2. Adding more independent variables to a regression model can only increase R^2 and never reduce it. This is because the error sum of squares (SSE) can never become larger with more independent variables and the total sum of squares (SST) is always the same for a given set of responses.

3. Regression models developed under conditions where the number of data points is roughly equal to the number of variables will yield high values of R^2 even though the model may not be useful. For example, for only two data points, the regression line will pass perfectly through the two points and r^2 will be one. Even though r^2 is one, there will be no useful prediction.

The strategy for using R^2 to evaluate regression models should not entirely focus on maximizing the R^2 value. Rather, the intent should be to find the point where adding more independent variables is not worthwhile in terms of overall effectiveness of the regression model. For most practical situations, R^2 values greater than 0.62 are considered acceptable.

Since R^2 can often be made larger by including a large number of independent variables, it is sometimes suggested that a modified measure be used, which recognizes the number of independent variables in the model. This modified measure is referred to as adjusted coefficient of multiple determination, R_a^2. It is defined mathematically as

$$R_a^2 = 1 - \left(\frac{n-1}{n-p} \right) \frac{SSE}{SST}$$

where
   n is the number of observations used to fit the model
   p is the number of coefficients in the model (including the constant term)
   $p-1$ is the number of independent variables in the model

R_a^2 may actually become smaller when another independent variable is introduced into the model. This is because the decrease in SSE may be more than offset by the loss of a degree of freedom in the denominator, $n-p$.

The *coefficient of multiple correlation* is defined as the positive square root of R^2. That is,

$$R = \sqrt{R^2}$$

Thus, the higher the value of R^2, the higher the correlation in the fitted model.

Residual Analysis

A residual is the difference between the predicted value computed from the fitted model and the actual value from the data. The ith residual is defined as

$$e_i = Y_i - \hat{Y}_i$$

where
   Y_i is the actual value
   $\hat{Y}_i$ is the predicted value

The SSE and the mean square error (MSE) are computed as

$$SSE = \sum_i e_i^2$$

$$\sigma^2 \approx \frac{\sum_i e_i^2}{n-2} = MSE$$

where n is the number of data points. A plot of residuals versus predicted values of the dependent variable can be quite revealing. The plot for a good regression model will have a random pattern. A noticeable trend in the residual pattern indicates a problem with the model. Some possible reasons for an invalid regression model are as follows:

1. Insufficient data.
2. Important factors not included in model.
3. Inconsistency in data.
4. No functional relationship exists.

Graphical analysis of residuals is important for assessing the appropriateness of regression models. The different possible residual patterns are presented in Figures 9.1–9.6.

When we plot the residuals versus the independent variable, the result should appear ideally as shown in Figure 9.1. Figure 9.2 shows a residual pattern indicating nonlinearity of the regression function. Figure 9.3 shows a pattern suggesting nonconstant variance (i.e., variation in σ^2). Figure 9.4 presents a residual pattern implying interdependence of error terms. Figure 9.5 shows a pattern depicting the presence of outliers. Figure 9.6 represents a pattern suggesting omission of independent variables.

Time Series Analysis

Time series analysis is a technique that attempts to predict the future using historical data. The basic principle of time series analysis is that the sequence of observations is based on jointly distributed random variables. The time series observations denoted by $Z_1, Z_2, ..., Z_T$ are assumed to be drawn from some joint probability density function (pdf) of the form

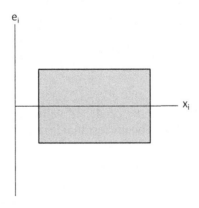

FIGURE 9.1
Ideal residual pattern

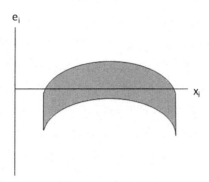

FIGURE 9.2
Residual pattern for nonlinearity

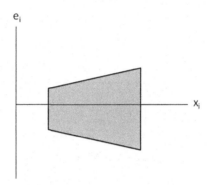

FIGURE 9.3
Residual pattern for nonconstant variance

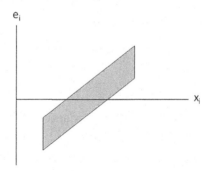

FIGURE 9.4
Residual pattern for the interdependence of error terms.

$$f_{1,\ldots,T}\left(Z_{1,\ldots,T}\right)$$

The objective of time series analysis is to use the joint density to make probability inferences about future observations. The concept of stationarity implies that the distribution of the time series is invariant with regard to any time displacement. That is,

$$f\left(Z_t,\ldots,Z_{t+k}\right) = f\left(Z_{t+m},\ldots,Z_{t+m+k}\right)$$

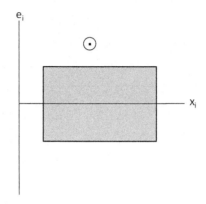

FIGURE 9.5
Residual pattern for the presence of outliers.

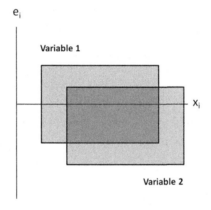

FIGURE 9.6
Residual pattern for the omission of independent variables.

where
    t is any point in time
    k and m are any pair of positive integers

A stationary time series process has a constant variance and remains stable around a constant mean with respect to time reference. Thus,

$$E(Z_t) = E(Z_{t+m})$$

$$V(Z_t) = V(Z_{t+m})$$

$$\mathrm{Cov}(Z_t, Z_{t+1}) = \mathrm{Cov}(Z_{t+k}, Z_{t+k+1})$$

Nonstationarity in a time series may be recognized in a plot of the series. A widely scattered plot with no tendency for a particular value is an indication of nonstationarity.

Stationarity and Data Transformation

In some cases where nonstationarity exists, some form of data transformation may be used to achieve stationarity. For most time series data, the usual transformation tried is differencing. Differencing involves the creation of a new series by taking differences between successive periods of the original series. For example, first regular differences are obtained by

$$w_t = Z_t - Z_{t-1}$$

To develop a time series forecasting model, it is necessary to describe the relationship between a current observation and previous observations. Such relationships are described by the sample autocorrelation function defined as shown in the following formula:

$$r_j = \frac{\sum_{t=1}^{T-j}(Z_t - \bar{Z})(Z_{t+j} - \bar{Z})}{\sum_{t=1}^{T}(Z_t - \bar{Z})^2}, \quad j = 0,1,\ldots,T-1$$

where
   T is the number of observations
   Z_t is the observation for time t
   $\bar{Z}$ is the sample mean of the series
   j is the number of periods separating pairs of observations
   r_j is the sample estimate of the theoretical correlation coefficient

The coefficient of correlation between two variables Y_1 and Y_2 is defined as

$$\rho_{12} = \frac{\sigma_{12}}{\sigma_1 \sigma_2}$$

where
   σ_1 and σ_2 are the standard deviations of Y_1 and Y_2, respectively
   σ_{12} is the covariance between Y_1 and Y_2

The standard deviations are the positive square roots of the variances defined as

$$\sigma_1^2 = E\left[(^{Y_1}]\right]$$

$$\sigma_2^2 = E\left[(^{Y_2}]\right]$$

The covariance, σ_{12}, is defined as

$$\sigma_{12} = E[(Y_1 - \mu_1)(Y_2 - \mu_2)]$$

which will be zero if Y_1 and Y_2 are independent. Thus, when $\sigma_{12} = 0$, we also have $\rho_{12} = 0$. If Y_1 and Y_2 are positively related, then σ_{12} and ρ_{12} are both positive. If Y_1 and Y_2 are negatively related, then σ_{12} and ρ_{12} are both negative. The correlation coefficient is a real number between −1 and +1:

$$1.0 \le \rho_j \le +1.0$$

Time series modeling procedure involves the development of a discrete linear stochastic process in which each observation, Z_t, may be expressed as

$$Z_t = \mu + \mu_t + \Psi_1 u_{t-1} + \Psi_2 u_{t-2} + \cdots$$

where
  μ is the mean of the process
  Ψ_i's are model parameters, which are functions of autocorrelations

Note that this is an infinite sum indicating that the current observation at time t can be expressed in terms of all previous observations from the past. In a practical sense, some of the coefficients will be zero after some finite point q in the past. The u_t's form the sequence of independently and identically distributed random disturbances with mean zero and variance sigma sub u^2. The expected value of the series is obtained by

$$E(Z_t) = \mu + E(u_t + \Psi_1 u_{t-1} + \Psi_2 u_{t-2} + \cdots)$$
$$= \mu + E(u_t)[1 + \Psi_1 + \Psi_2 + \cdots]$$

Stationarity of the time series requires that the expected value be stable. That is, the infinite sum of the coefficients should be convergent:

$$\sum_{i=0}^{\infty} \Psi_i = c$$

where
  $\Psi_0 = 1$
  c is a constant

The theoretical variance of the process, denoted by γ_0, can be derived as

$$\gamma_0 = E\left[Z_t - E(Z_t) \right]^2$$
$$= E\left[\left(\mu + u_t + \Psi_1 y_{t-1} + \Psi_2 u_{t-2} \cdots \right) - \mu \right]^2$$
$$= E\left[u_t + \Psi_1 y_{t-1} + \Psi_2 u_{t-2} \cdots \right]^2$$
$$= E\left[u_t^2 + \Psi_1^2 u_{t-1}^2 + \Psi_2^2 u_{t-2}^2 + \ldots \right] + E\left[cross - products \right]$$
$$= E\left[u_t^2 \right] + \Psi_1^2 E\left[u_{t-1}^2 \right] + \Psi_2^2 E\left[u_{t-2}^2 \right] + \ldots$$
$$= \sigma_u^2 + \Psi_1^2 \sigma_u^2 + \Psi_2^2 \sigma_u^2 + \ldots$$
$$= \sigma_u^2 \left(1 + \Psi_1^2 + \Psi_2^2 + \ldots \right)$$
$$= \sigma_u^2 \sum_{i=0}^{\infty} \Psi_i^2$$

where σ_u^2 represents the variance of the u_t's. The theoretical covariance between Z_t and Z_{t+j}, denoted by γ_j, can be derived in a similar manner to obtain

$$\gamma_j = E\left\{\left[Z_t - E(Z_t)\right]\left[Z_{t+j} - E(Z_{t+j})\right]\right\}$$

$$= \sigma_u^2\left(\Psi_j + \Psi_1\Psi_{j+1} + \Psi_2\Psi_{j+2} + \cdots\right)$$

$$= \sigma_u^2\sum_{i=0}^{\infty}\Psi_{i+j}\Psi$$

Sample estimates of the variances and covariances are obtained by

$$c_j = \frac{1}{T}\sum_{t=1}^{T-j}\left(Z_t - \bar{Z}\right)\left(Z_{t+j} - \bar{Z}\right), \quad j = 0, 1, 2, \ldots$$

The theoretical autocorrelations are obtained by dividing each of the autocovariances, γ_j, by γ_0. Thus, we have

$$\rho_j = \frac{\gamma_j}{\gamma_0}, \quad j = 0, 1, 2, \ldots$$

and the sample autocorrelations is obtained by

$$r_j = \frac{c_j}{c_0}, \quad j = 0, 1, 2, \ldots$$

Moving Average Processes

If it can be assumed that $\Psi_i = 0$ for some $i > q$, where q is an integer, then our time series model can be represented as

$$z_t = \mu + u_t + \Psi_1 u_{t-1} + \Psi_2 u_{t-2} + \cdots + \Psi_q u_{t-q}$$

which is referred to as a moving-average process of order q, usually denoted as MA(q). For notational convenience, we will denote the truncated series as presented in the following:

$$Z_t = \mu + u_t - \theta_1 u_{t-1} - \theta_2 u_{t-2} - \cdots - \theta_q u_{t-q}$$

where $\theta_0 = 1$. Any MA(q) process is stationary since the condition of convergence for the Ψ_i's becomes

$$\left(1 + \Psi_1 + \Psi_2 + \cdots\right) = \left(1 - \theta_1 - \theta_2 - \cdots - \theta_q\right)$$

$$= 1 - \sum_{i=0}^{q}\theta_i$$

which converges since q is finite. The variance of the process now reduces to

$$\gamma_0 = \sigma_u^2 \sum_{i=0}^{q} \theta_i$$

We now have the autocovariances and autocorrelations defined, respectively, as shown in the following:

$$\gamma_j = \sigma_u^2 \left(-\theta_j + \theta_1\theta_{j+1} + \cdots + \theta_{q-j}\theta_q \right), \quad j = 1,\ldots,q$$

where $\gamma_j = 0$ for $j > q$

$$\rho_j = \frac{\left(-\theta_j + \theta_1\theta_{j+1} + \cdots + \theta_{q-j}\theta_q \right)}{\left(1 + \theta_1^2 + \cdots + \theta_q^2 \right)}, \quad j = 1,\ldots,q$$

where $\rho_j = 0$ for $j > q$.

Autoregressive Processes

In the preceding section, the time series, Z_t, is expressed in terms of the current distur-bance, ut, and past disturbances, $ut - i$. An alternative is to express Z_t, in terms of the cur-rent and past observations, Z_{t-i}. This is achieved by rewriting the time series expression as

$$u_t = Z_t - \mu - \Psi_1 u_{t-1} - \Psi_2 u_{t-2} - \cdots$$

$$u_{t-1} = Z_{t-1} - \mu - \Psi_1 u_{t-2} - \Psi_2 u_{t-3} - \cdots$$

$$u_{t-2} = Z_{t-2} - \mu - \Psi_1 u_{t-3} - \Psi_2 u_{t-4} - \cdots$$

Successive back substitutions for the u_{t-i}'s yields the following:

$$u_t = \pi_1 Z_{t-1} - \pi_2 Z_{t-2} - \cdots - \delta$$

where π_i's and δ are model parameters and are functions of Ψ_i's and μ. We can then rewrite the model as

$$Z_t = \pi_1 Z_{t-1} + \pi_2 Z_{t-2} + \cdots + \pi_p Z_{t-p} + \delta + u_t$$

which is referred to as an *autoregressive process of order p*, usually denoted as AR(p). For notational convenience, we will denote the AR process as shown in the following:

$$Z_t = \varphi_1 Z_{t-1} + \varphi_2 Z_{t-2} + \cdots + \varphi_p Z_{t-p} + \delta + u$$

Thus, AR processes are equivalent to MA processes of infinite order. Stationarity of AR processes is confirmed if the roots of the following characteristic equation lie outside the unit circle in the complex plane:

$$\left(1 - \varphi_1 x - \varphi_2 x^2 - \cdots - \varphi_p x^p \right) = 0$$

where x is a dummy algebraic symbol. If the process is stationary, then we should have

$$E(Z_t) = \varphi_1 E(Z_{t-1}) + \varphi_2 E(Z_{t-2}) + \cdots + \varphi_p E(Z_{t-p}) + \delta + E(u_t)$$

$$= \varphi_1 E(Z_t) + \varphi_2 E(Z_t) + \cdots + \varphi_p E(Z_t) + \delta$$

$$= E(Z_t)(\varphi_1 + \varphi_2 + \cdots + \varphi_p) + \delta$$

which yields

$$E(Z_t) = \frac{\delta}{(1 - \varphi_1 - \varphi_2 - \cdots - \varphi_p)}$$

Denoting the deviation of the process from its mean by $Z_t{}^d$, the following is obtained:

$$Z_t{}^d = Z_t - E(Z_t) = Z_t - \frac{\delta}{(1 - \varphi_1 - \varphi_2 - \cdots - \varphi_p)}$$

$$Z_{t-1}{}^d - Z_{t-1} - \frac{\delta}{(1 - \varphi_1 - \varphi_2 - \cdots - \varphi_p)}$$

Rewriting the previous expression yields

$$Z_{t-1} = Z_{t-1}{}^d + \frac{\delta}{(1 - \varphi_1 - \varphi_2 - \cdots - \varphi_p)}$$

$$\vdots$$

$$Z_{t-k} = Z_{t-k}{}^d + \frac{\delta}{(1 - \varphi_1 - \varphi_2 - \cdots - \varphi_p)}$$

If we substitute the AR(p) expression into the expression for $Z_t{}^d$, we will obtain

$$Z_t{}^d = \varphi_1 Z_{t-1} + \varphi_2 Z_{t-2} + \cdots + \varphi_p Z_{t-p} + \delta + u_t - \frac{\delta}{(1 - \varphi_1 - \varphi_2 - \cdots - \varphi_p)}$$

Successive back substitutions of Z_{t-j} into the preceding expression yields

$$Z_t{}^d = \varphi_1 Z_{t-1}{}^d + \varphi_2 Z_{t-2}{}^d + \cdots + \varphi_p Z_{t-p}{}^d + u_t$$

Thus, the deviation series follows the same AR process without a constant term. The tools for identifying and constructing time series models are the sample autocorrelations, r_j. For the model identification procedure, a visual assessment of the plot of r_j against j, called the sample correlogram, is used. Table 9.1 presents examples of *sample correlograms* and the corresponding time series models.

A wide variety of sample correlogram patterns can be encountered in time series analysis. It is the responsibility of the analyst to choose an appropriate model to fit the prevailing time series data. Several statistical computer programs are available for performing time series analysis. SPSS and STATGRAPHICS are two familiar examples.

TABLE 9.1

Identification of Some Time Series Models

Characteristics of Sample Correlogram	Model Type	Model
a 1.0 Spikes at lags 1 to q 0 1 2	MA(2)	$Z_t = u_t - \theta_1 u_{t-1} - \theta_2 u_{t-2}$
b 1.0 Exponential decay 0	AR(1)	$Z_t = u_t + \theta_1 u_{t-1}$
c 1.0 Damped sine wave form 0	AR(2)	$Z_t = u_t + \theta_1 u_{t-1} + \theta_2 u_{t-2}$

Classical Inventory Management

Inventoried items are an important component of any project, even in nonmanufacturing operations. Any resource can be viewed as an "inventoried" item for the purpose of managing enterprise projects. Consequently, inventory management strategies are essential for comprehensive project planning and control. Tracking activities is analogous to tracking inventory items. The important aspects of inventory management for systems project management are

1. Ability to satisfy work demands promptly by supplying materials from stock
2. Availability of bulk rates for purchases and shipping
3 Possibility of maintaining more stable and level resource or workforce

Some of the basic and classical inventory control techniques are discussed in the following.

Economic Order Quantity Model

The economic order quantity (EOQ) model determines the optimal order quantity based on purchase cost, inventory carrying cost, demand rate, and ordering cost. The objective is to minimize the total relevant costs (TRC) of inventory. For the formulation of the model, the following notations are used:

Q is the replenishment order quantity (in units)

A is the fixed cost of placing an order

v is the variable cost per unit of the item to be inventoried

r is the inventory carrying charge per dollar of inventory per unit time

D is the demand rate of the item

TRC is the total relevant cost per unit time

Figure 9.7 shows the basic inventory pattern with respect to time. One complete cycle starts from a level of Q and ends at zero inventory.

The TRC for order quantity Q is given by the expression as follows. Figure 9.8 shows the costs as functions of replenishment quantity.

$$TRC(Q) = \frac{Qvr}{2} + \frac{AD}{Q}$$

When the TRC(Q) function is optimized with respect to Q, we obtain the expression for EOQ:

$$EOQ = \sqrt{\frac{2AD}{vr}}$$

which represents the minimum TRC of inventory. The previous formulation assumes that the cost per unit is constant regardless of the order quantity. In some cases, quantity discounts may be applicable to the inventory item. The formulation for quantity discount situation is presented in the following.

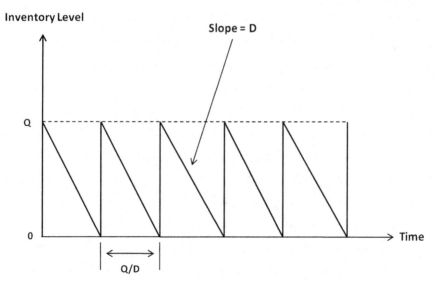

FIGURE 9.7
Basic inventory pattern.

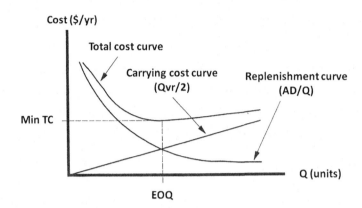

FIGURE 9.8
Inventory costs as functions of replenishment quantity.

Quantity Discount

A quantity discount may be available if the order quantity exceeds a certain level. This is referred to as the single breakpoint discount. Figure 9.9 presents the price breakpoint for quantity discount.

The unit cost is represented as shown in the following equation

$$v = \begin{cases} v_0, & 0 \le Q < Q_b \\ v_0(1-d), & Q_b \le Q \end{cases}$$

where
v_0 is the basic unit cost without discount
d is the discount (in decimals)
d is applied to all units when $Q \le Q_b$
Q_b is the breakpoint

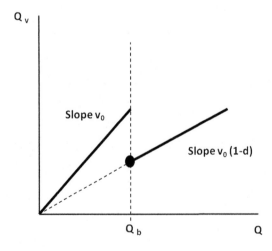

FIGURE 9.9
Price breakpoint for quantity discount.

Calculation of TRC

For $0 \leq Q < Q_b$, we obtain

$$\text{TRC}(Q) = \left(\frac{Q}{2}\right)v_0 r + \left(\frac{A}{Q}\right)D + Dv_0$$

For $Q_b \leq Q$, we have

$$\text{TRC}(Q)_{\text{discount}} = \left(\frac{Q}{2}\right)v_0(1-d)r + \left(\frac{A}{Q}\right)D + Dv_0(1-d)$$

Note that, for any given value of Q, $\text{TRC}(Q)_{\text{discount}} < \text{TRC}(Q)$. Therefore, if the lowest point on the $\text{TRC}(Q)_{\text{discount}}$ curve corresponds to a value of $Q^* > Q_b$ (i.e., Q is valid), then set $Q_{\text{opt}} = Q^*$.

Evaluation of the Discount Option

The trade-off between extra carrying costs and the reduction in replenishment costs should be evaluated to see whether the discount option is cost justified. Reduction in replenishment cost is composed of

1. Reduction in unit value
2. Fewer replenishments per unit time

   Case a: If reduction in acquisition costs > extra carrying costs, then set $Q_{\text{opt}} = Q_b$.

   Case b: If reduction in acquisition costs < extra carrying costs, then set $Q_{\text{opt}} = \text{EOQ}$ with no discount.

   Case c: If Q_b is relatively small, then set $Q_{\text{opt}} = \text{EOQ}$ with discount. The three cases are illustrated in Figure 9.10.

Based on the three cases shown in Figure 9.10, the optimal order quantity, Q_{opt}, can be found as follows:

1. *Step 1*: Compute EOQ when d is applicable:

$$\text{EOQ (discount)} = \sqrt{\frac{2AD}{v_0(1-d)r}}$$

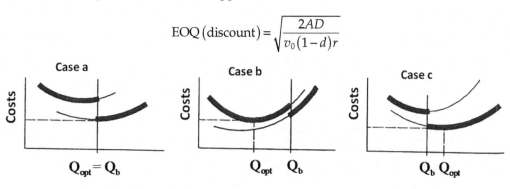

FIGURE 9.10
Cost curves for discount options.

2. *Step 2*: Compare EOQ(d) with Q_b:

    If EOQ(d) $\geq Q_b$, set $Q_{opt} =$ EOQ(d)

    If EOQ(d) $< Q_b$, go to Step 3

3. *Step 3*: Evaluate TRC for EOQ and Q_b:

$$\text{TRC}(\text{EOQ}) = \sqrt{2ADv_0r} + Dv_0$$

$$\text{TRC}(Q_b)_{discount} = \left(\frac{Q_b}{2}\right)v\,(1-d)r + \left(\frac{A}{Q_b}\right)D + Dv_0(1-d)$$

If $\text{TRC}(\text{EOQ}) < \text{TRC}(Q_b)$, set $Q_{opt} =$ EOQ(no discount):

$$\text{EOQ}(\text{no discount}) = \sqrt{\frac{2AD}{v_0r}}$$

If $\text{TRC}(\text{EOQ}) > \text{TRC}(Q_b)$, set $Q_{opt} = Q_b$

Suppose $d = 0.02$ and $Q_b = 100$ for the three items shown in Table 9.2.

Item 1 (case a)

Step 1: EOQ(discount) = 19 units < 100 units
Step 2: EOQ(discount) $< Q_b$, go to Step 3
Step 3: TRC values

$$\text{TRC}(\text{EOQ}) = \sqrt{2(1.50)(416)(14.20)(0.24)} + 416(14.20)$$

$$= \$5972.42/\text{year}$$

$$\text{TRC}(Q_b) = \frac{100(14.20)(0.98)(0.24)}{2} + \frac{(1.50)(416)}{100} + 416(14.20)(0.98)$$

$$= \$5962.29/\text{year}$$

Since TRC(EOQ) > TRC(Q_b), set $Q_{opt} = 100$ units.

Item 2 (case b)

Step 1: EOQ(discount) = 21 units < 100 units
Step 2: EOQ(discount) $< Q_b$, go to Step 3
Step 3: TRC values

TABLE 9.2

Items Subject to Quantity Discount

Item	D (Units/Year)	v_0 ($/Unit)	(A) ($)	r ($/$/Year)
Item 1	416	14.20	1.50	0.24
Item 2	104	3.10	1.50	0.24
Item 3	4160	2.40	1.50	0.24

$$TRC(EOQ) = \sqrt{2(1.50)(104)(3.10)(0.24)} + 104(3.10)$$

$$= \$337.64/\text{year}$$

$$TRC(Q_b) = \frac{100(3.10)(0.98)(0.24)}{2} + \frac{(1.50)(104)}{100} + 104(3.10)(0.98)$$

$$= \$353.97/\text{year}$$

$$TRC(EOQ) < TRC(Q_b), \quad \text{set } Q_{\text{opt}} = EOQ(\text{without discount}):$$

$$EOQ = \sqrt{\frac{2(1.50)(104)}{3.10(0.24)}}$$

$$= 20 \text{ units}$$

Item 3 (case c)

Step 1: Compute EOQ(discount)

$$EOQ(\text{discount}) = \sqrt{\frac{2(1.50)(4160)}{2.40(0.98)(0.24)}}$$

$$= 149 > 100 \text{ units}$$

Step 2: EOQ(discount) > Q_b. Set $Q_{\text{opt}} = 149$ units.

Sensitivity Analysis

Sensitivity analysis involves a determination of the changes in the values of a parameter that will lead to a change in a dependent variable. It is a process of determining how wrong a decision will be if assumptions on which the decision is based prove to be incorrect. For example, a "decision" may be dependent on the changes in the values of a particular parameter. For example, inventory cost may be the parameter on which the decision depends. Cost itself may depend on the values of other parameters as shown in the following:

$$\text{Subparameter} \rightarrow \text{Main parameter} \rightarrow \text{Decision}$$

It is of interest to determine what changes in parameter values can lead to changes in a decision. With respect to inventory management, we may be interested in the cost impact of deviation of actual order quantity from the EOQ. The sensitivity of cost to departures from EOQ is analyzed as presented in the following:

Let p represent the level of change from EOQ:

$$|p| \leq 1.0$$
$$Q' = (1-p)EOQ$$

Percentage cost penalty (PCP) is defined as

$$PCP = \frac{TRC(Q') - TRC(EOQ)}{TRC(EOQ)}(100)$$

$$= 50\left(\frac{p^2}{1+p}\right)$$

A plot of the cost penalty is shown in Figure 9.11.

It is seen that the cost is not sensitive to minor departures from EOQ. We can conclude that changes within 10% of EOQ will not significantly affect the TRC. Two special inventory control algorithms are discussed in the following.

Wagner–Whitin Algorithm

The Wagner–Whitin (W–W) algorithm is an approach to deterministic inventory model. It is based on dynamic programming technique. Dynamic programming is a mathematical procedure for solving sequential decision problems. The assumptions of W–W algorithm are

1. Expected demand are known for N periods into the future
2. The periods are equal in length
3. No stockout or backordering is allowed
4. All forecast demand will be met
5. Ordering cost A may vary from period to period
6. Orders are placed at the beginning of a period
7. Order lead time is zero
8. Inventory carrying cost is $(v)(r)$/unit/period
9. Inventory carrying cost is charged at the beginning of a period for the units carried forward from the previous period

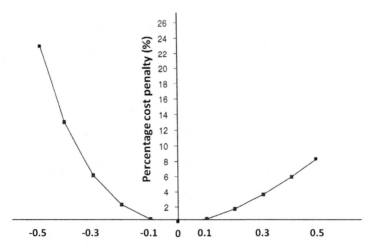

FIGURE 9.11
Sensitivity analysis based on PCP.

The optimality properties of W–W algorithm are as follows:

1. *Property 1.* Replenishment takes place only when inventory level is zero.
2. *Property 2.* There is an upper limit to how far back before period j we would include D_j in a replenishment quantity.

If the requirements for period j are so large that

$$D_j(vr) > A,$$

$$\text{i.e., } D_j > \frac{A}{vr}$$

then the optimal solution will have a replenishment at the beginning of period j. That is, inventory must go to zero at the beginning of period j.

The earliest value of j, where $D_j > A/vr$, is used to determine the horizon (or number of periods) to be considered when calculating the first replenishment quantity.

Notations and Variables

N is the number of periods in the planning horizon ($j = 1, 2, …, N$)

t is the number of periods considered when calculating first replenishment quantity ($j = 1, 2, …, t$),

where $t \leq N$

vr is the inventory carrying cost/unit/period of inventory carried forward from period j to period $j + 1$
A is the ordering cost per order
I_j is the inventory brought into period j
D_j is the demand for period j
Q_j is the order quantity in period j

$$\delta_j = \begin{cases} 0 & \text{if } Q_j = 0 \\ 1 & \text{if } Q_j > 1 \end{cases}$$

$$I_{j+1} = I_j + \delta_j Q_j - D_j$$

Let $F(t)$ = total inventory cost to satisfy demand for periods 1, 2, …, t:

$$F(t) = \left[vrI_1 + A\delta_1 Q_1 \right]$$
$$+ \left[vr(I_1 + \delta_1 Q_1 - D_1) + A\delta_2 Q_2 \right]$$
$$+ \left[vr(I_1 + \delta_1 Q_1 - D_1 + \delta_2 Q_2 - D_2) + A\delta_3 Q_3 \right]$$
$$+ \cdots + \left[\cdots + A\delta_t Q_t \right]$$

The expression for $F(t)$ assumes that the initial inventory is known. The objective is to minimize total inventory cost.

Propositions for the W–W Algorithm

Proposition 1: $I_j Q_j = 0$. This is because if we are planning to place an order in period j, we will not carry an inventory I_j into period j. Since stockout is not permitted, an order must be placed in period j if the inventory carried into period j is zero. Either one of I_j or Q_j must be zero.

Proposition 2: An optional policy exists such that for all periods

$$Q_j = 0$$

or

$$Q_j = \sum_{i=1}^{t} Q_i, \quad j \le t \le N$$

This means that order quantity in period j is either zero or the total demand of some (or all) of the future periods.

Proposition 3: There exists an optimal policy such that if demand D_{j+t} is satisfied by Q_j^*, then demands $D_j, D_{j+1}, D_{j+2}, \ldots, D_{j+t-1}$ are also satisfied by Q_j^*.

Proposition 4: If an optimal inventory plan for the first t periods is given for an N-period model in which $I_j = 0$, for $j < t$, then it is possible to determine the optimal plan for periods $t + 1$ through N.

Proposition 5: If an optimal inventory plan for the first t periods of an N-period model is known in which $Q_t > 0$, it is not necessary to consider periods 1 through $t - 1$ in formulating the optimal plan for the rest of the periods. That is, we reinitialize the problem to start at time t and proceed forward.

Computational Example of W–W Algorithm

W–W algorithm is best explained by an example. Suppose we have an item that is to be inventoried over four periods. We want to determine the best inventory strategy covering all four periods.

$A = \$10/\text{order}$

$vr = \$0.04/\text{unit/period}$

Initial Inventory, $I_1 = 0$ (i.e., no starting inventory)

Periods (Month), j	1	2	3	4
D_j	40	70	90	60

Period 1: $D_1 = 40$				Optimal Solution	
I_2	$Q = 40$	$Q = 110$	$Q = 200$	$Q = 260$	Cost	Q_0
0	$0 + 10 = \$10.00$				$\$10.00$	40
70		$70 (0.04) + 10 = \$12.80$			$\$12.80$	110
160			$160 (0.04) + 10 = \$16.40$		$\$16.40$	200
220				$220 (0.04) + 10 = \$18.80$	$\$18.80$	260

The value of I_2 represents the possible inventory to be carried to period 2. The values in the first column are the possible inventories that can be carried from period 1 to period 2 based on the order quantity, Q, shown in the top row. The possible carryforward inventories are 0, 70, 160 (70 + 90), and 220 (70 + 90 + 60). The possible order quantities are 40, 110 (40 + 70), 200 (40 + 70 + 90), and 260 (40 + 70 + 60). The minimum cost in each row is entered in the column under cost, and the column corresponding to the minimum cost is used to determine the optimal Q, which is entered in the last column.

Period 2: $D_2 = 70$					Optimal Solution	
I_3	$Q = 0$	$Q = 70$	$Q = 160$	$Q = 220$	Cost	Q_0
0	12.8 + 0 = $12.80	10 + 10 = $20.00			$12.80	0
90	16.40 + 90 (0.04) = $20		10 + 10 + 90 (0.04) = $23.60		$20.00	0
150	18.80 + 150 (0.04) = $24.80			10 + 10 + 6.0 = $26.00	$24.80	0

For the $12.80 in the second column, if $Q_2 = 0$ and $I_3 = 0$, then we must have carried 70 ($D_2 = 70$) from period 1, which leads us to the $I_2 = 70$ row and a minimum cost of $12.80 in that row. For the $16.40 is the second row, if 90 is to be carried from period 2 to period 3 and $Q_2 = 0$, then it means we must have ordered $Q = 200$ in period 1, which corresponds to a cost of $16.40.

Period 3: $D_3 = 90$				Optimal Solution	
I_4	$Q = 0$	$Q = 90$	$Q = 150$	Cost	Q_0
0	20 + 0 = $20.00	12.80 + 10 = $22.80		$20.00	0
60	24.80 + (60) (0.04) = $27.20		12.80 + (60) (0.04) + 10 = $25.20	$25.20	150

Period 4: $D_4 = 60$			Optimal Solution	
I_5	$Q = 0$	$Q = 60$	Cost	Q_0
0	$25.20	20 + $10 = $30.00	$25.20	0

For the $25.20 in the second column, if $Q_4 = 0$, then we must have carried 60 from period 3, which leads us to the minimum cost of $25.20 in the $I_4 = 60$ row. Therefore, the optimal policy is to place an order for 110 units at the beginning of period 1 and an order of 150 units at the beginning of period 3.

$Q_1 = 110$

$Q_3 = 150$

Total cost = $25.20

This is obtained by backtracking from the final table:

$$\$25.20 \rightarrow Q_4 = 0 \rightarrow \$25.20 + Q_3 = 150 \rightarrow \$12.80 \rightarrow Q_2 = 0 \rightarrow \$12.80 + Q_1 = 110$$

Although there are modern computationally efficient methods of accomplishing what this algorithm does, it is still a classic approach to inventory control, and there is valuable learning advantage from going through the steps of the algorithm. It is through the understanding of manual computational process that students can learn to develop validated computer programming of a technique.

Silver-Meal Heuristic

Silver-meal heuristic is a simple inventory control technique that is recommended for items that have significantly variable demand pattern. Its objective is to minimize TRC per unit time for the duration of the replenishment quantity:

$$TRC = A + \text{Carrying costs}$$

where A is the cost of placing an order

Procedure. Evaluate the following expression for increasing values of T:

$$TRCUT(T) = \frac{TRC(T)}{T}$$

$$= \frac{A + \text{Carrying costs}}{T}$$

where TRCUT = TRCs per unit time.

Stop when TRCUT(T + 1) > TRCUT(T).

Set optimal T: $T^* = T$.

The silver-meal heuristic is recommended for use when demand variability coefficient (VC) ≥ 0.2.

$$VC = \frac{\text{Variance of demand per period}}{\text{Square of average demand per period}}$$

$$= \frac{N \sum\limits_{j=1}^{N} \left[D_j \right]^2}{\left[\sum\limits_{j=1}^{N} D_j \right]^2} - 1$$

If VC < 0.2, use basic EOQ formula.

If VC $\geq$ 0.2, use silver-meal heuristic.

Example 9.1

Suppose we have an inventory situation characterized by the following data:

12-month planning horizon
$A = \$35$ (cost of placing an order)

$v = \$4.75/\text{unit}$ (dollar value of one unit of inventory)
$r = \$0.02/\$/\text{month}$ (dollar cost of carrying one dollar value of inventory for one period)

Period, j	1	2	3	4	5	6	7	8	9	10	11	12
Demand, D_j	18	31	23	95	29	37	50	39	30	88	22	36

The silver-meal heuristic will proceed as presented in the following:
Iteration 1 (Month $j = 1$)

D_j	Q	T	A	D_2vr	$2D_3vr$	$3D_4vr$	$4D_5vr$	Row Sum	Cumulative Sum	Cumulative Sum/T
18	18	1	35					35	35	35
31	49	2		2.95				2.95	37.95	18.98
23	72	3			4.37			4.37	42.32	14.11
95	167	4				27.08		27.08	69.4	17.35
29	196	5					11.02	11.02	80.42	16.08

$T_1^* = 3, Q_1^* = 72$ units, $\text{TRCUT}(T_1^*) = 14.11$
The \$2.95 is calculated as $31(\$4.75)(0.02)$ while the \$4.37 is calculated as $2\{23(\$4.75)(0.02)\}$. The cost calculations for the periods are as follows:

Period 1: A
Period 2: $A + D_2$
Period 3: $A + D_2 + 2D_3$
Period 4: $A + D_2 + 2D_3 + 3D_4$

Iteration 2 (Month $j = 4$)

D	Q	T	A	D_5vr	$2D_6vr$	$3D_7vr$	$4D_8vr$	Row Sum	Cumulative Sum	Cumulative Sum/T
95	95	1	35					35	35	35
29	124	2		2.76				2.76	37.76	18.88
37	161	3			7.03			7.03	44.79	14.93
50	211	4				14.25		14.25	59.04	14.76
39	250	5					14.92	14.92	73.96	14.79

$T_2^* = 4, Q_4^* = 211$ units, $\text{TRCUT}(T_2^*) = 14.76$.
Note that inventory carried forward from period $j = 4$ will start at $j = 5$.
Iteration 3 (Month $j = 8$)

D	Q	T	A	D_8vr	$2D_9vr$	$3D_{10}vr$	$4D_{11}vr$	Row Sum	Cumulative Sum	Cumulative Sum/T
39	39	1	35					35	35	35
30	69	2		2.85				2.85	37.85	18.93
88	157	3			16.72			16.72	54.57	18.19
22	179	4				4.18		4.18	58.75	14.69
36	215	5					13.68	13.68	72.43	14.49

$T_3^* = 5$, $Q_8^* = 215$ units, $\text{TRCUT}(T_3^*) = 14.49$

Period, *j*	1	2	3	4	5	6	7	8	9	10	11	12	Total
Starting inventory	0	54	23	0	116	87	50	0	176	146	58	36	—
Q_j	72	—	—	211	—	—	—	215	—	—	—	—	498
D_j	18	31	23	95	29	37	50	39	30	88	22	36	498
Ending inventory	54	23	0	116	87	50	0	176	146	58	36	0	746

$$\text{Total Inventory carrying} = 1\big[31(4.75)(0.02)\big] + 2\big[23(4.75)(0.02)\big] + 1\big[29(4.75(0.02))\big]$$

$$+ 2\big[37(4.75)(0.02)\big] + 3\big[50(4.75)(0.02)\big] + 1\big[30(4.75)(0.02)\big]$$

$$+ 2\big[88(4.75)(0.02)\big] + 3\big[22(4.75)(0.02)\big] + 4\big[36(4.75)(0.02)\big]$$

$$= (4.75)(0.02)[31 + 46 + 29 + 74 + 150 + 30 + 176 + 66 + 144]$$

$$= (0.095)[746]$$

$$= \$70.87,$$

$$\text{Total ordering cost} = 3(\$35) = \$105,$$

$$\text{Total Cost} = \$70.87 + \$105 = \$175.87$$

Note that D_2 is carried for one month, D_3 is carried for two months, and D_5 is carried for one month. Inventory analysis is needed to support procurement goals as outlined in the project management body of knowledge. Many computerized commercial and proprietary systems are available for tracking inventory. Some of these can be adapted for inventory analysis in project management.

Seasonal Pattern Modeling

There are numerous applications of statistical distributions to practical, real-world problems. Several distributions have been developed and successfully utilized for a large variety of problems. Despite the large number of distributions available, it is often confusing to determine which distribution is applicable to which real-world random variable. In many applications, the choices are limited to a few familiar distributions due either to a lack of better knowledge or computational ease. Such familiar distributions include the *normal, exponential,* and *uniform* distributions. To accommodate cases where the familiar distributions do not adequately represent the variable of interest, some special purpose distributions have been developed. Examples of such special distributions are *Pareto, Rayleigh, log-normal,* and *Cauchy* distributions. Consider the *cyclic pdf* (also called *periodic distribution* or *seasonal distribution*) for special cases involving random variables that are governed by cyclic processes. These special cases include seasonal inventory control and time series processes. Instabilities in a system can be caused by erratic materials supply, cyclic inventory, and seasonal work patterns. A time series is a collection of observations that are drawn from a periodic or cyclic process. Examples of operating time series include monthly cost, quarterly revenue, and seasonal energy consumption. The standard cyclic pdf is a continuous wave-form periodic function defined on the interval [0, 2π)]. The standard function is transformed into a general cyclic function defined over any time series interval [*a*, *b*]. One possible application of the cyclic distribution is the estimation of time-to-failure for components or equipment that undergoes periodic maintenance. Another application may be in the statistical analysis of the peak levels in a time series process.

Modeling Approach

The trigonometric sine function provides the basis for the cyclic distribution. The basic sine function is given by

$$y = \sin\theta, \quad -\infty < \theta < \infty$$

where θ is in radians. The sine function is shown in Figure 9.12.

The function is cyclic with period 2π. The sine function has its maxima and minima (1 and −1 respectively) at the points $\theta = \pm n\pi/2$, where n is an odd positive integer. The function also satisfies the relationship

$$\sin(\theta + 2n\pi) = \sin\theta$$

for any integer n. In general, any function $f(x)$ satisfying the relationship

$$f(x + nT) = f(x)$$

is said to be cyclic with period T, where T is a positive constant and n is an integer. A graph of $f(x)$ truncated to an interval [(a, a + T) or (a, a + T)] is called *one cycle of the function*.

Standard Cyclic PDF

The cyclic pdf is defined as

$$f(x) = \begin{cases} \dfrac{1}{\pi}\sin^2 mx & 0 \le x \le 2\pi, \quad m = 1,2,3,\ldots \\ 0 & \text{otherwise} \end{cases}$$

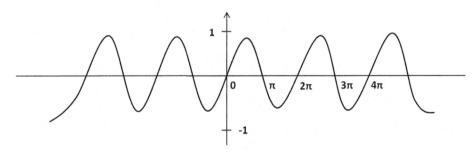

FIGURE 9.12
Basic sine function.

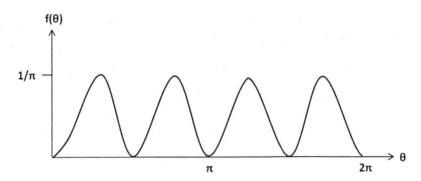

FIGURE 9.13
Standard cyclic distribution for $m = 2$.

where

x is in radians

m (positive integer) is the number of peaks (or modes) associated with the function on the interval $[0, \pi]$, which is half of the period

Thus, $f(x)$ represents a multimodal distribution. A graph of the standard cyclic pdf is presented in Figure 9.13.

To be a legitimate pdf, $f(x)$ must satisfy the following conditions:

1. $f(x) \geq 0$ for all x in $[0, 2\pi]$.
2. $\int_0^{2\pi} f(x)dx = 1$.
3. $P(x_1 < X < x_2) = \int_{x_1}^{x_2} f(x)dx$.

The first condition can be seen to be true by inspection. Each value of x yields a probability value equal to or greater than zero. The second condition can be verified by examining the definite integral

$$\int_0^{2\pi} f(x)dx = \frac{1}{\pi} \int_0^{2\pi} \sin^2(mx)dx$$

$$= \frac{1}{\pi}\left[\frac{x}{2} - \frac{\sin(2mx)}{4m}\right]_0^{2\pi}$$

$$= \frac{1}{\pi}\left[\pi - \frac{\sin(4mx)}{4m}\right]$$

$$= 1$$

The third condition is easily verified by computing the probability that X will fall between x_1 and x_2 as the area under the curve for $f(x)$ from x_1 to x_2.

Expected Value

The expected value of X is given by

$$E[X] = \int_0^{2\pi} xf(x)dx$$

$$= \frac{1}{\pi}\int_0^{2\pi} x\sin^2(mx)dx$$

$$= \frac{1}{\pi}\left[\frac{x^2}{4} - \frac{x\sin(2mx)}{4m} - \frac{\cos(2mx)}{8m^2}\right]_0^{2\pi}$$

$$= \pi$$

Variance

The variance of X is calculated from the theoretical definition of variance, which makes use of the definite integral from 0 to 2π. This is given by:

$$V[X] = \int_0^{2\pi} x^2 f(x)dx - \{E[X]\}^2$$

$$= \frac{1}{\pi}\int_0^{2\pi} x^2 \sin^2(mx)dx - \pi^2$$

$$= \frac{1}{\pi}\left[\frac{x^3}{6} - \left(\frac{x^2}{4m} - \frac{1}{8m^3}\right)\sin(2mx) - \frac{x\cos(2mx)}{4m^2}\right]_0^{2\pi} - \pi^2$$

$$= \frac{1}{3}\pi^2 - \frac{1}{2m^2}$$

Cumulative Distribution Function

The cumulative distribution function of $f(x)$ is given by

$$F(x) = \int_0^x f(r)dr$$

$$= \int_0^x \frac{1}{\pi}\sin^2(mr)dr$$

$$= \frac{2mx - \sin(2mx)}{4m\pi}$$

General Cyclic PDF

For practical applications, a general form of the cyclic pdf defined over any real interval $[a, b]$ will be of more interest. This general form is given by

$$f(x) = \begin{cases} \dfrac{1}{\pi}\sin^2 mr & 0 \le r \le 2\pi, \quad m = 1,2,3,\dots \\ 0 & \text{otherwise} \end{cases}$$

where
  m is half of the number of peaks expected over the interval [a, b]
  r is the radian equivalent of the real number x in the interval [a, b]

The transformations from real units to radians and vice versa are accomplished by the expression as follows:

$$r = 2\pi \frac{x-a}{b-a}, \quad 0 \le r \le 2\pi, \quad a \le x \le b$$

$$x = a + (r)\frac{b-a}{2\pi}, \quad 0 \le r \le 2\pi, \quad a \le x \le b$$

Using the earlier transformation relationships and the expressions derived previously for the standard cyclic random variable, the mean and variance of the general cyclic random variable are derived to be

$$E[X] = a + \frac{b-a}{2\pi}(E[r])$$

$$= a + \frac{b-a}{2\pi}(\pi)$$

$$= \frac{a+b}{2}$$

$$V[X] = 0 + \left(\frac{b-a}{2\pi}\right)^2 V[r]$$

$$= \frac{(b-a)^2}{4\pi^2}\left(\frac{\pi^2}{3}\right)$$

$$= \frac{(b-a)^2}{12}$$

Note that these expressions are identical to the expressions for the mean and variance of the uniform distribution.

Application Examples

Let us suppose that a system contains a certain type of component that undergoes cyclic maintenance. The cyclic distribution will be an appropriate model for the time-to-failure analysis for the system. Assume that the useful life of the component is estimated to be 4 years. The component will undergo maintenance three times during the 4-year period. The failure times of the component after its first installation and between maintenance will have identical bell shapes. This is represented graphically in Figure 9.14.

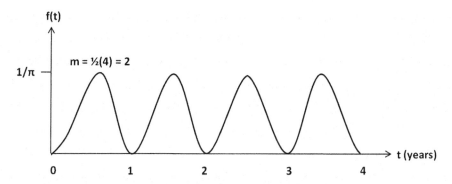

FIGURE 9.14
Cyclic distribution for component time-to-failure.

Now, suppose we are interested in the probability that the component will fail no later than 2.8 years after its first installation. The calculations, using the cyclic pdf, will proceed as follows:

$$a = 0, \quad b = 4, \quad m = 2, \quad x = 2.8$$

$$r = \frac{(2\pi)(2.8 - 0)}{(4 - 0)} = 1.4\pi = 4.3982 \text{ rad}$$

$$P(0 \le X \le 2.8) = P(0 \le r \le 4.3982) = \int_0^{4.3982} \frac{1}{\pi} \sin^2(2r)\,dr = 0.7378$$

For another example, suppose we are interested in analyzing the peak points in the distribution of monthly energy consumption. Suppose the times of occurrence of consumption peaks are periodically distributed and evenly spaced from month to month. The peak times are modeled as shown in Figure 9.15.

The graphical representation shows that on a month-to-month basis, the most likely peak point is the middle of the month. Any other peak point (e.g., end of the month) can be accommodated by shifting the time reference to the distribution.

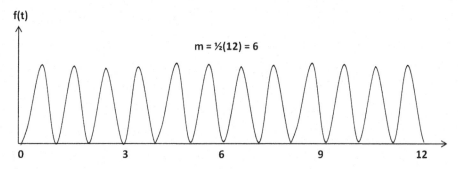

FIGURE 9.15
Cyclic distribution for peak occurrence times

Now, suppose we are interested in the probability that there will be a peak between $t = 1$ and $t = 4.1$ months. The analysis will be done as follows:

$$a = 0, \quad b = 12, \quad m = 6, \quad x_1 = 1.0, \quad x_2 = 4.1$$

$$r_1 = \frac{2\pi}{12} = 0.1667\pi$$

$$r_2 = \frac{2\pi}{12(4.1)} = 0.6833\pi$$

$$P(1 \le X \le 4.1) = P(0.1667\pi \le r \le 0.6833\pi)$$

$$= \int_{0.1667\pi}^{0.6833\pi} \frac{1}{\pi} \sin^2(6r)\,dr$$

$$= F(0.6833\pi) - F(0.1667\pi)$$

$$= 0.2506$$

The cyclic distribution can be modeled for seasonal data collected for project parameters that fit the assumptions of the distribution. It is interesting to note the similarity between the cyclic distribution and other distributions over the interval $[0, 2\pi]$. The parameter m determines how many cycles (or "waves") will be in the $[0, \pi]$ interval. The higher the value of m, the greater the number of waves. Figure 9.16 shows the cyclic distribution for $m = 1/2$, 1, and 2.

It can be seen that as m goes to infinity, the cyclic distribution approaches the uniform distribution because the number of waves become so large and so close that they are practically indistinguishable. Recall that the mean and variance of the cyclic distribution are π and $\pi^2/3 - 1/(2m^2)$, respectively, while the mean and variance of the uniform distribution are $(a + b)/2$ and $(b - a)^2/12$, respectively. For a uniform distribution over the interval $[0, 2\pi]$, the mean and variance reduce to π and $\pi^2/3$, respectively. Now,

$$\lim_{m \to \infty} V[x]_{\text{cyclic}} = \lim_{m \to \infty} \left\{ \frac{\pi^2}{3} - \frac{1}{2m^2} \right\} = \frac{\pi^2}{3}$$

The cyclic distribution is intrinsically symmetric while the beta distribution is symmetric if and only if its shape parameters, α_1 and α_2, are equal. The standard beta distribution is given by

$$g(x) = \begin{cases} \dfrac{x^{\alpha_1 - 1}(1 - x)^{\alpha_2 - 1}}{B(\alpha_1, \alpha_2)}, & 0 < x < 1 \\ 0, & \text{otherwise} \end{cases}$$

where

$$B(\alpha_1, \alpha_2) = \int_0^1 t^{\alpha_1 - 1}(1 - t)^{\alpha_2 - 1}\,dt$$

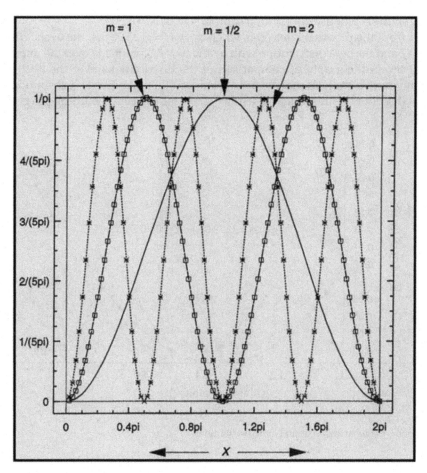

FIGURE 9.16
Cyclic distribution for $m = 1/2$, 1, and 2

We will compare the cyclic distribution with $m = 1/2$ to a general beta distribution defined over the interval $[0, 2\pi]$. To obtain a general beta distribution over the interval $[0, 2\pi]$, we will perform a function transform as shown next:

$$h(x) = \begin{cases} \dfrac{g(x/2\pi)}{2\pi}, & 0 < x < 2\pi \\ 0, & \text{otherwise} \end{cases}$$

Recall that the beta distribution is symmetric if and only if $\alpha_1 = \alpha_2$. Now, we define an error square function, $error\,(x)$, as the square of the difference between the general beta distribution and the cyclic distribution:

$$\text{Error}(x) = \sum_{k=1}^{N} \left\{ f\left(2\pi \frac{k}{N}\right) - h\left(2\pi \frac{k}{N}\right) \right\}^2, \quad m = \frac{1}{2}$$

Figure 9.17 shows a plot of *error* (*x*) for a sample size of $N = 1,000$.

It is seen from the plot that *error* (*x*) is when $\alpha_1 = \alpha_2 = 3.35$. Thus, the cyclic distribution with $m = 1/2$ and the beta distribution with α_1 and $\alpha_2 = 3.35$ coincide over the interval $[0, 2\pi]$. Figure 9.18 shows the graphical comparison of the distributions over the $[0, 2\pi]$ interval. Similar comparative analysis may be conducted for other symmetric distributions such as the normal distribution for modeling activity times for project scheduling purposes.

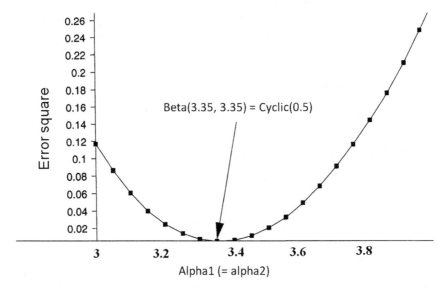

FIGURE 9.17
Plot of error square comparing cyclic and beta distributions.

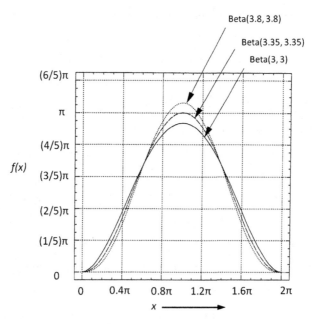

FIGURE 9.18
Graphical comparison of cyclic and beta distributions.

Exercises

9.1 For the *T*-period moving average forecast method, show that

$$f_{n+1} = \frac{\sum\limits_{t=n-T+1}^{n} d_t}{T}$$

$$= f_n + \left[\frac{d_n - d_{n-T}}{T}\right]$$

9.2 Develop a numeric example of exponential smoothing forecast. Justify the value of the smoothing factor you use in the example.

9.3 Use weighted average forecast approach to generate the forecast in Exercise 9.2. Compare and discuss the result with the forecast generated in Exercise 9.2.

9.4 Given that the cost of inventory is described by a continuous cash flow function defined as

$$f(t) = vc(Q - \lambda t), \quad t \le T$$

where

  v is the inventory carrying charge ($/$ of inventory per unit time)

  c is the unit cost of inventory items

  Q is the reorder quantity

  λ is the demand rate per unit time

  T is the length of one complete inventory cycle $f(T) = 0$

  a. If the annual nominal interest rate is r%, use direct integration to derive an expression for the present value of inventory cost for one complete inventory cycle.

  b. If $v = \$0.15$ per $\$$ inventory per month, $r = 12\%$, $c = \$10/$unit, $Q = 150$ units, and $\lambda = 20$ units/month, use your formula in part (a) to find the present value of the inventory cost for one complete cycle.

9.5 Repeat Exercise 9.4 for the case where the cost of inventory is described by a cyclic distribution over the interval $[0, 2\pi]$ with $m = 4$.

9.6 Suppose the monthly demand for an inventoried item is a random variable having the following distribution,

$$f(x) = \begin{cases} \dfrac{1}{50}\sin^2\left(\dfrac{\pi x}{100}\right), & 0 \le x \le 100 \\ 0, & \text{otherwise} \end{cases}$$

Find the probability that the demand will exceed 75 units in a particular month.

9.7   In a store, the time (in years) when a product is sold is an independent random variable having the following distribution:

$$f(x) = \begin{cases} 2\sin^2(2\pi x), & 0 \le x \le 1 \\ 0, & \text{otherwise} \end{cases}$$

The price of the product, $P(x)$, is time-dependent and is represented as follows:

$$P(x) = \begin{cases} 2 - \sin^2(2\pi x), & 0 \le x \le 1 \\ 0, & \text{otherwise} \end{cases}$$

What is the expected revenue from the product?

9.8   The price of an item in a project supply store is a random variable having the following distribution:

$$f(x) = \begin{cases} \dfrac{1}{20}\sin^2\left(\dfrac{\pi(x-35)}{10}\right), & 35 \le x \le 75 \\ 0, & \text{otherwise} \end{cases}$$

Which of the following price ranges has the highest probability? [35, 36]; [50, 51]; [68, 69]

9.9   The 10-year historical demand data for a part is presented as follows:

Year	1	2	3	4	5	6	7	8	9	10
Demand	2150	1900	2100	1850	2050	1900	2200	1900	2100	1850

Determine the forecast for year 11 using the following forecasting approaches:

a.   Simple average

b.   Simple 4-period moving average

c.   Weighted 4-period moving average with: $\alpha_1 = 0.4$, $\alpha_2 = 0.3$, $\alpha_3 = 0.2$, $\alpha_4 = 0.1$

d.   Exponentially weighted moving average with $\alpha = 0.75$

e.   Regression model

10

Multiresource Scheduling

Scarcity of Project Resources

In many complex projects, sequence-based scheduling and function-based scheduling are used so that the project can be responsive to the prevailing scenarios within the project environment. It is in this context that a generic multiresource project scheduling methodology is presented in this chapter. The approach incorporates the following:

1. Resource characteristics, such as preferences, time-effective capabilities, costs, and availability of project resources
2. Performance interdependencies among different resource groups

Based on the previous considerations, the methodology maps the most adequate resource units to each newly scheduled project activity as presented by Milatovic and Badiru (2004). The major challenge in this generic model development is to make it applicable to realistic project environments, which often involve multifunctional resources, whose capabilities or other characteristics may cross activities, as well as within a single activity relative to specific interactions among resources themselves. The scope of this research challenge further increases when the actual duration, cost, and successful completion of a project activity is assumed as resource driven and dependent on the choice of particular resource units assigned to it.

The proposed methodology dynamically executes two alternative procedures: the *activity scheduler* and *resource mapper*. The *activity scheduler* prioritizes and schedules activities based on their attributes and may also attempt to centralize selected resource loading graphs based on activity resource requirements. The *resource mapper* considers resource characteristics, incorporates interdependencies among resource groups or types, and maps the available resource units to newly scheduled activities according to a project manager's prespecified mapping (objective) function.

Notations Used in the Methodology

i is the project activity i, such that $i = 1, ..., I$.

I is the number of activities in project network.

t_c is the decision instance, i.e., time moment at which one or more activities qualify to be scheduled since their predecessor activities have been completed.

$PR(i)$ is the set of predecessor activities of activity i.

$Q(t_c)$ is the set of activities qualifying to be scheduled at t_c, i.e., $Q(t_c) = \{i \mid PR(i) = \emptyset\}$.

j is the resource type j, $j = 1, \ldots, J$.

J is the number of resource types involved in the project.

R_j is the number of units of resource type j available for the project.

$\langle j, k \rangle$ is the notation for kth unit of type j.

$\rho_i^{\,j}$ is the number of resource units type j required by activity i.

$u_{t_c}^{\,j,k}$ is a binary variable with a value of one, if kth unit of type j is engaged in one of the project activities that is in progress at the decision instance t_c, and zero otherwise. All $u_{t_c}^{\,j,k}$'s are initially set to zero.

$t_i^{\,j,k}$ is the time-effective executive capability of kth unit of resource type j if assigned to work on activity i.

$p_i^{\,j,k}$ is the preference of kth unit of resource type j to work on activity i.

$c_i^{\,j,k}$ is the estimated cost of kth unit of resource type j if assigned to work on activity i.

$\alpha_i^{\,j,k}(t_c)$ is the desired start time or interval availability of kth unit of type j to work on activity i at the decision instance t_c. In many cases, this parameter is invariant across activities, and the subscript i may often be dropped.

Analysis of Project Resources

Project resources, generally limited in quantity, are the most important constraints in scheduling of activities. In cases when resources have prespecified assignments and responsibilities toward one or more activities, their allocation is concurrently performed with the scheduling of applicable activities. In other cases, an activity may only require a certain number of (generic) resource units of particular type(s), which are assigned after the scheduling of the particular activity. These two approaches coarsely represent the dominant paradigms in project scheduling. The objective of this research is to propose a new strategy that will shift these paradigms to facilitate a more refined guidance for allocation and assignment of project resources. In other words, there is a need for tools that will provide for more effective resource tracking, control, interaction, and, most importantly, resource-activity mapping.

The main assumption in the methodology of this paper is that project environments often involve multicapable resource units with different characteristics. This is especially the case in knowledge-intensive settings and industries, which are predominantly staffed with highly trained personnel. The specific characteristics considered were resource preferences, time-effective capabilities, costs, and availability. Each resource unit's characteristics may further vary not only across project activities but also within a single activity relative to interaction among resource units. Finally, resource preferences, cost, and time-effective capabilities may also independently vary with time due to additional factors, such as learning, forgetting, weather, type of work, etc. Therefore, although we do not exclude a possibility that an activity duration is independent of resources assigned to it, in this research, we assume that it is those resource units assigned to a particular activity that determine how long it will take for the activity to be completed.

The scheduling strategy as illustrated earlier promotes a more balanced and integrated activity-resource mapping approach. Mapping the most qualified resources to each project activity, and thus preserving the values of resource, is achieved by proper consideration or resource time-effective capabilities and costs. By considering resource preferences and availability, which may be entered in either crisp or fuzzy form, the model enables consideration of personnel's voice and its influence on a project schedule and quality. Furthermore, resource-interactive dependencies may also be evaluated for each of the characteristics and their effects incorporated into resource-activity mapping. Finally, by allowing flexible and dynamic modifications of scheduling objectives, the model permits managers or analysts to incorporate some of their tacit knowledge and discretionary input into project schedules.

Resource Modeling Background

Literature presents extensive work on scheduling projects by accounting for worker (resource) preferences, qualifications, and skills as decisive factors to their allocation. Yet, a survey of some 400 top contractors in construction showed that 96.2% of them still use basic critical path method (CPM)s for project scheduling (Mattila and Abraham, 1998). Roberts (1992) pointed out that information sources for project planners and schedulers are increasingly nonhuman and stressed, and that planners must keep computerized tools for project management and scheduling in line and consider the perspectives of human resources used by projects. In other words, the author warns that too much technicalities may prompt and mislead managers into ignoring human aspects of project management.

Franz and Miller (1993) considered a problem of scheduling medical residents to rotations and approached it as a large-scale multiperiod staff assignment problem. The objective of the problem was to maximize residents' schedule preferences while meeting hospital's training goals and contractual commitments for staffing assistance.

Gray et al. (1993) discussed the development of an expert system to schedule nurses according to their scheduling preferences. Assuming consistency in nurses' preferences, an expert system was proposed and implemented to produce feasible schedules considering nurses' preferences, besides accounting for overtime needs, desirable staffing levels, patient acuity, etc. A similar problem was also addressed by Yura (1994), where the objective was to satisfy worker's preferences for time off as well as overtime, but under due date constraints.

Campbell (1999) further considered allocation of cross-trained resources in multidepartmental service environment. Employers generally value more resource units with various skills and capabilities for performing greater number of jobs. It is in those cases when managers face challenges of allocating these workers such that the utility of their assignment to a department is maximized. The results of experiments showed that the benefits of cross-training utilization may be significant. In most cases, only a small degree of cross-training captured the most benefits, and tests also showed that beyond a certain amount, the additional cross-training adds little additional benefits. Many companies face problems of continuous reassignment of people to facilitate new projects. Cooprider (1999) suggests a seven-step procedure to help companies consider a wide spectrum of parameters when distributing people within particular projects or disciplines.

Badiru (1993) proposed *critical resource diagramming* (CRD), which is a simple extension to traditional CPM graphs. In this approach, criticalities in project activities may also be reflected on project resources. Different resource types or units may vary in skills, supply, availability, or cost. This discrimination in resource importance should be accounted for when carrying out their allocation in scheduling activities. Unlike activity networks, the CRDs use nodes to represent each resource units. Also, unlike activities, a resource unit may appear more than once in a CRD network, specifying all different tasks for which a particular unit is assigned to. Similar to CPM, the same backward and forward computations may be performed to CRDs.

Multiresource Methodology

The methodology of this paper represents an analytical extension of CRDs presented by Badiru (1993). As previously mentioned, the design considerations of the proposed model consist of two distinct procedures: activity scheduling and resource mapping. At each decision instance during a scheduling process, the *activity scheduler* prioritizes and schedules some or all candidate activities, and then the *resource mapper* iteratively assigns the most adequate resource units to each of the newly scheduled activities.

Representation of Resource Interdependencies and Multifunctionality

This methodology is primarily focused on renewable resources. In addition, resources are not necessarily categorized into types or groups according to their similarities (i.e., into personnel, equipment, space, etc.), but more according to hierarchy of their interdependencies. In other words, we assume that time-effective capabilities, preferences, or even cost of any particular resource unit assigned to work on an activity may be dependent on other resource units also assigned to work on the same activity. Some or all of these other resource units may, in similar fashion, be also dependent on a third group of resources, and so on. Based on the concluded assumptions, we model competency of project resources in terms of the following four resource characteristics: time-effective capabilities, preferences, cost, and availability. Time-effective capability of a resource unit with respect to a particular activity is the amount of time the unit needs to complete its own task if assigned to that particular activity. Preferences are relative numerical weights that indicate personnel's degree of desire to be assigned to an activity, or a manager's perception on assigning certain units to particular activities. Similarly, each resource unit may have different costs associated with it relative to which activities it gets assigned to. Finally, not all resource units may be available to some or all activities at all times during project execution. Thus, the times during which a particular unit is available to some or all activities are also incorporated into the mapping methodology. Each of the characteristics described may vary across different project activities. In addition, some or all of these characteristics (especially time-effective capabilities and preferences) may also vary within a particular activity relative to resource interaction with other resources that are also assigned to work on the same activity.

Those resources, whose performance is totally independent of their interaction with other units, are grouped together and referred to as the type or group "one" and allocated first to scheduled activities. Resource units whose performance or competency is affected by their interaction with the type or group "one" units are grouped into type or group "two" and assigned (mapped) next. Resource units whose competency or performance is a function of type "two" or both types "one" and "two" are grouped into type "three" and allocated to scheduled activities after the units of the first two types have been assigned to them.

A project manager may consider one, more than one, or all four characteristics when performing activity-resource mapping. For example, a manager may wish to keep project costs as low as possible, while at the same time attempting to use resources with the best time-effective capabilities, consider their availability, and even incorporate their voice (in case of humans) or his/her own perception (in cases of human or nonhuman resources) in the form of preferences. This objective may be represented as follows:

$$u_i^{j,k} = f\left(t_i^{j,k}, c_i^{j,k}, p_i^{j,k}, \alpha_i^{j,k}(t_c)\right)$$

Mapping units of all resource types according to the same mapping function may often be impractical and unrealistic. Cost issues may be of greater importance in mapping some, while inferior to time-effective capabilities of other resource types. To accommodate the need for a resource-specific mapping function as mapping objective, we formulated the mapping function as additive utility function. In such a case, each of its components pertains to a particular resource type and is multiplied by a *Kronecker's delta* function. Kronecker's delta then detects resource type whose units are currently being mapped and filters out all mapping function components, except the one that pertains to the currently mapped resource type.

As an example, consider again a case where all resource types would be mapped according to their time-effective capabilities, except in the case of resource types "two" and "three," where costs would also be of consideration, and in the case of type "five," resource preferences and availabilities would be considered:

$$U_i^{j,k} = f\left(t_i^{j,k}\right) + f_2\left(c_i^{j,k}\right) \cdot \delta(j,2) + f_3\left(c_i^{j,k}\right) \cdot \delta(j,3) + f_5\left(p_i^{j,k}, \alpha_i^{j,k}(t_c)\right) \cdot \delta(j,5)$$

The previous example illustrates a case where mapping of resource units is performed according to filtered portions of a manager's objective (mapping) function, which may in turn be dynamically adaptive and varying with project scheduling time. As previously indicated, some resource characteristics may be of greater importance to a manager in the early scheduling stages of a project rather than in the later stages. Such a mapping function may be modeled as follows:

$$U_i^{j,k} = f_g\left(t_i^{j,k}, c_i^{j,k}, p_i^{j,k}, a_i^{j,k}(t_c)\right) + \sum_{s \in T} f_s\left(t_i^{j,k}, c_i^{j,k}, p_i^{j,k}, a_i^{j,k}(t_c)\right) \cdot w\left(t_{LO}^s, t_{HI}^s, t_c\right)$$

where
f_g is the component of the mapping function that is common to all resource types
f_s is the component of the mapping function that pertains to a *specific* project scheduling interval

$t_{LO}{}^{s}, t_{HI}{}^{s}$ is the specific time interval during which resource mapping must be performed according to a unique function

T is the set of previously defined time intervals for a particular project

$w\left(t_{LO}{}^{s}, t_{HI}{}^{s}, t_{c}\right)$ is the window function with a value of one if t_{c} falls within the interval $\left(t_{LO}{}^{s}, t_{HI}{}^{s}\right)$, and zero otherwise

Finally, it is also possible to map different resource types according to different objectives *and* at different times simultaneously by simply combining the two concepts already mentioned. For example, assume again that a manager forms his objective in the early stage of the project based on resources' temporal capabilities, costs, and preferences. Then at a later stage, the manager wishes to drop the costs and preferences and consider only resource capabilities, with the exception of resource type "three," whose costs should still remain in consideration for mapping. An example of a mapping function that would account for this scenario may be as follows:

$$U_i^{j,k} = f\left(c_i^{j,k}, p_i^{j,k}, t_i^{j,k}\right) \cdot w(0,30,t_c) + \left(f\left(t_i^{j,k}\right) + f\left(c_i^{j,k}\right) \cdot \delta(j,3)\right) \cdot w(30,90,t_c)$$

Modeling of Resource Characteristics

For resource units whose performance on a particular activity is independent of their interaction with other units, that is, for the *drivers*, $t_i^{j,k}$ is defined as the time, t, it takes kth unit of resource type j to complete its own task or process when working on activity i. Thus, different resource units, if multicapable, can be expected to perform differently on different activities. Each *dependent* unit, on the other hand, instead of $t_i^{j,k}$, generally has a set of interdependency functions associated with it.

In this research, we consider two types of interactive dependencies among resources, which due to their simplicity, are expected to be the most commonly used ones: *additive* and *percentage*. *Additive* interaction between a *dependent* and each of its *driver* resource unit indicates the amount of time that the *dependent* will need to complete its own task if assigned to work in conjunction with a particular driver. This is in addition to the time the driver itself needs to spend working on the same activity:

$$\left(T_i^{j,k}\right)_z = \left(t_i^{j_D,k_D} + \tilde{t}_i^{j,k}\right) \cdot y_i^{j_D,k_D}$$

where

$\langle j_D, k_D \rangle \in D^{j,k}$, where $D^{j,k}$ is a set of *driver* units (each defined by an indexed pair $\langle j_D, k_D \rangle$) for a particular resource unit $\langle j,k \rangle$

$\left(T_i^{j,k}\right)_z$ is the zth interactive time-effective dependency of kth unit of type j on its *driver* $\langle j_D, k_D \rangle$, $z = 1, \ldots,$ size $(D^{j,k})$. The actual number of these dependencies will depend on a manager's knowledge and familiarity with their resources

$\tilde{t}_i^{j,k}$ is the time needed in addition to $t_i^{j_D,k_D}$ for kth *dependent* unit of type j to complete its task on activity i if it interacts with its *driver* unit j_D, k_D

$y_i^{j_D,k_D}$ is the binary (zero-one) variable, indicating mapping status of the *driver* unit $\langle j_D, k_D \rangle$. It equals one if the unit $\langle j_D, k_D \rangle$ is assigned to activity i, and zero if the unit

$\langle j_D, k_D \rangle$ has been assigned to activity i. Therefore, each $\left(T_i^{j,k}\right)_z$ will have a nonzero value only if $y_i^{jD,kD}$ is also nonzero (i.e., if the *driver* resource unit $\langle j_D, k_D \rangle$ has been previously assigned to activity i)

The percentage interactive dependency is similarly defined as

$$\left(T_i^{j,k}\right)_z = t_i^{jD,kD}\left(1 + \tilde{t}_i^{j,k}\%\right) \cdot y_i^{jD,kD}$$

where $\tilde{t}_i^{j,k}\%$ is the percentage of time by which $t_i^{jD,kD}$ will be prolonged if the unit k of type j interacts with its *driver* $\langle j_D, k_D \rangle$.

Modeling cost characteristics follows a similar logic used for representation of temporal capabilities and interdependencies. In place of $t_i^{j,k}$, we now define a variable $c_i^{j,k}$, which represents the cost (say, in dollars) of kth unit of resource type j if it gets assigned to work on activity i. This value of $c_i^{j,k}$ may be invariant regardless of a unit's interaction with other resources, or it may vary relative to interaction among resources, and thus, implying cost interdependencies, which need to be evaluated before any mapping is performed (provided that the cost considerations are, indeed, a part of a manager's utility or objective for mapping).

In cases when a cost of a resource unit for an activity varies depending on its interaction with units of other (lower indexed) types, we define cost dependencies as

$$\left(C_i^{j,k}\right)_z = \tilde{c}_i^{j,k} \cdot y_i^{jD,kD}$$

where
$y_i^{jD,kD}$ is a binary variable, which indicates the status of the particular resource, named *driver* $\langle j_D, k_D \rangle$, as defined in the previous section
$\tilde{c}_i^{j,k}$ is the interactive cost of kth unit of type j on its *driver* $\langle j_D, k_D \rangle$, with respect to activity i
$\left(C_i^{j,k}\right)_z$ is the zth evaluated interactive cost dependency of kth unit of type j on its *driver* $\langle j_D, k_D \rangle$, $z = 1, ..., \text{size}(D^{j,k})$. The values of each $\left(C_i^{j,k}\right)_z$ equal $\tilde{c}_i^{j,k}$ when $y_i^{jD,kD}$ equals one, and zero otherwise. The actual number of these interactive cost dependencies will again depend on a manager's knowledge and information about available resources

Given a set of cost dependencies, we compute the overall $c_i^{j,k}$ as a sum of all evaluated $\left(C_i^{j,k}\right)_z$'s as follows:

$$c_i^{j,k} = \sum_{z=1}^{\left|D^{j,k}\right|}\left(C_i^{j,k}\right)_z$$

In many instances, due to political, environmental, safety, or community standards, esthetics, or other similar nonmonetary reasons, pure monetary factors may not necessarily prevail in decision making. It is those other nonmonetary factors that we wish to capture by introducing preferences in resource mapping to newly scheduled activities. The actual representation of preferences is almost identical to those of the costs

$$\left(P_i^{j,k}\right)_z = \tilde{p}_i^{j,k} \cdot y_i^{j_D,k_D}$$

where

$\tilde{p}_i^{j,k}$ is an interactive preference of kth unit of type j on its *driver* $\langle j_D, k_D \rangle$, with respect to activity i

$\left(P_i^{j,k}\right)_z$ is zth evaluated interactive preference dependency of kth unit of type j, with respect to activity i

Finally, again identically to modeling costs, $p_i^{j,k}$ is computed as

$$p_i^{j,k} = \sum_{z=1}^{\left|D^{j,k}\right|} \left(P_i^{j,k}\right)_z$$

Having certain number of resource units of each type available for a project does not necessarily imply that all of the units are available all the time for the project or any of its activities in particular. Due to transportation, contracts, learning, weather conditions, logistics, or other factors, some units may only have *time preferences* for when they are available to start working on a project activity or the project as a whole. Others may have *strict time intervals* during which they are allowed to start working on a particular activity or the project as a whole. The latter, strictly constrained availability, may be easily accommodated by the previously considered *window* function, $w(t_{\mathrm{LO}}, t_{\mathrm{HI}}, t_c)$.

In many cases, especially for humans, resources may have a desired or "ideal" time when to start their work or be available in general. This flexible availability can simply be represented by using the fuzzification calculation as follows:

$$\alpha_i^{j,k}(t_c) = \frac{1}{1 + a\left(t_c - \tau_i^{j,k}\right)^b}$$

where

$\tau_i^{j,k}$ is the desired time for kth unit of resource type j to start its task on activity i. This desirability may either represent the voice of project personnel (as in the case of preferences), or manager's perception on resource's readiness and availability to take on a given task

$\alpha_i^{j,k}(t_c)$ is the fuzzy membership function indicating a degree of desirability of $\langle j, k \rangle$th unit to start working on activity i, at the decision instance t_c

a is the parameter that adjusts for the width of the membership function

b is the parameter that defines the extent of *start time* flexibility

Resource Mapper

At each scheduling time instance, t_c, available resource units are mapped to newly scheduled activities. This is accomplished by solving J number of zero-one linear integer problems (i.e., one for each resource type), where the coefficients of the decision vector correspond to evaluated mapping function for each unit of the currently mapped resource type

$$\max \sum_{h \in \Omega(t_c)} \sum_{k=1}^{R_j} U_h^{j,k} \cdot y_h^{j,k} \quad \text{for } j = 1, \dots, J$$

where
  $y_h^{j,k}$ is the binary variable of the decision vector
  $\Omega(t_c)$ is the set of newly scheduled activities at decision instance t_c

A $y_i^{j,k}$ resulting in a value of one would mean that kth unit of resource type j is mapped to ith ($i \in \Omega(t_c)$) newly scheduled activity at t_c. The aforementioned objective in each of J number of problems is subjected to four types of constraints, as illustrated hereafter.

1. The first type of constraint ensures that each newly scheduled activity receives its required number of units of each project resource type

$$\sum_{k=1}^{R_j} y_i^{j,k} = \rho_i^j \quad \text{for } i \in \Omega(t_c) \quad \text{for } j = 1, \dots, J$$

2. The second type of constraint prevents mapping of any resource units to more than one activity at the same time at t_c

$$\sum_{i \in \Omega(t_c)} y_i^{j,k} \le 1 \quad \text{for } k = 1, \dots, R_j \quad \text{for } j = 1, \dots, J$$

3. The third type of constraint prevents mapping of those resource units that are currently in use by activities in progress at time t_c

$$\sum_{k=1}^{R_j} u_{t_c}^{j,k} \cdot y_i^{j,k} = 0 \quad \text{for } i \in \Omega(t_c) \quad \text{for } j = 1, \dots, J$$

4. The fourth type of constraint ensures that the variables in the decision vector $y_i^{j,k}$ take on binary values

$$y_i^{j,k} = 0 \text{ or } 1 \quad \text{for } k = 1, \dots, R_j, i \in \Omega(t_c) \quad \text{for } j = 1, \dots, J$$

Therefore, in the first of the total of J runs at each decision instance t_c, available units of resource type "one" compete (based on their characteristics and prespecified mapping function) for their assignments to newly scheduled activities. In the second run, resources of type "two" compete for their assignments. Some of their characteristics, however, may vary depending on the "winners" from the first run. Thus, the information from the first run is used to refine the mapping of type or group "two" resources. Furthermore, the information from either or both of the first two runs is then used in tuning the coefficients of the objective function for the third run when resources of type "three" are mapped.

  Due to the nature of linear programming, zeros in the coefficients of the objective do not imply that corresponding variables in the solution will also take the value of zero. In our case, that would mean that although we flagged off a resource unit as unavailable, the solution may still map it to an activity. Thus, we need to strictly enforce the interval (un) availability by adding information into constraints. Thus, we perturbed the third mapping constraint that was previously set to prohibit mapping of resource units at time t_c which are in use by activities in progress at that time. The constraint was originally defined as

$$\sum_{k=1}^{R_j} u_{t_c}{}^{j,k} \cdot y_i{}^{j,k} = 0 \quad \text{for } i \in \Omega(t_c) \quad \text{for } j = 1, \ldots, J$$

To now further prevent mapping of resource units whose $\alpha_i{}^{j,k}(t_c)$ equals zero at t_c, we modify the previous constraint as follows:

$$\sum_{k=1}^{R_j} \left(u_{t_c}{}^{j,k} + \left(1 - \alpha_i{}^{j,k}(t_c)\right)\right) \cdot y_i{}^{j,k} = 0 \quad \text{for } i \in \Omega(t_c) \quad \text{for } j = 1, \ldots, J$$

This modified constraint not only filters out those resource units that are engaged in activities in progress at t_c but also those units that were flagged as unavailable at t_c due to any other reasons.

Activity Scheduler

Traditionally, a project manager estimates duration of each project activity first, and then assigns resources to it. In this study, although we do not exclude a possibility that an activity duration is independent of resources assigned to it, we assume that it is those resource units and their skills or competencies assigned to a particular activity that determine how long it will take for the activity to be completed. Normally, more capable and qualified resource units are likely to complete their tasks faster and vice versa. Thus, activity duration in this research is considered a *resource-driven activity attribute*.

At each decision instance t_c (in resource-constrained nonpreemptive scheduling as investigated in this study), activities whose predecessors have been completed enter the set of qualifying activities, $Q(t_c)$. In cases of resource conflicts, we often have to prioritize activities in order to decide which ones to schedule. In this methodology, we prioritize activities based on two (possibly conflicting) objectives:

1. Basic *activity attributes*, such as the *current amount of depleted slack*, number of successors, and initially estimated optimistic activity duration, d_i.
2. Degree of manager's desire to *centralize* (or balance) *the loading* of one or more preselected project resource types.

Amount of depleted slack, $S_i(t_c)$, is defined in this research as a measure of how much total slack of an activity from unconstrained CPM computations has been depleted each time the activity is delayed in resource-constrained scheduling due to lack of available resource units. The larger the $S_i(t_c)$ of an activity, the more it has been delayed from its unconstrained schedule, and the greater probability that it will delay the entire project.

Before resource-constrained scheduling of activities (as well as resource mapping that is performed concurrently) starts, we perform a single run of CPM computations to determine initial unconstrained *latest finish time, LFT$_i$* of each activity. Then, as the resource-constrained activity scheduling starts, at each decision instance t_c, we calculate $S_i(t_c)$ for each candidate activity (from the set $Q(t_c)$) as follows:

$$S_i(t_c) = \frac{t_c + d_i}{LFT_i} = \frac{t_c + d_i}{LST_i + d_i} \quad i \in Q(t_c)$$

$S_i(t_c)$, as a function of time, is always a positive real number. The value of its magnitude is interpreted as follows:

When $S_i(t_c) < 1$, the activity i still has some slack remaining and it may be safely delayed.

When $S_i(t_c) = 1$, the activity i has depleted all of its resource-unconstrained slack and any further delay to it will delay its completion as initially computed by conventional unconstrained CPM.

When $S_i(t_c) > 1$, the activity i has exceeded its slack and its completion will be delayed beyond its unconstrained CPM duration.

Once calculated at each t_c, the current *amount of depleted*, $S_i(t_c)$, is then used in combination with the other two activity attributes for assessing activity priority for scheduling. (These additional attributes are *the number of activity successors*, as well as its *initially estimated duration d_i*.) The number of successors is an important determinant in prioritizing because, if an activity with many successors is delayed, chances are that any of its successors will also be delayed, thus eventually prolonging the entire project itself. Therefore, the prioritizing weight, $w_p{}^t$, pertaining to basic activity attributes is computed as follows:

$$w_i{}^p = S_i(t_c) \left(\frac{\varsigma_i}{\max(\varsigma_i)} \right) \left(\frac{d_i}{\max(d_i)} \right)$$

where
$w_i{}^p$ is the activity prioritizing weight that pertains to basic activity attributes
ς_i is the number of successors activities of current candidate activity i
$\max(\varsigma_i)$ is the maximum number of activity successors in project network
$\max(d_i)$ is the maximum of the most optimistic activity durations in a project network

The second objective that may influence activity prioritizing is a manager's desire for a somewhat centralized (i.e., balanced) resource loading graph for one or more resource groups or types. This is generally desirable in cases when a manager does not wish to commit all of the available project funds or resources at the very beginning of the project, or to avoid frequent hiring and firing or project resources needlessly.

In this methodology, an attempt is made to balance (centralize) loading of prespecified resources by scheduling those activities whose resource requirements will minimize the increase in loading graph's stairstep size of the early project stages, and then minimize the decrease in the step size in the later stages. A completely balanced resource loading graph contains no depression regions as defined by Konstantinidis (1998), i.e., it is a nondecreasing graph up to a certain point at which it becomes nonincreasing.

The activity prioritizing weight that pertains to attempting to centralize resource loading is computed in this research as follows:

$$w_i{}^r = \sum_{j=1}^{J} \frac{\rho_i{}^j}{R_j}$$

where

$w_i{}^r$ is the prioritizing weight that incorporates activity resource requirements
$\rho_i{}^j$ is the number of resource type j units required by activity i
R_j is the total number of resource type j units required for the project

Notice that $w_i{}^p$ and $w_i{}^r$ are weights of possibly conflicting objectives in prioritization of candidate activities for scheduling.

To further limit the range of $w_i{}^r$ between zero and one, we scale it as follows:

$$w_i{}^r = \frac{w_i{}^r}{\max\left(w_i{}^r\right)}$$

With the two weights $w_i{}^p$ and $w_i{}^r$ defined and computed, we further use them as the coefficients of activity scheduling objective function

$$\max\left(\sum_{i \in Q(t_c)} w_i{}^p \cdot x_i \right) + W\left(\sum_{i \in Q(t_c)} \left(1 - w_i{}^r \cdot x_i\right) \right)$$

where

x_i is the binary variable whose value becomes one if a candidate activity $i \in Q(t_c)$ is scheduled at t_c, and zero if the activity i is not scheduled at t_c
W is the decision maker's supplied weight that conveys the importance of resource centralization (balancing) in project schedule

Notice that W is a parameter that allows a manager to further control the influence of $w_i{}^p$. Large values of W will place greater emphasis on the importance of resource balancing. However, to again localize the effect of W to the early stages of a project, we dynamically decrease its value at each subsequent decision instance, t_c, according to the following formula:

$$W_{new} = W_{old}\left(\frac{\displaystyle\sum_{i=1}^{I} d_i - \sum_{i \in H(t_c)} d_i}{\displaystyle\sum_{i=1}^{I} d_i} \right)$$

where

$\displaystyle\sum_{i=1}^{I} d_i$ is the sum of all the most optimistic activity durations (as determined by conventional resource unconstrained CPM computations) for all activities in project network
$H(t_c)$ is the set of activities that have been so far scheduled by the time t_c

Figure 10.1 shows a Gantt chart and resource loading graphs of sample project with seven activities and two resource types. Clearly, neither of the two resource types is balanced. The same project has been rerun using the previous reasoning, and shown in Figure 10.2.

Notice that the loading of resource type 2 is now fully balanced. The loading of resource type 1 still contains depression regions, but to a considerably lesser extent than in Figure 10.1.

Previously, it was proposed that one way of balancing resource loading was to keep minimizing the increase in the stairstep size of the loading graph in the early project stages, and then minimize the decrease in the step size in the later stages. The problem with this reasoning is that a continuous increase in the loading graph in early stages may eventually lead to scheduling infeasibility due to limiting constraints in resource availability. Therefore, an intelligent mechanism is needed that will detect the point when resource constraints become binding and force the scheduling to proceed in a way that will start the decrease in resource loading, as shown earlier in Figure 10.1. In other

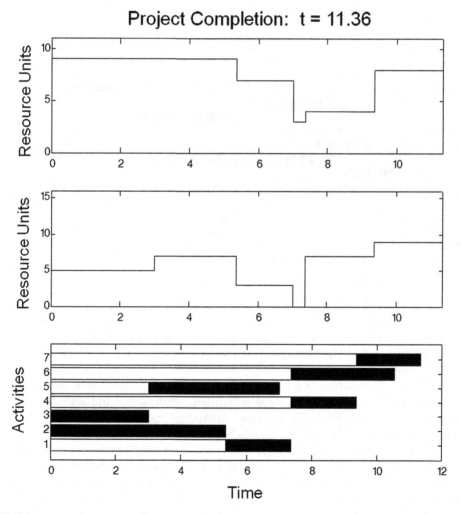

FIGURE 10.1
Gantt chart and resource loading graphs.

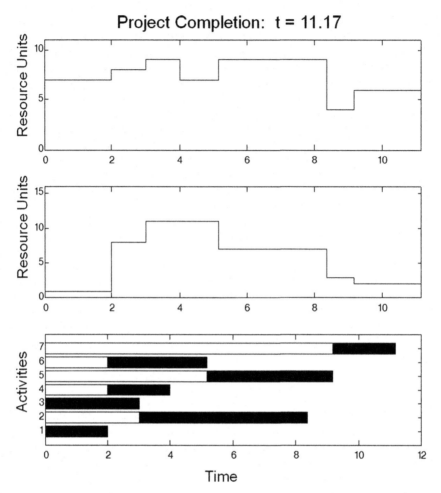

FIGURE 10.2
Rerun of Gantt chart and resource loading graphs.

words, we need to formulate a linear programming model whose constraints will drive the increase in resource stairstep shaped loading function up to a point when limits in resource availability are reached. At that point, the model must adjust the objective function and modify (relax) the constraints, to start minimizing the stairstep decrease of resource loading.

To ensure this, the constraints are formulated such that at each decision instance t_c, maximal number of candidate activities are scheduled, while satisfying activity precedence relations, preventing the excess of resource limitations, and most importantly, flag off the moment when resource limitations are reached. To facilitate a computer implementation and prevent the strategy from crashing, we introduce an auxiliary zero-one variable, $\hat{x}$, in this study, referred to as the *peak flag*. The value of $\hat{x}$ in the decision vector is zero as long as current constraints are capable of producing a feasible solution. Once that is impossible, all variables in the decision vector must be forced to zero, except $\hat{x}$, which will then take a value of one and indicate that the peak of resource loading is reached. At that moment, the constraints that force the increase in resource loading are relaxed (eliminated).

The *peak flag* is appended to the previous objective function as follows:

$$\max\left(\sum_{i\in Q(t_c)} w_i^{\,p}\cdot x_i\right)+W\left(\sum_{i\in Q(t_c)}\left(1-w_i^{\,r}\cdot x_i\right)\right)-b\hat{x}$$

where b is the arbitrary large positive number (in computer implementation of this study, b was taken as $b=\sum_{i=1}^{I} d_i$).

There are two types of constraints associated with the preceding objective of scheduling project activities. The first type simply serves to prevent scheduling of activities, which would overuse available resource units

$$\sum_{i\in Q(t_c)}\rho_i^{\,j}\cdot x_i+\left(R_j-\sum_{i\in G(t_c)}\rho_i^{\,j}\right)\hat{x}\le\left(R_j-\sum_{i\in G(t_c)}\rho_i^{\,j}\right),\quad j=1,\dots,J$$

where
x_i is the candidate activity qualified to be scheduled at t_c
$G(t_c)$ is the set of activities that are in progress at time t_c

$\left(R_j-\sum_{i\in G(t_c)}\rho_i^{\,j}\right)$ is the difference between the total available units of resource type j (denoted as R_j) and the number of units of the same resource type being currently consumed by the activities in progress during the scheduling instant t_c

The second type of constraints serves to force the gradual increase in the stairstep resource loading graphs. In other words, at each scheduling instant t_c, this group of constraints will attempt to force the model to schedule those candidate activities whose total resource requirements are greater than or equal to the total requirements of the activities that have just finished at t_c. The constraints are formulated as follows:

$$\sum_{i\in Q(t_c)}\rho_i^{\,j}x_i+\left(\sum_{i\in F(t_c)}\rho_i^{\,j}\right)\hat{x}\ge\left(\sum_{i\in F(t_c)}\rho_i^{\,j}\right),\quad j\in D$$

where
$F(t_c)$ is the set of activities that have been just completed at t_c
D is the set of manager's preselected resource types whose loading graphs are to be centralized (i.e., balanced)

$\left(\sum_{i\in F(t_c)}\rho_i^{\,j}\right)$ is the total resource type j requirements by all activities that have been completed at the decision instance t_c

Finally, to ensure an integer zero-one solution, we impose the last type of constraints as follows:

$$x_i=0\text{ or }1\quad\text{for }i\in Q(t_c)$$

As previously discussed, once $\hat{x}$ becomes unity, we adjust the objective function and modify the constraints that will, from that point on, allow a decrease in resource loading graph(s). Objective function for activity scheduling is modified such that the product $w_i^r \cdot x_i$ is not being subtracted from one any more, while the second type of constraints is eliminated completely

$$\min\left(-\sum_{i \in Q(t_c)} w_i^t \cdot x_i\right) - W\left(\sum_{i \in Q(t_c)} w_i^r \cdot x_i\right)$$

subject to

$$\sum_{i \in Q(t_c)} \rho_i^j \cdot x_i \leq \left(R_j - \sum_{i \in G(t_c)} \rho_i^j\right), \quad j = 1, \ldots, J$$

$$x_i = 0 \text{ or } 1$$

Since the second type of constraints is eliminated, resource loading function is now allowed to decrease. The first type of constraints still remains in place to prevent any overuse of available resources.

Model Implementation and Graphical Illustrations

The model as described in the preceding text has been implemented in a software prototype *project resource mapper* (PROMAP), with its code, input format, and sample outputs illustrated in the appendices. The output consists of five types of charts. The more traditional ones include project *Gantt chart* (Figure 10.3) and *resource loading graphs* (Figure 10.4) for all resource groups or types involved in a project. More specific graphs include *resource-activity mapping grids* (Figure 10.5), *resource utilization* (Figure 10.6), and *resource cost* (Figure 10.7) bar charts.

Based on the imported resource characteristics, their interdependencies, and the form of the objective, the *resource-activity mapping grid* provides a decision support in terms of which units of each specified resource group should be assigned to which particular project activity. Therefore, the *resource-activity grids* are, in effect, the main contributions of this study. *Unit utilization charts* track the resource assignments and provide a relative resource usage of each unit relative to the total project duration. The bottom (darker shaded) bars indicate the total time it takes for each unit to complete all of its own project tasks. The upper (lighter shaded) bars indicate the total additional time a unit may be locked in or engaged in an activity by waiting for other units to finish their tasks. In other words, the upper bars indicate the total possible resource idle time during which it cannot be reassigned to other activities because it is blocked waiting for other units to finish their own portions of work. This information is useful in nonpreemptive scheduling, as assumed in this study, as well as in contract employment of resources. *Resource cost* charts compare total project resource expenditures for each resource unit.

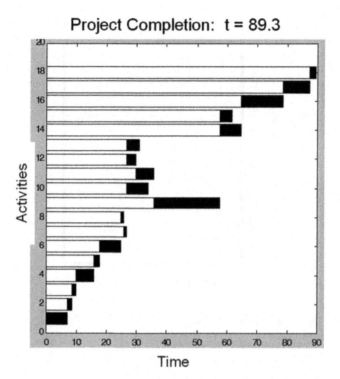

FIGURE 10.3
Project completion at time 89.3.

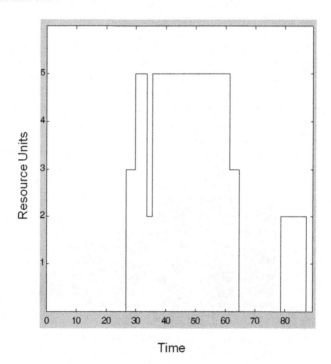

FIGURE 10.4
Resource type 3 loading graph.

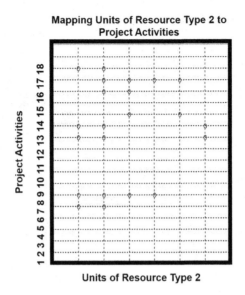

FIGURE 10.5
Mapping of resource type 2 units to project activities.

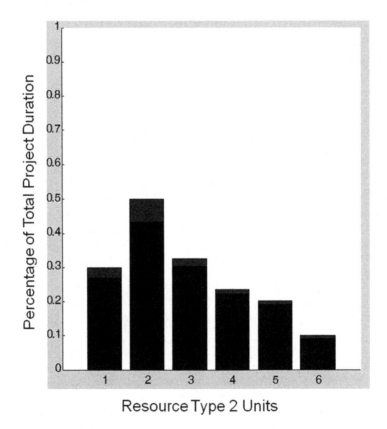

FIGURE 10.6
Time percentage of resource type 2 units engagement versus total project duration.

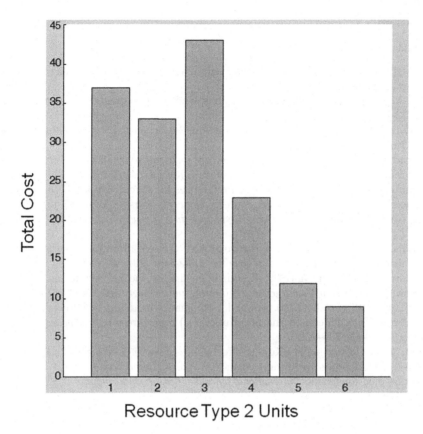

FIGURE 10.7
Project cost for type 2 resource units.

Sequence-Based Scheduling and Function-Based Scheduling

The model presented here represents an initial step toward a more comprehensive resource-activity integration in project scheduling and management. The methodology makes project scheduling more adaptive to what is currently going on in the project. Sequence-based scheduling is responsive to what has already been scheduled and the implications on subsequent scheduling of activities. Function-based scheduling is responsive to the level and type of resource functions already allocated in a project and the implications for subsequent resource allocation decisions.

The multiresource methodology presented in this chapter provides for both effective activity scheduling, based on dynamically updated activity attributes, as well as intelligent iterative mapping of resources to each activity, based on resource characteristics and preselected shape of project manager's objectives. The model consists of two complementary procedures: an *activity scheduler* and *resource mapper*. The procedures are alternatively being executed throughout the scheduling process at each newly detected decision instance, such that the final output is capable of providing decision support and recommendations with respect to both scheduling project activities and resource assignments. This approach allows human, social, as well as technical resources to interact and be utilized in value-creating ways, while facilitating effective resource tracking and job distribution control. Other rigorous scheduling models can be found in Jozefowska and Weglarz (2006).

References

Badiru, A.B. (1993). Activity resource assignments using critical resource diagramming, *Project Management Journal*, 14(3), 15–21.

Campbell, G.M. (1999). Cross-utilization of workers whose capabilities differ, *Management Science*, 45(5), 722–732.

Cooprider, C. (1999). Solving a skill allocation problem, *Production and Inventory Management Journal*, Third Quarter, 34(3), 1–6.

Franz, L.S. and Miller, J.L. (1993). Scheduling medical residents to rotations: Solving the large scale multiperiod staff assignment problem, *Operations Research*, 41(2), 269–279.

Gray, J.J., McIntire, D., and Doller, H.J. (1993). Preferences for specific work schedulers: Foundation for an expert-system scheduling program, *Computers in Nursing*, 11(3), 115–121.

Jozefowska, J. and Weglarz, J. (eds.) (2006). *Perspectives in Modern Project Scheduling*, Springer, New York.

Konstantinidis, P.D. (1998). A model to optimize project resource allocation by construction of a balanced histogram, *European Journal of Operational Research*, 104, 559–571.

Mattila, K.G. and Abraham, D.M. (1998). Resource leveling of linear schedules using integer linear programming, *Journal of Construction Engineering and Management*, 124(3), 232–244.

Milatovic, M. and Badiru, A.B. (2004). Applied mathematics modeling of intelligent mapping and scheduling of interdependent and multifunctional project resources, *Applied Mathematics and Computation*, 149(3), 703–721.

Roberts, S.M. (1992). Human skills—Keys to effectiveness, *Cost Engineering*, 34(2), 17–19.

Yura, K. (1994). Production scheduling to satisfy worker's preferences for days off and overtime under due-date constraints, *International Journal of Production Economics*, 33, 265–270.

11

Case Examples and Applications

Case Example: Project Systems View of World Economy

This case example emphasizes that, using a world systems viewpoint and a project management structure, world economic interdependency can be leveraged to achieve strategic deterrent in pursuit of world peace and harmony. Military and political approaches are often the default approaches to pursuing deterrent. The premise of this case example is that other approaches, such as economic interdependence and collaborative global project execution, can be equally effective. The basic questions of interest are as follows:

1. Do we pursue mutual assured destruction or mutual assured economic advancement?
2. Can we formulate strategic deterrent of conflict through economic interdependency?

This example addresses strategic deterrence from a nonmilitary project systems perspective rather than a military approach, capitalizing on the concept of economic interdependency of nations. Everything is economically interconnected in the modern shrunk world of fast travel and instantaneous communication. It is a systems world, and all nations must demonstrate a systems view of the world when approaching deterrent strategies. This requires a multilateral view where each nation sees the same color of national development. Teamwork and mutual understanding take on a different flavor when different cultures are involved. A structural understanding of the differing cultures can be formulated to become a unifying call for mutual assured economic advancement. The chapter uses a global systems framework for arguing the case for collaborative deterrent focused on mutual assured advancement. The chapter also advances the notion that "Together, we can accomplish deterrence and united, we can all flourish economically." It is hoped that the perspectives expressed herein can better move us toward working together to build a world that all of the world can be proud of.

Global teamwork is the basis for succeeding and flourishing in a diverse world system. In his magazine article, "Twin Fates," Badiru (2009a) argued that global partnering is essential for economic survival. Bhargava (2006) addresses many of the global issues involved in the development challenges faced by nations. Not only must we think globally, but also we must act globally through culturally sensitive hierarchy of the needs of each nation. Maslow's hierarchy of needs (Maslow, 1943) is applicable not only for individuals but also to the collective needs of nations. This chapter uses a global systems framework for arguing the case for collaborative deterrent focused on mutual assured economic advancement. The chapter also advances the notion that "Together, we can accomplish deterrence and united, we can all flourish economically." Sustaining each military enterprise (Mathaisel, 2008) through operational transformation provides a mechanism for achieving mutually desired results among cooperating nations. It is hoped that the perspectives expressed herein can better move us toward working together to build a world that all of the world can be proud of regardless of national orientation or perceived favorable alliances.

It is impossible for any nation to run its defense system without some level of dependency on some other nation's collaborative interest. In a marketing context, Libert and Spector (2008) demonstrate the power of collective thinking in achieving global goals. Moffat (2003) presents a complexity theory associated with nonlinearity of interdependency for self-organization in dynamic systems. He views economic systems as natural systems with a high level of interconnected complexity. When we talk of strategic deterrence in one nation, there is an implicit expectation that some other nation will be involved in the execution of deterrent actions. While a visible and nimble military power could be an effective deterrent, it is still essential that the subdued nation be nurtured to a vibrant economic outlook. Even where one nation is subjugated to another, would we rather preside over a destitute nation or an economically thriving nation? This assessment is particularly important when addressing evolving trends in command and control processes of advanced nations. Alberts et al. (2007) contains several intriguing perspectives of the future of command and control, which can be related to accomplishing mission objectives, such as deterrent. Garrett and Rendon (2007) present templates for adopting best practices at the local operational levels to a more global level, whereby one good turn deserves another, particularly from a risk mitigation standpoint. The pursuit of deterrent is intended to eliminate threats from one nation toward the security and welfare of other nations. Deterrence, in this context, focuses on nonproliferation of weapons of mass destruction. While the topic of strategic deterrence is very broad, it is prudent to focus on issues that have mutual symbiotic benefits for the nations involved on either side of the pursuit. This chapter advocates using economic interdependence as a platform for an alternate strategy for deterrence and global peace. Several issues impinge upon the pursuits of peace and deterrent from a global operational perspective. Many of these issues are addressed in the literature as can be found in Smith (2006), Kent (2008), Potts (2003), Kass (2006), Drew and Snow (2006), Spires (1998), Grossman and Christensen (2007), Alberts and Hayes (2005, 2007).

Given the choice of picking between mutual assured destruction and economic advancement, any rational nation would pick the latter, granted that some rationality may have to be imbued into erstwhile nations through nonmilitary options. The premise here is that economic-based strategic deterrence, from a nonmilitary perspective, offers a more sustainable positive result because it capitalizes on the concept of economic interdependency of nations. Everything is economically interconnected in the modern shrunk world typified by fast travel and instantaneous communication. It is a systems world, and all nations must demonstrate a systems view of the world when approaching deterrent strategies. This requires a multilateral view where each nation pursues the same path of national development, and economic successes are shared in win–win alliances. The concept of sharing success at the command level (Goldfein, 2001) can be expanded to the larger world stage. Teamwork and mutual understanding take on a different flavor when different cultures are involved. A structural understanding of the differing cultures can be formulated to become a unifying call for mutual assured economic advancement.

Global Systems View of the World

International involvement requires a global perspective. This is aptly articulated by Brungess (1994) in his caution about suppression of the enemy and joint war fighting in an uncertain world. Systems management is the pursuit of organizational goals within the constraints of time, cost, and performance expectations. The approach of systems thinking (Boardman and Sauser, 2008) allows a comprehensive assessment of all the parts in relation to the whole. Systems expectations can be in terms of physical products, service capability, and/or generation of desired results. Global interactions of business and industry imply that global situational awareness be instituted for collaborative deterrent measures. Global situational awareness is the process of recognizing and appreciating unique factors that define the operating characteristics in given geographical locations. These factors can range widely from one country to another. What are believed to be normal operating conditions in home–country operations may be taboo, illegal, restricted, forbidden, or condescending in another country. Far too often, nations do not pay enough attention to this fact despite the public rhetoric. Training programs, briefings, seminars, and sensitivity role plays are often used to prepare personnel for global interactions. But the fact remains that global awareness failure points still exist among nations when national defense issues are being deliberated. World news headlines on any given day reveal national interconnections in oil, autos, industrial credits, construction, and so on. Such mega economic deals are not limited to the G8, G10, G12, or G20. Even the little non-G economies are in the mix of international economic affairs. The common thread that connects all of these nations is global mutual economic advancement goal. This can be leveraged to pursue deterrent objectives.

Economic Interdependence Model

Economic vitality is the foundation of national development. Economic development is one of the primary means of improving the standard of living in a nation. History indicates the profound effect that the industrial revolution had on world development. Industrial development will continue to play an active role in the economic strategies of rational nations as they strive to cater to the basic needs of their peoples. A nation that cannot institute and sustain economic development will be politically delinquent and industrially stymied, and, thus, be more prone to militarily destructive tendencies. A solid mutually assured economic foundation can positively drive the political and military processes in any nation.

To achieve and sustain mutual development, both the political and economic aspects must come into play in formulating cooperative deterrent measures. The model presented in Figure 11.1 illustrates this interdependence graphically. As economic outlook improves along its axis and political system advances along its own axis, the prospect for a mutual intergovernmental deterrent improves along the axis of its dependency. Politics determines the crux of economics, economics determines the crux of politics, and both can be leveraged for the pursuit of sustainable deterrent. Laying a solid foundation for industrial development, as opined by Badiru (1993), can facilitate a lasting coexistence of political, economic, and deterrent activities. The mutual role of super powers must be as facilitators

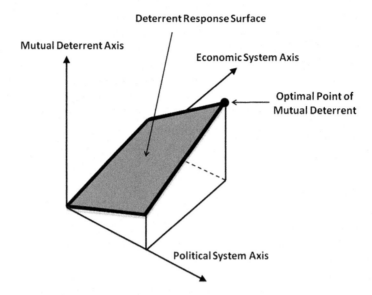

FIGURE 11.1
Mutual deterrent as a function of economic and political systems.

for mutual economic advancement for all nations. The concepts, strategies, tools, and techniques of national governance must be brought to bear in moving deterrent positively along the axes of its independence on economic and political systems.

Hybridization of Cultures

Economic interdependency is intertwined with hybridization of cultures because a flow of products often affect the flow of ideology. The increased interface of cultures through international project outsourcing is gradually leading to the emergence of hybrid cultures in many developing countries. A hybrid culture derives its influences from diverse factors, where there are differences in how the local population views education, professional loyalty, social alliances, leisure pursuits, and information management. A hybrid culture poses a big challenge to managing internationally focused deterrent programs. But the problem can be mitigated by a strategic program of situational awareness (Badiru, 2009b). Table 11.1 presents some of the pros and cons of cultural enmeshing at the global level within the context of deterrent.

TABLE 11.1

Pros and Cons of Global Hybrid Cultures

Pros	Cons
• Global awareness for competition	• Subjugation of one culture to the other
• Culturally diverse workforce	• Cultural differences in work ethics
• Access of international labor force	• Loss of cultural identity
• Intercultural peace and harmony	• Bias and suspicion
• Facilitation of social interaction	• Confusion about cultural boundaries
• Closing of trade gaps	• Nonuniform business vocabulary

Market Integration as a Factor of Deterrent

Many western industries cannot compete globally on the basis of labor cost, which is where the pursuit of competitiveness is often directed. The competitive advantage for many manufacturers will come from appropriate infusion of technology into the enterprise. Strategic research, development, and implementation of technological innovations will give manufacturers the edge needed to successfully compete globally. In spite of the many decades of lamenting about the future of manufacturing, very little has been accomplished in terms of global market integration as a function of achieving competitiveness. Part of the problem is the absence of a unified strategic systems approach. Managing global and distributed production teams requires a fundamental systems approach.

The recommendation offered here is to pursue more integrative linkages of technical issues of production and the operational platforms available in industry. Many concepts have been advanced on how to bridge the existing gaps. But what is missing appears to be a pragmatic systems-oriented road map that will create a unified goal that adequately, mutually, and concurrently addresses the business-oriented focus of practitioners in industry and the knowledge-oriented pursuits of researchers in academia. The problems embody both scientific and management issues. Many researchers have not spent adequate time in industry to fully appreciate the operational constraints of industry. Hence, there is often a disconnection between what research dictates and what business practice requires. An essential need is the development of a global systems road map. One aspect that is frequently ignored in this respect is the set of human–cultural factors operating in globally distributed work teams. A culturally sensitive systems management approach will enable an appreciation of this crucial component of global deterrent from terrorist tendencies.

Pursuit of Global Awareness for the Sake of Deterrent

To enhance global awareness (Badiru, 2009b), the preparation approach that can be more effective and sustainable is to address specific sets of questions without making any prior assumptions in each and every global engagement. Even previous international experience can become out of phase because local situations can change at any time. The questions categorized in the following can be helpful.

Questions for Overseas Economic Engagement

1. What is the national mission in the overseas location?
2. Are the personnel aware of the global situation as it relates to the country or region of operation?
3. Will the assignment be as a team, individual operation, or organizational attachment?
4. What legal status will be associated with the operation during the overseas assignment?

5. What country locations are involved in the assignment?

6. What healthcare situations and medical services are available during the assignment?

7. What personal family arrangements and precautions are relevant for the overseas assignment?

8. What prudent precautions are possible in terms of health insurance and medical records?

9. What is the currency exchange rate? Location and access to *"bureau de change?"*

Questions for Assessing Local Conditions

1. What kind of local environment will be in effect at the assignment location? Urban city? Rural? Jungle? Mountain? Desert?

2. What is the typical weather pattern? Seasons? Cold? Hot? When, where, and how long?

3. What are the general living conditions? Access to clean water? Bathroom facilities?

4. What are the domestic transportation options? Cost? Reliability? Access? Rivers? Trails? Trains? Shuttle buses? Commuter planes?

5. What is the level of environmental consciousness?

6. What is the local social hierarchy?

Questions for Economic–Cultural Nuances

1. Is there a compilation of acceptable and unacceptable behavior in the assignment location?

2. What are the political beliefs?

3. What are the typical customs?

4. What are the prevailing religious beliefs and restrictions?

5. Are there guidelines for the use of body language and gestures?

6. What are the offensive symbols and words?

Questions for Geographic Awareness

1. What are the bordering regions, areas, and/or countries to the assignment location?

2. What is the operating terrain? Flat? Hilly? Road system?

3. Is there access to rivers, oceans, beaches, etc.?

4. What is the vegetation like?

5. What are the major industries? Manufacturing factories? Service industry?

6. Are there farming establishments? Access to fresh foods?

Questions for General Assessment

1. What is the social environment?

2. Is there night life? Is it accessible? Is it safe?

3. What is the crime statistics?

4. What is the mode of law enforcement?

5. What is the restriction on currency importation and exportation?

6. What is the local manpower level? Skills? Availability? Cost? Dependability?

7. What is the educational opportunity? Access? Affordability? Level?

8. What international support organizations are available? Red Cross?

9. What is the human rights record of the locality?

If the preceding questions are addressed forthrightly, collaborative deterrent programs can run more effectively, efficiently, and successfully. Typical sources of barriers to a comprehensive global awareness include the following:

1. *Denial*: This presents the danger of not fully recognizing the problem of global awareness. A person may dismiss the problem or minimize the gravity of the problem.

2. *Past experience*: This has the danger of luring the personnel into an erroneous decision based on a past or similar experience.

3. *Complacency*: This has the danger of feeling of overconfidence about knowing what obtains in the assignment location, thereby dismissing the necessity for proactive preparation.

4. *Insufficient information*: This has the danger of misinterpretation due to limited or unreliable information about the assignment location.

Labor as a Vehicle of Mutual Development

Some cultures, by their inherent nature, offer lower labor costs for international operations. Consequently, the search for a competitive operating site might take an organization to a culture that is totally different from the base station. The pervasiveness of fast routing of information makes it easy for a business to find a seemingly receptive overseas site to relocate operations. The consequence of such relocation is a transfer of culture in either direction, often with limited situational awareness. Many organizations do not fully appreciate the differences in the operating cultures. They pay attention only to the physical and economic aspects of their operations. But more often than not, the cultural shock and unsuccessful assimilation (from either end of the culture transfer) can lead to failure of economic-based deterrent agreements.

Many of the economically underserved countries in Asia, Africa, Oceania, and Latin America are frequent targets for international market development. Those countries have common characteristics, such as highly dependent economies (devoted to producing primary products for the developed world), traditional, rural social structures, high population growth, and widespread poverty. The following characteristics may exist to constitute barriers to successful global deterrent. Some of these are summarized as follows:

1. Limited access to information (substandard telecommunications infrastructure)
2. Politically induced trade barriers
3. Cultural norms that impede free flow of information
4. Existence and abundance supply of cheap, albeit untrained, workforce
5. Orientation of manpower toward artisan and apprenticeship labor

Awareness of Overseas Workforce Constraints

Most industrial outsourcing points are located in developing and underdeveloped nations. These locations often have repressive cultures that are replete with norms that the Western world would find unacceptable. A cultural bridge usually is missing between the developed nations and the developing nations with respect to workforce capabilities that can help support deterrent. Thus, there are increasing cultural and economic disparities between global business partners. Some of the local issues to be factored into global situational awareness programs include

1. Poverty
2. Pollution
3. Disease
4. Inferior health services
5. Political oppression
6. Gender biases
7. Economic and financial scams
8. Wealth inequities
9. Social permissiveness among the elite

Western personnel posted to these regions are shocked by the level of cultural differences that they experience. In some cases, they maintain a "laissez-faire" and hands-off attitude. But there have also been cases where some of the personnel take advantage of the loose culturally acceptable social contacts that could prove to be detrimental to international cooperation.

Hierarchy of National Needs

The psychology theory of "hierarchy of needs" postulated by Maslow (1943) still governs how different cultures respond along the dimensions of global economic expectations. A culturally induced disparity in the hierarchy of needs implies that we may not be able to fulfill market responsibilities along the global spectrum of international engagement. In a culturally different workforce, the specific levels and structure of the needs may be drastically different from the typical mode observed in the western culture. Maslow's hierarchy of needs consists of five stages:

1. *Physiological needs*: These are the needs for the basic necessities of life, such as food, water, housing, and clothing (i.e., survival needs). This is the level where access to money is most critical.

2. *Safety needs*: These are the needs for security, stability, and freedom from physical harm (i.e., desire for a safe environment).

3. *Social needs*: These are the needs for social approval, friends, love, affection, and association (i.e., desire to belong). For example, social belonging may bring about better economic outlook that may enable each individual to be in a better position to meet his or her social needs.

4. *Esteem needs*: These are the needs for accomplishment, respect, recognition, attention, and appreciation (i.e., desire to be known).

5. *Self-actualization needs*: These are the needs for self-fulfillment and self-improvement (i.e., desire to arrive). This represents the stage of opportunity to grow professionally and be in a position to selflessly help others.

What have these needs got to do with strategic deterrent? Well, ultimately, the need for and commitment to deterrent boil down to each person's perception based on his or her location on the hierarchy of needs. In Figure 11.2, this case example expands the hierarchy

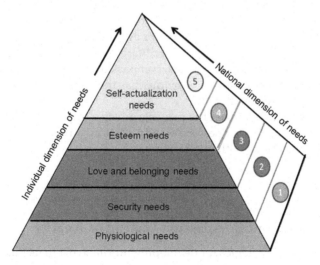

FIGURE 11.2
Hierarchy of needs pyramid expressing stages of national needs.

of needs pyramid to generate a 3D rendition that incorporates national hierarchy of needs. The location of each nation along its hierarchy of needs will determine how the nation perceives and embraces deterrent overtures from other nations.

In an economically underserved culture, most workers will be at the basic level of physiological needs, and there may be cultural constraints on moving from one level to the next higher level. This fact has an implication on how cultural interfaces may fail between host and guest nations involved in a global deterrent program. In terms of national hierarchy of needs, this chapter offers the following characterization:

1. *Level 1 of national needs*: This is the national need for basic essentials of economic vitality to support the provision of food and housing for citizens. Can the nation feed itself?

2. *Level 2 of national needs*: This is need for national defense. Can the nation feel safe from external attack? Can the nation protect itself?

3. *Level 3 of national needs*: This is the need for a nation to belong to some national alliances. Can the nation be invited to join economic groups, e.g., G8, G12, or Gxx? Does the nation have affinity on some world stage?

4. *Level 4 of national needs*: This is the level of having national respect. Is the nation esteemed in some aspect of national pride? What positive thing is the country known for?

5. *Level 5 of national needs*: This is the level of being classified as a "power" in some respect. Does the nation have a recognized niche on the World stage?

Prevention Is Better than Correction

At the community level of social ills, preemption is far better than incarceration. On the world stage of deterrent, prevention is far better than correction. Proactive programs that help to forestall conflict are often cheap, subtle, and innocuous, such as offering social and economic support to the less fortunate, providing a basis for optimism in developing nations, creating an atmosphere of belonging for everyone, offering encouragement, showing empathy where and when needed, and facilitating educational advancement opportunities. For underdeveloped nations, support, compassionate discipline, and comfort are as much a responsibility of the national government as they are of all nations that are linked in the intertwined global economy.

No nation is too far removed from the possible adverse impacts of political delinquency of militant nations. Education is one sure way to advance a nation and minimize villainous tendencies in interactions with other nations.

Empathy, Sympathy, and Compassion Instead of Apathy

There are three potential strategies for developing mutual trust to facilitate mutual pursuit of deterrent. Instead of apathy, the more advanced nations must exhibit empathy, sympathy, and compassion toward the less developed nations through the following strategies:

Educate

An important aspect of mutual deterrent is education. An intellectually astute partner is more likely to be a rational discussant in deterrent issues. Mutual partners must make a commitment for the following:

1. Inform positively.
2. Create educational opportunities.
3. Facilitate intellectual dialog.

Engage

Mutual partners must constructively engage the adversary, not necessarily through military options. Possible approaches include the following:

1. Address deterrent violations promptly.
2. Engage offender directly with logic predicated on education.
3. Explain mutual consequences.

Empower

An empowered partner is a more viable contributor to the mutual welfare of all partners. Some issues that can be addressed include the following:

1. Encourage self-determination to contribute to deterrent expectations.
2. Provide avenues for self-sufficiency.
3. Facilitate workforce development.

Sustainability of Global Alliances

Sustainability—it is not just for the environment. Deterrent sustainability on the global front is as much a need as the traditional components of systems management that spans planning, organizing, scheduling, and control. For global deterrent and alliances, preemption of cultural conflicts is far better than postoccurrence remedies. Proactive pursuit of deterrent best practices can pave the way for the success of deterrent measures. This chapter emphasizes that deterrent can be more sustainable if the participating nations recognize the mutual economic dependencies that exist among them.

Summary of the Case Example

Cultural infeasibility is one of the major impediments to economic outsourcing in the emerging global economy. The business climate of today is very volatile. This volatility, coupled with cultural limitations, creates problematic operational elements for global economic alliances involving a developing country. Global situational awareness and its best practices are essential for any nation engaged in negotiating deterrent programs. In these days of globalization, nothing happens without international cooperation, and cooperation cannot succeed without effective situational awareness on each side of the cooperating partners. For an overall safer world, nations must focus on mutual commerce rather than conflict. We all share the world and should work together to protect and preserve it for future generations. The more powerful nations are encouraged to "speak softly, but carry a big stick," that may never need to be used militarily on account of mutual economic interdependency.

Case Application: Coordination of Multinational Projects

The following case study illustrates the importance of a robust organizational structure for multinational projects. The case study involves a large-scale international project that consists of the engineering, procurement, and construction of seven liquid gas tanks together with their ancillary system and control building in Persian Gulf in the early 1980s. The use and capacity of the tanks are as follows:

1. Three liquefied natural gas (LNG) tanks of $80,000\,m^3$ each
2. Two liquefied petroleum gas (LPG) propane tanks of $50,000\,m^3$ each
3. Two LPG butane tanks of $50,000\,m^3$ each

The ancillary systems that are part of the project include

1. A propane and butane vapor recovery system
2. A low-pressure flare system
3. An offsite control room

Project Setting

The construction is to be done on DAX Island in the Persian Gulf (Middle East), approximately 300 miles from the city of Abu Dhabi (the capital of the United Arab Emirates). DAX is a small island of 2 square miles and has a population of 5,000, all of whom work in the only two plants on the island. Due to the large project size, additional 3,000 people will be working and living on the island during the construction phase, which is a considerable increase in the population of the island. The island has salty soil and cannot sustain any vegetation. Also, there isn't any source of fresh water except for desalinated sea water. Desalination is a very slow process and can cause considerable delays when it comes to pouring concrete.

Technical Requirements

Each of the seven tanks will consist of two separate and structurally independent liquid containers: a primary inner metallic container and a secondary outer concrete container. Each of the containers will be constructed of a material suitable for the low-temperature liquid.

Each container should be capable of holding the required volume of the stored liquid for an indefinite period without any deterioration of the container or its surroundings.

The secondary concrete container should be capable of withstanding the effect of fire exposure from the adjacent tanks and an external impact of 21 tons traveling at high speed without loss of structural integrity. Adequate insulation should be provided between the primary and secondary tanks to limit the heat inleak. All the tanks should be provided with the necessary pumping and piping systems for the receipt, storage, and loading of the LNG and LPG from the nearby plant to the different tankers. A blast-proof offsite control room should be provided. All storage and loading operations should be controlled from this control room. The project total cost is estimated at U.S. $600 million. The detailed project network consists of around 48,000 activities.

Organizational Relationships

GASC company is the owner of the project. Their main offices are located in Abu Dhabi city. GASC is an operating company that has never managed any engineering or construction project. GASC turned to its mother company, NOCC, for help and signed an agreement with them for the management of this project.

NOCC is the national oil company of the Emirate of Abu Dhabi with main offices in Abu Dhabi city. It is completely owned by the government of Abu Dhabi and plays an extremely important role in the national economy. It is headed by a general manager responsible for the eight different directorates, one of which is the project directorate. One of the divisions of the project's directorate is the gas projects division, headed by a manager. The gas projects division is selected to manage the project.

GASC is one of the several operating companies whose majority of shares is owned by NOCC. NOCC's share in GASC is 60%, and the remaining shares are owned by TOTAC of France, SHELC of Holland, PBC of Great Britain, and MITSC of Japan. While it is considered convenient to have the major owning company manage the project for GASC, such an arrangement has drawbacks, mainly the fact that the client company (who should have the final say in its own project) is a subsidiary of the hired management company.

The gas projects division is staffed with a skeleton staff. Although capable, this staff can only oversee the management of the project and cannot perform all the actual engineering, procurement, and construction management. The Houston-based KELLC company is selected to provide the services under the direct supervision of the gas projects division. KELLC's scope of work includes the basic design of the tanks, the detailed design of all the piping and ancillary systems, the procurement of all free issue materials, and the management of construction. All the engineering and procurement activities are to be performed out of KELLC's regional office in London. KELLC, a well-known process engineering firm, has limited experience with concrete tanks. Their selection was conditional on their acceptance to hire the Belgian civil engineering firm TRACT as a consultant to help them in the critical civil engineering problems. NOCC will also hire the American-based consulting firm DMRC to do the soil investigation and testing work.

The construction work is packaged into 15 small contracts and 1 large (75% of the total construction work) contract. All the small contracts were awarded to local construction companies, while the large main contract, which included the tanks, piping, and ancillary systems, was awarded to the Chicago-based CBAIC company. CBAIC will have to open three new offices. One office will be located in London, next to KELLC's regional office, during the engineering phase. Another office will be in Abu Dhabi city for the construction management. A third office will be on DAX island for the construction operations. CBAIC is a reputable tanking contractor. However, its experience in concrete tanks is quite limited. They will have to hire the French civil engineering firm SBC as a subcontractor.

SBC's experience in low-temperature concrete is limited. They will have to hire the specialized Belgian firm CBC as a consultant.

Safety and Environmental Considerations

The project's safety requirements are very high. The safety of those living on such a small island in case of an accident is of major concern to the owning company. To ensure that the required quality and safety standards are achieved, NOCC will hire the French third party inspection company BVC as a consultant.

Since the engineering office is in London, the French and the Belgian engineers are to commute to London as necessary to provide their inputs to the project. This will continue for the whole engineering phase, which is expected to last for 2 years.

Procurement activities will be handled out of KELLC's London office. Materials and equipment are to be delivered to the construction site. Supplies will be purchased in the open market at the most competitive prices. Steel will be purchased from Japan and Belgium; pipe from Germany, France, and Japan; valves from Sweden and France; pumps from the United States; compressors from Switzerland and Japan; and vessels from Italy. A total of around 600 purchase orders will be issued.

Analysis of Project Scenario

The organizational setup has to be quite flexible and should change in accordance with the project requirements. Since it is expected to be a fast-track project, the most active period will be the second half of the engineering phase, which corresponds to the first half of the construction phase. The organizational setup for the project stretches across national boundaries, practices, and regulations. It is, thus, very important that efforts of the project staff be closely coordinated.

The existing scenario suggests that all the planning, organizing, and staffing functions are fixed. However, the existing organizational setup and overall project plan pose some possible problems. The possible problems associated with the existing setup are discussed along with suggestions on how they could be alleviated by applying sound management principles to the planning, organizing, and staffing functions. Management should especially consider and forecast possible future problems and make provisions for handling the problems.

Understanding the scope and limitations of the project is vital due to the fact that it is a multinational high-tech project and of such large size. Special management practices suitable for high-tech and multinational projects should be applied.

Sources of Possible Problems and Solutions

The core of possible problems associated with this project lies in the complex organizational structure and the fact that it is a large multinational high-tech project. Next are listed the problem sources, associated possible problems that could arise, and recommended solutions or alternatives to alleviate the problems.

1. GASC turned to its mother company NOCC for help and subcontracted with NOCC for management of the project. There could be several drawbacks associated with this situation. NOCC may take the attitude of overcontrolling, since GASC is its subsidiary and begins to generate and implement decisions without contacting GASC for approval. For example, NOCC may overlook the performance standards specified

in the blueprints and set new performance standards that are lower than those specified by GASC. If NOCC feels that their way of managing jobs is best, and since GASC
has to come to them for help, NOCC may feel that making their own changes can
only benefit the project. However, this will only create a poor final product.

    a. GASC knows the expectations of the project and has planned for that in the blue-
prints and specifications. However, GASC is playing no part in the internal decision making as the project evolves. This is going to produce conflicts that will slow
down the project and create many more scheduling problems, financial problems,
and other problems that were not planned for initially. If GASC carefully covers the planning function, they will foresee the potential problems and realize
that their participation in decisions is of utmost importance for the project to be
streamlined and for the final system to perform its intended function efficiently.
Thus, GASC could either hire a more qualified internal staff with more expertise
in management or they could take a team of their top managers to coordinate with
NOCC on the project to ensure things will go as planned.

2. Too many companies are involved in the project. This results in several levels of
responsibility. The situation is compounded by the fact that the companies are so
globally widespread.

    a. Too many levels of responsibility create unnecessary red tape and can cause much
delay in the project's phases along with employee frustration and disinterest. First,
NOCC is not equipped to handle such a large-scale project and, thus, should be
eliminated from the organizational setup. This decision should have been made
in the planning stages as mentioned previously. However, in the organizing stage,
NOCC should also be eliminated because of its poor managing abilities which are
clearly shown in the poorly planned organizational setup.

    b. The organizational structure does not give a logical design of the interfaces needed
among personnel to assure a dedicated pursuit of the common goals in association with the project nor has it been approached carefully so as to facilitate coordination. Although this large project requires much expertise in various fields, a
better approach would be to simplify the levels of responsibilities by contracting
with fewer companies that are more specialized. The benefits of this approach
outweigh the costs. If the foundation of responsibility level and organizational
hierarchy is not set up adequately from the start, then the directing and controlling functions will be limited and project success may not be possible.

3. Several communication problems could arise due to the fact that the project is high
tech and multinational. The difference in cultural backgrounds and the complex organizational structure of this project will influence how communication is approached
and how it will be perceived by different levels of personnel. The management should
form an environment where a formal and informal exchange of ideas can take place.
Due to the numerous levels of responsibility and cultural and language differences
among countries, any form of communication that must pass through these several
levels should be concisely formatted. This will ensure that the original information
is clearly communicated and not be altered as it passes through the various levels.
Clear communication is very important in preventing project problems.

    a. The difference in time zones presents several communication obstacles to over-
come. Thus, to help resolve these problems, a global time frame should be determined, during which all the project offices in the following cities or countries can
communicate on any given day (24 h) by teleconference: London, Rome, Chicago,
Abu Dhabi, Tokyo, Geneva, Bonn, Paris, Sweden, and Belgium. In approaching

the problem of determining the time window, the two locations with the greatest time difference (Japan and Chicago) and the major project offices (London, Abu Dhabi, and DAX) should be jointly considered. The optimal time window should be 1 h each day in which Chicago can communicate from 6:00 a.m. to 7:00 a.m. and Japan from 9:00 p.m. to 10:00 p.m. Thus, this time frame will give the major project offices a practical communication time during their daily business hours. This is important since most project interactions will originate from these offices.

b. Another problem in communication is the language barriers among the various countries. To ensure that information is streamlined and correctly perceived, interpreters who are fluent in various foreign languages should be employed at the various project offices. Effective communication will ensure that coordination will be increased within all units, thus enhancing cooperation and creating satisfaction among members of the project.

4. A considerable increase in island population during the construction phase will have to be planned for. An additional 3,000 people will increase DAX's population by 60%. It is known that the island is currently starved of natural resources. Thus, high priority should be given to conserve these resources. Upper management should ensure that sufficient facilities be provided for the employees without polluting or destroying the natural environment.

5. Ordering supplies from so many different countries creates problems in logistics. Due to geographical diversity of the incoming supplies, delays and backlogs are likely to occur. A sound inventory system should be implemented to minimize any logistic problems or delays so that the project deadlines are accomplished.

In dealing with multinational high-tech projects, the management culture must be versatile and congenial. Several factors must be considered, including how to control the uncertainty in high-tech operations and how to streamline information among the various personnel. Multinational projects require management practices designed specifically to meet the unique challenges. Several factors to consider and specific suggestions for this project are presented next.

Strategic Planning

Since a frequent problem with technology is the extension of useful life well beyond the period of obsolescence, it is vital that NOCC defines the long-range purpose and useful life of the project so that replacement of the system can be planned for, financially.

Cost/Benefit Analysis

The bottom line in most operations is combined profit/benefit and performance. NOCC should analyze the cost of the high technology required versus the benefits to see if this project could be profitable.

Technology Assessment

The technology itself should be assessed in detail. GASC should determine whether there are alternative technologies that can perform the required objective of the project.

Acquisition

In the planning stage, NOCC should evaluate the policy of procuring and implementing the high-tech system. NOCC should evaluate the hardware and software components of the project, communication site selection and installation, and development and implementation along with training and postimplementation evaluation of the project.

Trade Agreements

Several contracts on procurement of supplies will need to be closely regulated to ensure that all governmental policies of the various countries are strictly followed.

Different Labor Practices

Labor practices vary from country to country in terms of the number of hours of work per day, the number of days to work per week, unions, relationships between personnel, payment practices, and so on. A compromise should be developed among the various workers and contracting companies to address these various factors.

Different Governmental and Political Ideologies

Differences among countries on politics could create several conflicts, especially if top management allows political ambitions to influence their actions toward work on the project. Thus, these differences in governments must be discussed and accounted for in the initial planning stage of the project.

Information Management

Information is a critical resource in managing high-tech operations. The organizational efficiency of the project will be affected by the quality of information generated and how the information is used. A multinational project manager may be bogged down with the problem of receiving too much information. Sophisticated data processing and information handling mechanisms will be needed to govern the mass of information that may be associated with the project.

In coordinating the activities of each phase of the project, a central division should be set up to monitor all the activities that go on and keep all the members of the project informed of the current situation. The organizational structure should also be streamlined to make use of the systems approach in which project teams interact both horizontally and vertically.

Undertaking such a vast multinational high-tech project requires a management with special expertise and capabilities. The project manager should have broad authority over all elements of the project and engage in all necessary managerial and technical actions required to complete the project successfully. Also, the project manager should have appropriate authority in making technical and design decisions and should be able to control funds, schedule, and quality of product.

Due to the large size of this international project, a team of managers is recommended to successfully control and direct the project. Cooperation, communication, and coordination are vital to the interactions among managers. The management team should be experienced and well educated in management principles as applied to high-tech and multinational projects in order for harmony to exist throughout the decision-making processes. The case study presented here gives insight to a real-world situation in which documented management principles can be applied for optimal project outcome.

A conducive organization structure must be formulated. Since there are interdependencies and subsidiary relationships among many of the companies involved in the project, the conventional organization structures may not be adequate. A sophisticated form of the mixed organization structure will, thus, be required. In summary, the potential problems that may adversely affect the project organization are

1. Shortage of food supply
2. Security concern (high-tension zone)
3. Potential cost overruns
4. Potential for catastrophic accidents
5. Foreign rules and regulations
6. Trade restrictions
7. Communication delays
8. Language barriers
9. Lack of fresh water

Recommendations

The following recommendations are offered for the case study project:

Simplify the levels of responsibilities by contracting with fewer companies.

Minimize the interdependencies of companies by using nonsubsidiary companies.

Predetermine communication time zones that are compatible with the timing of project decisions.

Reevaluate the level of technology required for the project.

Arrange for special favorable trade exceptions to facilitate movement of personnel and materials.

References

Alberts, D.S. and Hayes, R.E. (2005). *Power to the Edge: Command and Control in the Information Age*, CCRP Publication Series, Washington, DC.

Alberts, D.S. and Hayes, R.E. (2007). *Planning: Complex Endeavors*, CCRP Publication Series, Washington, DC.

Alberts, D.S. et al. (2007). The future of C2, special issue, *C2 Journal*, 1(1): 1–210.

Badiru, A.B. (1993). *Managing Industrial Development Projects: A Project Management Approach*, John Wiley & Sons, New York.

Badiru, A.B. (November 2009a). Global situational awareness using project management, *Industrial Engineer*, 41(12): 22–26.

Badiru, A.B. (March 2009b). Twin fates: Partnerships will keep manufacturers' doors open, *Industrial Engineer*, 41(3): 40–41.

Bhargava, V. (ed.) (2006). *Global Issues for Global Citizens: An Introduction to Key Development Challenges*, The World Bank, Washington, DC.

Boardman, J. and Sauser, B. (2008). *Systems Thinking: Coping with 21st Century Problems*, Taylor & Francis Group, Boca Raton, FL.

Brungess, J.R. (1994). *Setting the Context: Suppression of Enemy Air Defenses and Joint War Fighting in an Uncertain World*, Air University Press, Maxwell Air Force Base, Montgomery, AL.

Drew, D.M. and Snow, D.M. (2006) *Making Twenty-First-Century Strategy: An Introduction to Modern National Security Processes and Problems*, Air University Press, Maxwell Air Force Base, Montgomery, AL.

Garrett, G.A. and Rendon, R.G. (2007). *U.S. Military Program Management: Lessons Learned and Best Practices*, Management Concepts, Vienna, VA.

Goldfein, D.L. (2001). *Sharing Success Owning Failure: Preparing to Command in the Twenty-First Century Air Force*, Air University Press, Maxwell Air Force Base, Montgomery, AL.

Grossman, D. and Christensen, L.W. (2007). *On Combat: The Psychology and Physiology of Deadly Conflict in War and Peace*, PPCT Research Publications, Belleville, IL.

Kass, R.A. (2006). *The Logic of Warfighting Experiments*, CCRP Publication Series, Washington, DC.

Kent, G.A. (2008). *Thinking about America's Defense: An Analytical Memoir*, RAND, Santa Monica, CA.

Libert, B. and Spector, J. (2008). *We Are Smarter Than Me: How to Unleash the Power of Crowds in Your Business*, Wharton School Publishing/Pearson Education, Inc., Upper Saddle River, NJ.

Maslow, A.H. (1943). A theory of human motivation, *Psychology Review*, 1(3): 93–100.

Mathaisel, D.F.X. (2008). *Sustaining the Military Enterprise: An Architecture for a Lean Transformation*, Auerbach Publications, Boca Raton, FL.

Moffat, J. (2003). *Complexity Theory and Network Centric Warfare*, CCRP Publication Series, Washington, DC.

Potts, D. (ed.) (2003). *The Big Issue: Command and Combat in the Information Age*, CCRP Publication Series, Washington, DC.

Smith, E.A. (2006). *Effects Based Operations: Applying Network Centric Warfare in Peace, Crisis, and War*, CCRP Publication Series, Washington, DC.

Spires, D.N. (1998). *Beyond Horizons: A Half Century of Air Force Space Leadership*, Air University Press, Maxwell Air Force Base, Montgomery, AL.

12

Emerging Roles of Artificial Intelligence in Project Management

Artificial Intelligence in Project Management

Project management has always taken advantage of whatever technology emerges from science, technology, engineering, and mathematics research laboratories. The field of artificial intelligence (AI) continues to grow rapidly, fueled by the emergence of more advanced computers. The more computer power is available, the more can AI do for the benefit of project management. Project management is data-dependent. AI has the ability to handle and manipulate large amounts of data. The technique of "deep learning" that has developed in AI can enhance the reasoning process associated with project management. Thus, in the coming years, project management planning, organizing, scheduling, and control cannot escape the influence of AI. Consequently, project managers and analysts should be aware of the capabilities of AI and how to leverage it for project management functions.

AI is defined as the intelligence exhibited by machines or software tools. It relates to the field of study that studies how to create computers and computer software that are capable of intelligent behavior. Intelligence is needed by project management. AI can provide additional "smarts" for better management of projects. Deep learning in AI systems is the ability of the system to use massive amounts of data and number crunching to self-generate intelligent actions, which is a desirable function in complex project management.

Background of AI

The background of AI has been characterized by controversial opinions and diverse approaches. Despite the controversies that have ranged from the basic definition of intelligence to questions about the moral and ethical aspects of pursuing AI the technology continues to generate practical results. With increasing efforts in AI research, many of the prevailing arguments are being resolved with proven technical approaches. Expert systems, the main subject of this book, is the most promising branch of AI.

AI is a controversial name for a technology that promises much potential for improving human productivity. The phrase seems to challenge human pride in being the sole creation capable of possessing real intelligence. All kinds of anecdotal jokes about AI have been offered by casual observers. A speaker once recounted his wife's response when he told her that he was venturing into the new technology of AI. "Thank God, you are finally realizing how dumb I've been saying you were all these years," was alleged to have been the wife's words of encouragement. One whimsical definition of AI is "Artificial Insemination of knowledge into a machine." Despite the derisive remarks, serious embracers of AI may yet have the last laugh. It is being shown again and again that AI may hold the key to improving operational effectiveness in many areas of applications. Some observers have suggested changing the term AI to a less controversial one such as intelligent applications. This refers more to the way that computer and software are used innovatively to solve complex decision problems.

Natural intelligence involves the capability of humans to acquire knowledge, reason with the knowledge, and use it to solve problems effectively. By contrast, AI is defined as the ability of a machine to use simulated knowledge in solving problems.

The definition of intelligence had been sought by many great philosophers and mathematicians over the ages, including Aristotle, Plato, Copernicus, and Galileo. They attempted to explain the process of thought and understanding. The real key that started the quest for the simulation of intelligence did not occur, however, until the English philosopher Thomas Hobbes put forth an interesting concept in the 1650s. Hobbes believed that thinking consists of symbolic operations and that everything in life can be represented mathematically. These beliefs led directly to the notion that a machine capable of carrying out mathematical operations on symbols could imitate human thinking. This is the basic driving force behind the AI movement. For that reason, Hobbes is sometimes referred to as the grandfather of AI.

While the term "AI" was coined by John McCarthy relatively recently (1956), the idea had been considered centuries before. As long ago as 1637, Rene Descartes was conceptually exploring the ability of a machine to have intelligence when he said:

> For we can well imagine a machine so made that it utters words and even, in a few cases, words pertaining specifically to some actions that affect it physically. However, no such machine could ever arrange its words in various different ways so as to respond to the sense of whatever is said in its presence-as even the dullest people can do.

Descartes believed that the mind and the physical world are on parallel planes that cannot be equated. They are of different substances following entirely different rules and can thus not be successfully compared. The physical world (i.e., machines) cannot imitate the mind because there is no common reference point.

The 1800s saw advancement in the conceptualization of the computer. Charles Babbage, a British mathematician, laid the foundation for the construction of the computer, a machine defined as being capable of performing mathematical computations. In 1833, Babbage introduced an analytical engine. This computational machine incorporated two unprecedented ideas that were to become crucial elements in the modem computer. First, it had operations that were fully programmable, and second, it could contain conditional branches. Without these two abilities, the power of today's computers would be inconceivable. Due to a lack of financial support, Babbage was never able to realize his dream of building an analytic engine. However, his dream was revived through the efforts of later researchers. Babbage's basic concepts can be observed in the way that most computers operate today.

Another British mathematician, George Boole, worked on issues that were to become equally important. Boole formulated the laws of thought that set up rules of logic for representing thought. The rules contained only two-valued variables. By this, any variable in a logical operation could be in one of only two states: yes or no, true or false, all or nothing, 0 or 1, on or off, and so on. This was the birth of digital logic, a key component of AI effort.

In the early 1900s, Alfred North Whitehead and Bertrand Russell extended Boote's logic to include mathematical operations. This not only led to the formulation of digital computers but also made possible one of the first ties between computers and thought process. However, there was still no acceptable way to construct such a computer.

In 1938, Claude Shannon demonstrated that Boolean logic consisting of only two-variable states (e.g., on–off switching of circuits) can be used to perform logic operations. Based on this premise, Electronic Numerical Integrator and Computer (ENIAC) was built in 1946 at the University of Pennsylvania. ENIAC was a large-scale, fully operational electronic computer that signaled the beginning of the first generation of computers. It could perform calculations 1,000 times faster than its electromechanical predecessors. It weighed 30 tons, stood two stories high, and occupied 1,500 square feet of floor space. Unlike today's computers, which operate in binary codes (0s and 1s), ENIAC operated in decimal (0, 1, 2, …, 9) and required ten vacuum tubes to represent one decimal digit. With over 18,000 vacuum tubes, ENIAC needed a great amount of electrical power, so much so that it was said that it dimmed the lights in Philadelphia whenever it operated.

Two of the leading mathematicians and computer enthusiasts between 1900 and 1950 were Alan Turing and John von Neumann. In 1945, von Neumann insisted that computers should not be built as glorified adding machines, with all their operations specified in advance. Rather, he suggested computers should be built as general-purpose logic machines capable of executing a wide variety of programs. Such machines, von Neumann proclaimed, would be highly flexible and capable of being readily shifted from one task to another. They could react intelligently to the results of their calculations, could choose among alternatives, and could even play checkers or chess. This represented something unheard of at that time: a machine with built-in intelligence, able to operate on internal instructions.

Prior to von Neumann's concept, even the most complex mechanical devices had always been controlled from the outside, for example, by setting dials and knobs. Von Neumann did not invent the computer, but what he introduced was equally significant: computing by use of computer programs, the way it is done today. His work paved the way for what would later be called AI in computers.

Alan Turing also made major contributions to the conceptualization of a machine that can be universally used for all problems based only on variable instructions fed into it. Turing's universal machine concept, along with von Neumann's concept of a storage area containing multiple instructions that can be accessed in any sequence, solidified the ideas needed to develop the programmable computer. Thus, a machine was developed that could perform logical operations and could do them in varying orders by changing the set of instructions that were executed. Due to the fact that operational machines were now being realized, questions about the intelligence of machines began to surface. Turin's other contribution to the world of AI came in the area of defining what constitutes intelligence. In 1950, he designed the Turing test for determining the intelligence of a system. The test utilized the conversational interaction between three players to try to verify computer intelligence.

The test is conducted by having a person (the interrogator) in a room that contains only a computer terminal. In an adjoining room, hidden from view, a man (person A) and a woman (person B) are located with another computer terminal. The interrogator communicates with the couple in the other room by typing questions on the keyboard. The questions appear on the couple's computer screen and they respond by typing on their own keyboard. The interrogator can direct questions to either person A or B, but without knowing which is the man and the woman.

The purpose of the test is to distinguish between the man and the woman merely by analyzing their responses. In the test, only one of the people is obligated to give truthful responses. The other person deliberately attempts to fool and confuse the interrogator by giving responses that may lead to an incorrect guess. The second stage of the test is to substitute a computer for one of the two persons in the other room. Now the human is obligated to give truthful responses to the interrogator while the computer tries to fool the interrogator into thinking that it is human. Turing's contention is that if the interrogator's success rate in the human/computer version of the game is not better than his success rate in the man/woman version, then the computer can be said to be "thinking." That is, the computer possesses "intelligence." Turing's test has served as a classical example for AI proponents for many years.

By 1952, computer hardware had advanced far enough that actual experiments in writing programs to imitate thought processes could be conducted. The team of Herbert Simon, Allen Newell, and Cliff Shaw was organized to conduct such an experiment. They set out to establish what kinds of problems a computer could solve with the right programming. Proving theorems in symbolic logic, such as those set forth by Whitehead and Russell in the early 1900s, fits the concept of what they felt an intelligent computer should be able to handle.

It quickly became apparent that there was a need for a new higher-level computer language than was currently available. First, they needed a language that was more user-friendly and could take program instructions that are easily understood by a human programmer and automatically convert them into machine language that could be understood by the computer. Second, they needed a programming language that changed the way in which computer memory was allocated. All previous languages would preassign memory at the start of a program. The team found that the type of programs they were writing would require large amounts of memory and would function unpredictably. To solve the problem, they developed a list processing language. This type of language would label each area of memory and then maintain a list of all available memory. As memory became available it would update the list and when more memory was needed, it would allocate the amount necessary. This type of programming also allowed the programmer to be able to structure his or her data so that any information that was to be used for a particular problem could be easily accessed. The end result of their effort was a program called Logic Theorist. This program had rules consisting of axioms already proved. When it was given a new logical expression, it would search through all of the possible operations in an effort to discover a proof of the new expression. Instead of using a brute force search method, they pioneered the use of heuristics in the search method.

The Logic Theorist that they developed in 1955 was capable of solving 38 of 52 theorems that Whitehead and Russell had devised. It did them very quickly. What took Logic Theorist a matter of minutes would have taken years if it had been done by simple brute force on a computer. By comparison, the steps that it went through to arrive at a proof to those that human subjects went through showed that it had achieved a remarkable imitation of the human thought process. This system is considered the first AI program.

The First AI Conference

The summer of 1956 saw the first attempt to establish the field of machine intelligence into an organized effort. The Dartmouth Summer Conference, organized by John McCarthy, Marvin Minsky, Nathaniel Rochester, and Claude Shannon, brought together people whose work and interest formally established the field of AI. The conference, held at Dartmouth College in New Hampshire, was funded by a grant from the Rockefeller Foundation. It was at that conference that John McCarthy coined the term "AI." This was the same John McCarthy who developed the list processing (LISP) programming language, which later became a standard tool for AI development. In fact, in those days some computer makers developed what they called LISP machines. This author did buy one of those machines for his AI lab at the University of Oklahoma in the early 1990s (Badiru, 1990, 1993; Badiru and Cheung, 2002). In attendance at the first AI conference, in addition to the organizers, were Herbert Simon, Allen Newell, Arthur Samuel, Trenchard More, Oliver Selfridge, and Ray Solomonoff.

The Logic Theorist developed by Newell, Shaw, and Simon was discussed at the conference. Newell, Shaw, and Simon were far ahead of others in actually implementing AI ideas. The Dartmouth meeting served mostly as an avenue for the exchange of information and, more importantly, as a turning point in the main emphasis of work in the AI endeavor. Instead of concentrating on the hardware to imitate intelligence, the meeting set the course for examining the structure of the data being processed by computers, the use of computers to process symbols, the need for new languages, and the role of computers for testing theories.

The next major step in software technology came from Newell, Shaw, and Simon in 1959. The program they introduced was called General Problem Solver (GPS), which was intended to be a program that could solve many types of problems. It was capable of solving theorems, playing chess, or doing various complex puzzles. GPS was a significant step forward in AI. It incorporates several new ideas to facilitate problem-solving. The nucleus of the system was the use of means-end analysis, which involves comparing a present state with a goal state. The difference between the two states is determined, and a search is done to find a method to reduce this difference. This process is continued until there is no difference between the current state and the goal state.

To improve the search further, GPS contained two other features. The first is that, if while trying to reduce the deviation from the goal state, GPS finds that it has actually complicated the search process, it was capable of backtracking to an earlier state and exploring alternate solution paths. The second is that it was capable of defining subgoal states that, if satisfied, would permit the solution process to continue. In formulating GPS, Newell and Simon had done extensive work studying human subjects and the way they solved problems. They felt that GPS did a good job of imitating human subjects. They commented on the effort by saying:

> The fragmentary evidence we have obtained to date encourages us to think that the General Problem Solver provides a rather good first approximation to an information processing theory of certain kinds of thinking and problem-solving behavior. The processes of "thinking" can no longer be regarded as completely mysterious.

One criticism of GPS was that the only way the program obtained any information was through human input. The way and order in which the problems were presented was controlled by humans. Thus, the program was only doing what it was told to do. Newell and Simon argued the fact that the program was not just repeating steps and sequences but was actually applying rules to solve problems it had not previously encountered is indicative of intelligent behavior.

There were other criticisms as well. Humans are able to devise new shortcuts and improvise. GPS would always go down the same path to solve the same problem, making the same mistakes as before. It could not learn. Another problem was that while GPS was good when given a certain area or a specific search space to solve. In solving problems, it was difficult to determine what search space to use. Sometimes solving the problem is trivial compared with finding the search space. The problems posed to GPS were all of a specific nature. They were all puzzles or logical challenges; problems that could easily be expressed in symbolic form and operated on in a pseudomathematical approach. There are many problems that humans face that are not so easily expressed in symbolic form.

Also, in 1959, John McCarthy came out with a tool that was to greatly improve the ability of researchers to develop AI programs. He developed a new computer programming language called list processing. It was to become one of the most widely used languages in the field.

LISP is distinctive in two areas: memory organization and control structure. The memory organization is done in a tree fashion with interconnections between memory groups. Thus, it permits a programmer to keep track of complex structural relationships. The other distinction is the way the control of the program is done. Instead of working from the prerequisites to a goal, it starts with the goal and works backwards to determine what prerequisites are required to achieve the goal.

In 1960, Frank Rosenblatt did work in the area of pattern recognition. He introduced a device called PERCEPTRON that was supposed to be capable of recognizing letters and other patterns. It consisted of a grid of 400 photo cells connected with wires to a response unit that would produce a signal only if the light coming off the subject to be recognized crossed a certain threshold.

During the latter part of the 1960s, there were two efforts in another area of simulating human reasoning. Kenneth Colby at Stanford University and Joseph Weizenbaum at MIT wrote separate programs that were capable of interacting in a two-way conversation. Weizenbaum's program was called ELIZA. The programs were able to sustain realistic conversations by using very clever techniques. For example, ELIZA used a pattern-matching method that would scan for keywords like "I," "you," "like," and so on. If one of these words was found, it would execute rules associated with it. If no match was found, the program would respond with a request for more information or with a noncommittal response.

It was also during the 1960s that Marvin Minsky and his students at MIT made significant contributions towards the progress of AI. One student, T. G. Evans, wrote a program that would perform visual analogies. The program was shown two figures that had some relationship to each other and was then asked to find another set of figures from a set that matched the same relationship. The input to the computer was not done by a visual sensor (like the one worked on by Rosenblatt), but instead the figures were described to the system.

In 1968, another student of Minsky's, Daniel Bobrow, came out with a linguistic problem solver called STUDENT. It was designed to solve problems that were presented to it in a word-problem format. The key to the program was the assumption that every sentence was an equation. It would take certain words and turn them into mathematical operations. For example, it would convert "is," into "=," and "per" into "/."

Even though STUDENT responded very much the same way that a real student would, there was a major difference in the depth of understanding. While the program was capable of calculating the time, two trains would collide given the starting points and speeds of both, and it had no real understanding or even cared what a "train" or "time" was. Expressions like "perchance" and "this is it" could mean totally different things than what the program would assume. A human student would be able to discern the intended meaning from the context in which the terms were used.

In an attempt to answer the criticisms about understanding, another student at MIT, Terry Winograd, developed a significant program named SHRDLU. In setting up his program, he utilized what was referred to as a microworld or blocks world. This limited the scope of the world that the program had to try to understand. The program communicated in what appeared to be a natural language.

The world of SHRDLU consisted of a set of blocks of varying shapes (cubes, pyramids, etc.), sizes, and colors. These blocks were all set on an imaginary table. Upon request, SHRDLU would rearrange the blocks to any requested configuration. The program was capable of knowing when a request was unclear or impossible. For instance, if it was requested to put a block on top of a pyramid, it would request that the user specifies more clearly what block and pyramid are. It would also recognize that the block would not sit on top of the pyramid. Two other approaches that the program took that were new to programs were the ability to make assumptions and the ability to learn. If asked to pick up a larger block, it would assume that you meant a larger block than the one it was currently working on. If asked to build a figure that it did not know, it would ask for an explanation of what it was and, thereafter, it would recognize the object. One major sophistication that SHRDLU added to the science of AI programming was its use of a series of expert modules or specialists. There was one segment of the program that specialized in segmenting sentences into meaningful word groups, a sentence specialist to determine the relationship between nouns and verbs, and a scenario specialist that understood how individual scenes related to one another. This sophistication greatly enhanced the method in which instructions were analyzed.

As sophisticated as SHRDLU was at that time, other scholars were quick to point out its deficiencies. SHRDLU only responded to requests; it could not initiate conversations. It also had no sense of conversational flow. It would jump from performing one type of task to a totally different one if so requested. While SHRDLU had an understanding of the tasks it was to perform and the physical world in which it operated, it still could not understand very abstract concepts.

Branches of AI

The various attempts at formally defining the use of machines to simulate human intelligence led to the development of several branches of AI. Some of the subspecialties of AI include

1. Natura/language processing deals with various areas of research such as database inquiry systems, story understanders, automatic text indexing, grammar and style analysis of text, automatic text generation, machine translation, speech analysis, and speech synthesis.

2. Computer vision deals with research efforts involving scene analysis, image understanding, and motion derivation.

3. Robotics involves the control of effectors on robots to manipulate or grasp objects, locomotion of independent machines, and use of sensory input to guide actions.

4. Problem-solving and planning involves applications such as refinement of high-level goals into lower-level ones, determination of actions needed to achieve goals, revision of plans based on intermediate results, and focused search of important goals.

5. Learning deals with research into various forms of learning including rote learning, learning through advice, learning by example, learning by task performance, and learning by following concepts. In recent years, deep learning has become a key approach to getting AI to be more robust in independent thinking.

6. Expert systems deal with the processing of knowledge as opposed to the processing of data. It involves the development of computer software to solve complex decision problems.

Neural Networks

Neural networks, sometimes called connectionist systems, are networks of simple processing elements or nodes capable of processing information in response to external inputs. Neural networks were originally presented as models of the human nervous system. Just after World War II, scientists found out that the physiology of the brain was similar to the electronic processing mode used by computers. In both cases, large amounts of data are manipulated. In the case of computers, the elementary unit of processing is the bit, which is in either an "on" or "off" state. In the case of the brain, neurons perform the, basic data processing. Neurons are tiny cells that follow a binary principle of being either in a state of firing (on) or not firing (off). When a neuron is on, it fires a signal to other neurons across a network of synapses. In the late 1940s, Donald Hebb, a researcher, hypothesized that biological memory results when two neurons are active simultaneously. The synaptic connection of synchronous neurons is reinforced and given preference over connections made by neurons that are not active simultaneously. The level of preference is measured as a weighted value. Pattern recognition, a major strength of human intelligence, is based on the weighted strengths of the reinforced connections between various pairs of simultaneously active neurons. The idea presented by Hebb was to develop a computer model based on the way in which neurons form connections in the human brain. But the idea was considered to be preposterous at that time since the human brain contains 100 billion neurons, and each neuron is connected to 10,000 others by a synapse, unless it is a neuron from a moron, as quipped by Badiru (1990). Even with today's computing capability, it is still difficult to duplicate the activities of neurons. In 1969, Marvin Minsky and Seymour Pappert criticized existing neural network research as being worthless. It has been claimed that the pessimistic views they presented discouraged further funding for neural network research for several years. Funding was diverted instead to further research of expert systems, which Minsky and Pappert favored. Fortunately, neural networks made a strong comeback later on. Because neural networks are modeled after the operations of the brain, they hold considerable promise as building blocks for achieving the ultimate aim of AI. The present generation of neural networks uses artificial neurons. Each neuron is connected to at least one other neuron in a synapse-like fashion. The networks are based on some form of learning model. Neural networks learn by evaluating changes in input. Learning can be either supervised or unsupervised. In supervised learning, each response is guided by the given parameters. The computer is instructed to compare any inputs to ideal responses, and any discrepancy between the new inputs and ideal responses is recorded. The system then uses this data bank to guess how much the newly gathered data are similar to or different from the ideal responses. That is, how closely the pattern matches. Supervised learning networks are now commercially used for control systems, handwriting analysis, speech recognition, and other science and technology applications.

In unsupervised learning, input is evaluated independently and stored as a pattern. The system evaluates a range of patterns and identifies similarities and dissimilarities among them. However, the system cannot derive any meaning from the information without human assignment of values to the patterns. Comparisons are relative to other results, rather than to an ideal result. Unsupervised learning networks are used to discover patterns where a particular outcome is not known in advance, such as in physics research and the analysis of financial data. Several commercial neural network products have entered the market over the years, including NeuroShell from Ward Systems Group. The software is expensive but is relatively easy to use. It interfaces well with other software such as spreadsheets, as well as with C and other programming languages.

Despite the proven potential of neural networks, they drastically oversimplify the operations of the brain. The existing systems can only undertake elementary pattern-recognition tasks and are weak at deductive reasoning, math calculations, and other computations that are easily handled by conventional computer processing. The difficulty in achieving the promise of neural networks lies in our limited understanding of how the human brain functions. Undoubtedly, to model the brain accurately, we must know more about it. But a complete knowledge of the brain is still many years away.

Expert Systems

At some point in the past, expert systems were the rage of the AI movement. The rave of expert systems has waned in recent years in favor of other AI subfields. In the late 1960s to early 1970s, a special branch of AI began to emerge. The branch, known as expert systems, grew dramatically due to its ability to deliver results, albeit not in a thinking context. For years, expert systems represent the most successful demonstration of the capabilities of AI. Expert systems are the first truly commercial application of work done in the AI field and, as such, received considerable publicity for many years.

Unlike the desire to develop general problem-solving techniques that had characterized AI before, expert systems address problems that are focused. When Edward Feigenbaum developed the first successful expert system, DENDRAL, he had a specific type of problem that he wanted to be able to solve. The problem involved determining which organic compound was being analyzed in a mass spectrograph. The program was intended to simulate the work that an expert chemist would do in analyzing the data. This led to the term "expert system."

Between 1970 and 1980, numerous expert systems were introduced to handle several functions, from diagnosing diseases to analyzing geological exploration information. Of course, expert systems have not escaped the critics. Due to the nature of the system, critics argue that it does not fit the true structure of AI. Because of the use of only specific knowledge and the ability to solve only specific problems, some critics are apprehensive about referring to an expert system as intelligent. Proponents argue that if the system produces the desired results, it is of little concern whether it is intelligent or not. The author and his students at the University of Oklahoma developed several practical expert systems for various applications in the late 1980s and early 1990s (Badiru, 1991; Datar and Badiru, 1988; Sunku and Badiru, 1990; Arif and Badiru, 1992).

In 1972, Hubert Dreyfus initiated another debate of interest. Joseph Weizenbaum presented similar views in 1976. The issues that both authors raised touched on some of the basic questions that dated back to the time of Descartes. One of Weizenbaum's reservations concerned what should ethically and morally be handed over to machines. He maintained that the path that AI was pursuing was headed in a dangerous direction. Some aspects of human experience, such as love and morality, cannot be adequately imitated by machines.

While the debates were going on over how much AI could do, the work on getting AI to do more continued. In 1972, Roger Shrank introduced the notion of script, the set of familiar events that can be expected from a frequently encountered setting. This enables a program to assimilate facts quickly. In 1975, Marvin Minsky presented the idea of frames. Even though neither concept drastically advanced the theory of AI, they did help expedite research in the field.

In 1979, Minsky suggested a method that could lead to a better simulation of intelligence: the society of minds view, in which the execution of knowledge is performed by several programs working in conjunction simultaneously. This concept helped to encourage interesting developments such as present-day parallel processing.

During the 1980s, AI gained significant exposure and interest. AI, once restricted to the domain of esoteric research, has now become a practical tool for solving real problems. While AI is enjoying its most prosperous period, it is still plagued with disagreements, criticism, and skepticism. The emergence of commercial expert systems on the market has created both enthusiasm and skepticism. There is no doubt that more research and successful applications developments will help prove the potential of expert systems. It should be recalled that new technologies sometimes fail to convince all initial observers. IBM, which later became a giant in the personal computer business, hesitated for several years before getting into the microcomputer market because the company never thought that those little boxes called personal computers would ever have any significant impact on the society. Today's laptop market has proven otherwise.

Embedded Expert Systems

More expert systems are showing up, not as stand-alone systems, but as software tools (or apps) embedded in large software systems. This trend is bound to continue as systems integration takes hold in many software applications. Many conventional commercial packages, such as statistical analysis systems, data management systems, information management systems, project management systems, and data analysis systems, contain embedded heuristics that constitute expert systems components of packages. Even some computer operating systems now contain embedded expert systems designed to provide real-time systems monitoring and troubleshooting. With the success of embedded expert systems, the long-awaited payoffs from AI systems may be realized on a larger scope in the coming years. Because the technology behind expert systems has changed little over the past years, the issue is not whether the technology is useful but how to implement it more fruitfully. This is why the integrated systems approach of this book is very useful. The book focuses not only on the traditional principles of project management but also on the emerging technological tools that can aid project planning, scheduling, and control. Combining neural networks with expert systems, for example, will become more prevalent in the modern AI applications. In combination, the neural networks might be implemented as a tool for scanning and selecting data while the expert system would evaluate the data and present recommendations. More information about this can be found in Badiru et al. (1992), Milatovic et al. (2000), Milatovic and Badiru (2001), and Sieger and Badiru (1993).

Conclusion and Projection

The effort in AI is worthwhile as long as it increases the understanding that we have of intelligence and enables us to do things that we previously could not do. Due to the discoveries made in AI research, computers are now capable of things that were once beyond imagination. The field of project management will be hugely impacted as AI systems become more robust and more intelligent.

References

Arif, A.E. and Badiru, A.B. (1992). An integrated expert system with a fuzzy linguistic model for facilities layout, *Proceedings of the Sixth Oklahoma Symposium on Artificial Intelligence*, Tulsa, OK, November 11–12, pp. 185–194.

Badiru, A.B. (1990). Artificial intelligence applications in manufacturing, in *The Automated Factory Handbook: Technology and Management*, D.I. Cleland and B. Bidanda (eds.), TAB Professional and Reference Books, New York, pp. 496–526.

Badiru, A.B. (November 1991). OKIE-ROOKIE: An expert system for industry relocation assessment in Oklahoma, *Proceedings of the Fifth Oklahoma Symposium on Artificial Intelligence*, Norman, OK.

Badiru, A.B. (1993). The role of artificial intelligence and expert systems in new technologies, in *Management of New Technologies for Global Competitiveness*, C.N. Madu (ed.), Quorum Books, Greenwood Publishing Company, Westport, CT, pp. 301–317.

Badiru, A.B. and Cheung, J. (2002), *Fuzzy Engineering Expert Systems with Neural Network Applications*, John Wiley & Sons, New York.

Badiru, A.B., Gruenwald, L. and Trafalis, T. (1992). A new search technique for artificial intelligence systems, *Proceedings of the Sixth Oklahoma Symposium on Artificial Intelligence*, Tulsa, OK, November 11–12, pp. 91–96.

Datar, N.N. and Badiru, A.B. (November 1988). A prototype knowledge based expert system for robot consultancy – ROBCON, in *Proceedings of the 1988 Oklahoma Symposium on Artificial Intelligence*, Norman, OK, pp. 51–68.

Milatovic, M. and Badiru, A.B. (2001). Control sequence generation in multistage fuzzy control systems for design process, *AIEDAM (Artificial Intelligence for Engineering Design, Analysis, and Manufacturing)*, 15: 81–87.

Milatovic, M., Badiru, A. and Trafalis, T. (2000). Taxonomical analysis of project activity networks using competitive artificial neural networks, in *Intelligent Engineering Systems through Artificial Neural Networks*, Vol. 10, C.H. Dagli et al. (eds.), ASME Press, New York, pp. 437–442.

Sieger, D.B. and Badiru, A.B. (March 1993). An artificial neural network case study: Prediction versus classification in a manufacturing application, *Computers and Industrial Engineering*, 25(1–4): 381–384.

Sunku, R. and Badiru, A.B. (1990). ROBEX (robot expert): An expert system for manufacturing robot system implementation, *Proceedings of* 12th *Annual Conference on Computers and Industrial Engineering*, Orlando, FL, March 1990, *Computers & Industrial Engineering*, 19(1–4): 481–483.

Index

A

Abilene paradox, 91–93
Abstraction decision model, 30
Accomplishment cost procedure (ACP), 240
ACTIM, *see* Activity time (ACTIM)
Activity attributes, 462
Activity crashing, 134–140
Activity network, 119
Activity-on-arrow (AOA), 118, 164, 167
Activity-on-node (AON), 118, 123, 124, 164, 312
   after-work-hours chores, graph of, 168
   graphical representation for, 119, 120
Activity planning, 300–302
   takt time for, 191–193
Activity resource (ACTRES), 180
   vs. activity time and time resources,
      185–188
Activity scheduler, 453, 456, 462–468, 471
Activity time (ACTIM), 179–180, 184–185
   vs. activity resource and time resources,
      185–188
ACTRES, *see* Activity resource (ACTRES)
Additive utility model, 393–394
Administrative cooperation, 88
After-the-fact performance measurements, 236
After-work-hours chores
   AON graph of, 168
   project graph of, 167
   sample project for, 166
AGS, *see* Arithmetic gradient series (AGS)
AHP, *see* Analytic hierarchy process (AHP)
AI, *see* Artificial intelligence (AI)
Amortization schedules, of capital investment
      projects, 378–380
Analysis of variance (ANOVA), 418
Analytic hierarchy process (AHP), 406–411
Annual worth (AW), 374
ANOVA, *see* Analysis of variance (ANOVA)
Antecedent activity, network, 120
AOA, *see* Activity-on-arrow (AOA)
AON, *see* Activity-on-node (AON)
Arithmetic gradient series (AGS), 368–369,
      382–385
AR processes, *see* Autoregressive (AR)
      processes
Arrow, network, 119
Artificial intelligence (AI), 493–496
   branches of, 499–500
   expert systems, 501–502
      embedded, 502–503
   first conference, 496–499
   neural networks, 500–501
Assignment method, 189, 335–337
Assignment problem, 189
   in project optimization, 335–339
Associative cooperation, 88
Autocratic organization, 109
Autoregressive (AR) processes, 428–430
Average cost model, 207–211
AW, *see* Annual worth (AW)

B

Back arc, 165
Backward pass computations, 124–125
Badiru's rule, 12
Baker model, 235
Bandwagon effect, 33
Basic tent (BT) cash flow, 390
   computational formula analysis of, 384
Benefit–cost *(B/C)* ratio analysis, 373–374
Beta distribution, 143–144
Bicritical (BC) activity, 157
Blob organization, 106
Body of knowledge, 17–18
   components of, 18–20
Boilerplates, 70
Bottleneck resource node, 200
Bottom-up budgeting, 73
Brainstorming, 32
British Petroleum (BP) oil spill, 76
Brooks's algorithm, 184
BT cash flow, *see* Basic tent (BT) cash flow
Bubble organization, 106
Budget planning, 72–74
Bureaucratic processes, 97
Burgess's method, 203–204
Burst point network, 120
Burst resource node, 200

C

CAF, *see* Composite allocation factor (CAF)
Capability architecture, 14

Capitalized cost formula, 366–367
Carbon sequestration methods, 16
Cash-flow diagram (CFD), 361–362
Cause-and-effect diagram, 267–268
CBAIC company, 485
Central limit theorem, 146–147
Certain amount (CA), 392
Certainty equivalent (CE), 392
CFD, *see* Cash-flow diagram (CFD)
Chronological organization, 107
Cluster sampling, for project control, 263
Cobb–Douglas function, 215
Commitment cooperation, 89
Communication channel, 79–87
Communications management, 19
Compassion, 482–483
Complexity theory, 474
Component-by-component implementation,
    20–22
Composite allocation factor (CAF), 181–182
Compound amount factor, 362
    uniform series, 364–365
Compound interest rate, 355–357
Computational formula analysis, of basic tent
    cash flow, 384
Computer processing time, 162
Computer tools, 12
Connectionist systems, *see* Neural networks
Construction decision model, 30
Continuous compounding, interest
    rate for, 360
Continuous performance improvement (CPI),
    237–239
Control charts
    for budget tracking, 256–257
    for cumulative cost, 256
    graphical report, on task progress, 258
    monitoring revenues *vs.* expenses, 257
    for project monitoring, 255–256
    resource loading *vs.* project progress, 259
    task progress analysis, 258
    time *vs.* cost, 260, 261
    trade-off relationships, 260
Convenience sample, for project control, 262
Conventional network logic, 121
Cooperation, project management, 87–89
Coordination, project management, 89–90
Correlogram, 429
Cost benchmarking, 411–412
Cost/benefit analysis, 488
Cost conflict, 90
Cost estimate feasibility analysis, 68
Cost management, 19

Cost project control system, 239–240
CPI, *see* Continuous performance
    improvement (CPI)
CPM, *see* Critical path method (CPM)
Crash task duration, 135–137
CRD, *see* Critical resource diagram (CRD)
Critically dependent resource node, 200
Critical path method (CPM), 118, 121, 135, 141,
    151, 280, 335, 455, 463
    backward pass, 124–125
    crash task duration, 135–138
    data analysis, 123
    forward pass, 123–124, 128
    free slack, 126
    independent float, 126–128
    network scheduling, 273
    of partitioned and serial activities, 153
    precedence relationships in, 121
    subcritical paths, 128–130
    total slack, 125–126
Critical resource diagram (CRD), 198, 456
    computations, 199–200
    network development, 198–199
    node classifications, 200
    resource management, 198
    RS chart, 201–202
Cross arc, 165
Cultural change, 96
Cultural feasibility, 67
Cultural infeasibility, 484
Cumulative distribution function, 445
Cumulative production, 217
Cycle time, 191

D

Dartmouth Summer Conference, 496–497
Data and information requirements, 29
Decision-making models, *see* Group systems
    decision-making models
Decision model, 29–30
Decision tree analysis, 273–280
Decision variable, 299
Deductive statistics, 261–262
Deep learning, in artificial intelligence, 493
Defense Acquisition System, 15
Degradation–innovation cycle, 237
D-E-J-I model, *see* Design, evaluate, justify and
    integrate (D-E-J-I) model
Delphi method, 32–34
Department of Defense (DoD), 14, 15
Department of Defense Architecture
    Framework (DODAF), 14–15

Dependency cooperation, 88
Dependent resource node, 200
Depth-first search method, 165
Descendent activity, network, 120
Descriptive models, 30
Descriptive statistics, 261
Design, evaluate, justify and integrate (D-E-J-I)
    model
    design stage of, 41–43
    evaluation stage of, 44–45
    implementation, 44
    integration stage of, 49–53
    justification stage of, 47–48
    for project execution, 41
Dijkstra's algorithm, 342–343
Discounted payback period (DPP), 375–376
Distance learning (DL) educational project,
    80, 81
DL educational project, *see* Distance learning
    (DL) educational project
DoD, *see* Department of Defense (DoD)
DODAF, *see* Department of Defense
    Architecture Framework (DODAF)
DPP, *see* Discounted payback period (DPP)
Dual-use integration, 39
Dummy network, 119
Dynamic programming, 436
Dynamic resource integration, 39

E

Earliest completion (EC) time, 121–124, 126, 127
Earliest start (ES) time, 121–124, 126
    Gantt chart based on, 130
Economic analysis process, 355
Economic engagement, 477–478
Economic feasibility, 67
Economic interdependence model, 475–476
Economic order quantity (EOQ) model,
    430–432, 435
Effective interest rate, 359–361
Electronic Numerical Integrator and Computer
    (ENIAC), 495
ELIZA, 498
Embedded expert systems, 502–503
Empathy, 482–483
ENIAC, *see* Electronic Numerical Integrator and
    Computer (ENIAC)
EOQ model, *see* Economic order quantity (EOQ)
    model
Equity break-even point, 378, 380–382
Error sum of squares (SSE), 421, 422
Esteem needs, 64, 481

ET cash-flow profile, *see* Executive tent (ET)
    cash-flow profile
Executive tent (ET) cash-flow profile, 386
Expert systems, 501–502
Exponential smoothing forecast, 417
Ex-post-facto experimental research, 285
Extrapolation, 419
Extrinsic forecasting, 415

F

FC activity, *see* Finish critical (FC) activity
Feasibility analysis
    elements of, 68–69
    engineering and design, 68
    financial, 67, 68–69
Finish critical (FC) activity, 157
First artificial intelligence conference, 496–499
First-order half-life, 47
Fishbone diagram, 267–268
Fitted model, 213, 214
Fixed interest rate, 377–378
Flood Act (1936), 373
Forecasting techniques
    based on averages, 416–417
    coefficient of multiple correlation, 421
    determination coefficient, 420–421
    intrinsic and extrinsic, 415
    regression analysis, 418
        prediction and control, 419
        procedure for, 419–420
    residual analysis, 421–424
    stationarity and data transformation,
        424–427
        autoregressive processes, 428–430
        moving average processes, 427–428
    time series analysis, 422–424
Formal organization structure, 98–99
Formal project control system, 230–232
Forward arc, 165
Forward pass calculations, 123–124
Free slack (FS), 126
FS, *see* Free slack (FS)
Full-duplex communication, 84
Functional cooperation, 88
Functional organization, 100–101
Function-based scheduling, 471

G

Gantt chart, 129–131, 151, 232, 465, 466, 468, 469
    crashing and schedule compression, 134–140
    phase-based, 134

Gantt chart (*cont.*)
　progress-monitoring, 132
　variations, 131–135
GAO, *see* Government Accountability Office
　　(GAO)
GASC company, 485–487
General Problem Solver (GPS), 497–498
General tent equation (GTE), 382–383
　derivation of, 388–390
Geographic awareness, 478–479
Geometric series cash flow
　decreasing, 371–372
　increasing, 369–371
Global alliances sustainability, 483
Global awareness, 477
Global situational awareness, 475, 485
Global teamwork, 473
Goal programming, 344–349
"GO-NO-GO" chart, for project selection
　　analysis, 393
Government Accountability Office (GAO), 3
GPS, *see* General Problem Solver (GPS)
Greatest resource demand, 182
Greatest resource utilization rule, 183
Group systems decision-making models
　advantages, 31
　brainstorming, 32
　Delphi method, 32–34
　disadvantages, 32
　interviews, surveys and questionnaires,
　　35–36
　multivoting, 36
　nominal group technique, 34–35
　systems hierarchy, 36–40
GTE, *see* General tent equation (GTE)

H

Half-duplex communication, 84
Half-life computation, for learning curves,
　　45–47
Health-care projects, learning curve analysis in,
　　219–220
Heuristic scheduling, limitations of, 297
Hierarchy of needs, 481–482
Histogram, 268–269
Human-engineered fusion, 16
Human resource management, 19
Hungarian method, 190, 336–337
Hybrid cultures, 476
Hygiene factors, 65–66
Hypothesis testing

null and alternate hypothesis, 285–286
one-tailed *vs.* two-tailed, 286–288
producer's risk *vs.* the consumer's risk, 288

I

IF, *see* Independent float (IF)
Imposed cooperation, 88
Imposed precedence, 121
Independent float (IF), 126–128
Inductive statistics, 262
Industrial projects, 105
Inferential statistics, 261
Informal organization structure, 98–99
Informal project control system, 230–232
Information age technologies, 95
Information flow model
　cost and value of, 77–78
　for project decision making, 76–77
Information support plans, 15
Integer programming model, 306–308
Integrated systems implementation, 12–13
Integration management, 18, 40
Interest rates
　fixed and variable, 377–378
　nominal and effective, 359–361
　simple and compound, 355–357
Interfering slack (IS), 126
Internal rate of return (IRR), 372–373
Interpolation, 419
Interval measurement scale, 241
Intrinsic forecasting, 415
Inventory management, 430
　economic order quantity model, 430–432
　quantity discount, 432
Investment benchmark-feedback model, 412
Iron triangle model, 5
IRR, *see* Internal rate of return (IRR)
IS, *see* Interfering slack (IS)

J

Joint Capabilities Integration and Development
　　System, 14
Judgment sample, for project control, 262

K

KELLC company, 485, 486
Knapsack problem, 319–322
　activity scheduling, 322–324
Kronecker's delta, 457

L

Labor practices, 479–480, 489
Lateral cooperation, 88
Latest completion (LC) time, 126, 127, 154
Latest finish time (LFT), 463
Latest starting (LS) time, 121, 123–125, 126, 154
    Gantt chart based on, 131
Law of averages, 269
Leadership development, 95–96
Lead–lag relationships, in precedence
        diagramming method, 150, 154
Learning curve analysis, 44–45, 207
    graphical analysis, 212–214
    half-life computation for, 45–47
    in health-care projects, 219–220
    log-linear model
        average cost model, 207–211
        unit cost model, 211–212
    multivariate, 215–219
Learning process, interruption of, 219
Least-cost method, 328
Legal cooperation, 88
Legal systems considerations, 75–76
Limit theorem, 366
LINDO solution, for goal programming, 347
Linear constraints, 299
Linear programming (LP)
    for energy resource allocation, 302–304
    formulation, 298–300, 305–306
Linked bars, in Gantt chart, 131, 132
Liquefied natural gas (LNG), 484
Liquefied petroleum gas (LPG), 484
LISP, *see* List processing (LISP)
List processing (LISP), 497, 498
LNG, *see* Liquefied natural gas (LNG)
Local adaptation loop, for project transfer, 110,
    111
Local conditions assessment, 478
Logic Theorist program, 496, 497
Log-linear model, 45, 46
    average cost model, 207–211
    unit cost model, 211–212
LP, *see* Linear programming (LP)
LPG, *see* Liquefied petroleum gas (LPG)

M

Macrolevel planning, 60
MALU, *see* Minimum acceptable level of utility
        (MALU)
Management by exception (MBE), 66–67

Management by objective (MBO), 60–61, 66
Management by project (MBP), 11–12
Management conflict, 90
Managerial feasibility, 67
MA processes, *see* Moving average (MA)
        processes
Market integration, 477
Market organization, 106, 107
MARR, *see* Minimum attractive rate of return
        (MARR)
Maslow's hierarchy of needs, 64–65, 481
Mathematical scheduling, advantages of, 297
Matrix of reduced costs, 337
Matrix organization, 103–105
MBC, *see* Mid-bicritical (MBC)
MBE, *see* Management by exception (MBE)
MBO, *see* Management by objective (MBO)
MBP, *see* Management by project (MBP)
Means-end analysis, 497
Mean square error (MSE), 422
Merge point, network, 120
Microlevel planning, 60–61
Mid-bicritical (MBC), 157
Mid-normal critical (MNC), 157
Mid-reverse critical (MRC), 157
Milestone Gantt chart, 133
Military organization, 108
Minimum acceptable level of utility (MALU),
    399
Minimum attractive rate of return (MARR), 362
Mixed organization structure, 105
MNC, *see* Mid-normal critical (MNC)
Model testing, 418
Mortgage loans, 381
Motivation, 62–63
Motivators, 65–66
Moving average (MA) processes, 427–428
MRC, *see* Mid-reverse critical (MRC)
Multiattribute project selection, 390
Multinational projects
    acquisition, 488–489
    information management, 489–490
    organization, 112–115
    organizational relationships, 485–486
    problems and solutions, 486–488
    project setting, 484
    recommendations, 490
    safety and environmental considerations, 486
    strategic planning and technology
        assessment, 488
    technical requirements, 484–485
    trade agreements, 489

Multiple-boss organization, 104
Multiple-projects Gantt chart, 133, 134
Multiple resources, work rate tabulation for, 194
Multiplicative utility model, 395
Multiresource scheduling
    activity scheduler, 462–468
    characteristics modeling, 458–460
    interdependencies and multifunctionality,
        456–458
    loading graphs, 465, 466, 468, 469
    mapper, 453, 456, 460–462
    notations, 453–454
    project resources
        analysis of, 454–455
        scarcity of, 453
    resource-activity mapping grids, 470
    resource modeling, 455–456
    sequence-based and function-based
        scheduling, 471
    utilization, 470
Multistage sampling, 263–264
Multivariate learning curves, 215–219
Multivoting, 36
Murphy's law, 12

N

NAE, *see* National Academy of Engineering
    (NAE)
NASA, *see* National Aeronautics and Space
    Administration (NASA)
National Academy of Engineering (NAE), 15, 16
National Aeronautics and Space
    Administration (NASA), 75
Nearest neighbor algorithm, 340–342
Need feasibility analysis, 68
Net annual value (NAV), 367
Net FW (NFW), 372–373
Net PW (NPW), 372–373
Network analysis
    components of, 119–120
    control, 119
    planning, 118
    scheduling, 118–119
Neural networks, 500–501
Nitrogen cycle, 16
NOCC, 485–487
Node network, 119
Nominal group technique, 34–35
Nominal interest rate, 359–361
Nominal measurement scale, 241
Nonlinear cost–volume–profit model, 216

Nonnegativity constraint, 299
Nonsampling error, for project control, 263
Normal critical (NC), 156
Normal distribution, 270–273
Normalized solution time, 162
Normal task duration, 135–136
Northwest-corner method, 327–329, 330
Null hypothesis, 285–286

O

Objective function, 298, 299
OBS, *see* Organizational breakdown structure
    (OBS)
One cycle of the function, 443
One-tailed *vs.* two-tailed hypothesis testing,
    286
Operating characteristic curve, 287
Operational views (OVs), 15
Optimization models, 30
Ordinal measurement scale, 241
Organizational breakdown structure (OBS),
    97–98
Organizational performance, diversity of, 2–4
Overseas workforce constraints, awareness of,
    480
OVs, *see* Operational views (OVs)

P

Parallel priority scheduling process, 179
Parallel scheduling heuristic method, 118
Parametric cost analysis, 216
Pareto analysis, 128
    diagram, 266–267
    principle, 129
Parkinson's law, 12, 230
Payback period, 374–375
PBS, *see* Project breakdown structure (PBS)
PCP, *see* Percentage cost penalty (PCP)
pdf, *see* Probability density function (pdf)
PDM, *see* Precedence diagramming method
    (PDM)
Peak flag, 466, 467
Percentage cost penalty (PCP), 436
PERCEPTRON device, 498
Performance conflict, 90
Performance measure, 29
Performance project control system, 236–237
Period moving average forecast method, 416
Permanent investments formula, 367
Personality conflict, 91

PERT, *see* Program evaluation and review technique (PERT); Project Evaluation and Review Technique (PERT)
Peter's principle, 12
Phase-based Gantt chart, 134
Physiological needs, 64, 481
PMBOK®, *see* Project Management Body of Knowledge (PMBOK®)
PMI, *see* Project Management Institute (PMI)
PMIS, *see* Project management information system (PMIS)
PMO, *see* Project management office (PMO)
Polar plots, 400–406
Political feasibility, 68
Political organization, 109
Power conflict, 91
Power curve, 287
Precedence diagramming method (PDM), 118, 149
  anomalies in, 156–157
  complexity of, 157–162
  compressed network, 154–156
  lead–lag relationships in, 150, 152
  network example, 151
  resource-constrained network, 196–197
  scheduling heuristics, performance of, 163–164
  solution time, evaluation of, 162–163
Precedence network diagram, 120
Predecessor activity, 120
Predictive models, 30
Predictive scheduling, 117
Prescriptive models, 30
Present worth (PW) factor, 363, 366, 374, 383
  of increasing geometric series cash flow, 370
  uniform series, 363–364
Primary sampling unit, 263
Priority conflict, 91
Proactive programs, 482
Probabilistic decision analysis, 269–273
Probabilistic resource utilization, 204–207
Probability density function (pdf), 270, 395
Problem identification, 23
Problem statement, 28–29
Procedural precedence, 121
Process work feasibility analysis, 68
Procurement management, 19–20
Product organization, 101–102
Program evaluation and review technique (PERT), 177, 280, 335
Progress-monitoring Gantt chart, 132
Project analysis

documentation, 28
feasibility, 67–68
life cycle, 20–22
optimization
  assignment problem in, 335–339
  modeling for, 297
organization, 23, 26
  bubble, 106
  chronological, 107
  environmental factors in, 95
  formal and informal structures, 98–99
  functional, 100–101
  market, 106–107
  matrix, 103–105
  military, 108
  mixed, 105
  multinational, 112–115
  political and autocratic, 109
  product, 101–102
  sequential, 107–108
  social and cognitive domains, 95–96
  transferred project, 110–112
plan components, 61–62
planning, 23, 25–26
proposals
  incentives, 72
  preparation, 70–72
  request for proposal, 69–70
tracking and reporting, 24, 233–236
Project breakdown structure (PBS), 97
Project control system, 24, 27, 227
  axes of, 228
  continuous performance improvement, 237–239
  control charts (*see* Control charts)
  cost, 239–240
  data analysis and presentation, 245
    average deviation, 252
    average revenue, 247–250
    median revenue, 250
    mode, 251
    quartiles and percentiles, 251
    range of revenue, 251–252
    raw data, 245–246
    sample variance, 253–254
    standard deviation, 254–255
    total revenue, 247
  data determination and collection, 242–245
  decision tree analysis, 273–280
  experimentation
    information needs, 282
    personnel interactions, 281–282

Project control system (*cont.*)
    procedural steps, 283–285
    types of, 285
  flowchart, 231
  formal and informal, 230–232
  hypothesis testing, 285–288
  measurement scales, 241–242
  performance control, 236–237
  probabilistic decision analysis, 269–273
  procedural steps, 230
  project management information system, 240–241
  schedule, 232
  termination, 288–293
  through rescheduling, 280–281
  time, budget, and performance factors, 228–230
  tracking and reporting, 233–236
Project cost systems
  amortization of capitals, 378–380
  analytic hierarchy process, 406–411
  arithmetic gradient series, 368–369
  benchmarking, 411–412
  benefit–cost ratio, 373–374
  capitalized cost formula, 366–367
  cash-flow patterns and equivalence, 361–362
  compound amount factor, 362
  discounted payback period, 375–376
  economic analysis process, 355
  equity break-even point, 380–382
  general tent equation, derivation of, 388–390
  geometric series cash flow
    decreasing, 371–372
    increasing, 369–371
  interest rates
    fixed and variable, 377–378
    nominal and effective, 359–361
    simple and compound, 355–357
  internal rate of return, 372–373
  investment for multiple returns, 357–359
  multiattribute project selection, 390
  payback period, 374–375
  permanent investments formula, 367
  polar plots, 400–406
  present worth factor, 363
  tent cash flows analysis, 382–383
    design and analysis of, 383–388
  uniform series
    capital recovery factor, 364
    compound amount factor, 364–365
    PW factor, 363–364
    sinking fund factor, 365–366
  utility function, 395–400

    utility models, 390–393
      additive, 393–394
      multiplicative, 395
Project definition, 23
Project Evaluation and Review Technique (PERT), 118
  beta distribution, 143–144
  central limit theorem, 146–147
  complexity of, 158
  estimates and formulas, 140–142
  network example, 148–149
  probability calculation, 147–148
  project duration distribution, 146
  triangular distribution, 144–145
  uniform distribution, 145–146
Project graph, formulation of, 164
  activity-on-arc representation, 165–168
  procedure to detect cycles, 165, 166
Project impacts feasibility analysis, 69
Project management, 1
Project Management Body of Knowledge (PMBOK®), 17, 56, 58
Project management information system (PMIS), 240
Project Management Institute (PMI), 17, 55, 58
Project management office (PMO), 96
Project networks, complexity of, 157–162
Project personnel management systems, 96
Project resource mapper (PROMAP), 468
Project schedule optimization
  activity planning, 300–302
  formulation, 297–298
    linear programming, 298–300, 305–306
  goal programming, 344–349
  integer programming, 306–308
  resource combination, 302–304
  resource requirements analysis, 304–305
  time–cost trade-off model, 309–310
    assignment problem, in project optimization, 335–339
    knapsack activity scheduling, 322–324
    knapsack problem, 319–322
    maximum flow procedure, 310–311
    2-opt technique, 341
    procedural steps, 311–318
    sensitivity analysis for, 318–319
    shortest-path problem, 341–343
    transportation problem (*see* Transportation problem)
    transshipment formulation, 335
    traveling resource formulation, 339–341
Project scheduling, 24, 26–27

critical path method (*see* Critical path method (CPM))
Gantt chart, 129–131
  crashing and schedule compression, 134–140
  variations, 131–135
network analysis, fundamentals of, 117–120
Project Evaluation and Review Technique
  beta distribution, 143–144
  central limit theorem, 146–147
  estimates and formulas, 140–142
  network example, 148–149
  probability calculation, 147–148
  project duration distribution, 146
  triangular distribution, 144–145
  uniform distribution, 145–146
project graph, formulation of, 164
  activity-on-arc representation, 165–168
  procedure to detect cycles, 165, 166
Project-slippage-tracking Gantt chart, 135
Project systems
  implementation outline, 25–28
  logistics, 5
  structure, 22–25
Project termination, 24, 27, 288–289
  sensitivity analysis, for project control, 292–293
  system evaluation, 292
  verification and validation, 289–292
Project transfer organization, 110–112
PROMAP, *see* Project resource mapper (PROMAP)
Proportionate stratified sampling, for project control, 263
Proximity cooperation, 88
Pure project organization, 101
PW factor, *see* Present worth (PW) factor

Q

Qualitative tools, 12
Quality management, 19
Quantitative tools, 12
Quantity discount, 432

R

Random sample, for project control, 262
Ratio measurement scale, 241–242
Reactive scheduling, 117
Recurring data, 242, 283
Redundant dummy arcs, 167
Request for proposal (RFP), 69

Residual analysis, 421–424
Resource-activity mapping grid, 468, 470
Resource allocation systems, 23–24
  activity planning and scheduling, 191–193
  activity time, 179–180
  critical resource diagram, 198–200
  heuristics, 179–188
  idleness graph, 204, 205
  interruption of learning process, 219
  learning curve analysis (*see* Learning curve analysis)
  leveling, 203–204
  loading, 202–203
  longest-duration-first, 178–179
  and management strategies, 175–176
  PERT, 177
  probabilistic resource utilization, 204–207
  resource-constrained scheduling, 176–177
  RS chart, 200–202
  using simulated annealing, 347–349
  worker assignment, quantitative modeling of, 188–191
  work rate, 193–196
Resource conflict, 91
Resource-constrained scheduling, 176–177
  PDM network, 196–197
Resource cost charts, 468
Resource idleness graph, 204, 205
Resource leveling, 203–204
Resource loading, 202–203, 465, 466, 468, 469
Resource mapper, 453, 456, 460–462, 471
Resource over time (ROT), 180
Resource requirements analysis, 304–305
Resource schedule (RS) chart, 200–202
Resource scheduling method (RSM), 182
Resource utilization, 468, 470
Resource work rate, 193–196
Restriction network, 119
Reverse critical (RC), 156–157
Reversed STT (R-STT) cash-flow profile, 387–388
RFP, *see* Request for proposal (RFP)
Risk management, 19
"Rule of 72," 357–358
Run chart, 268
Russell's approximation method, 328–329

S

Safety feasibility, 67
Safety needs, 64, 481
Sampling bias, 263
Sampling error, 263
Satisficing models, 30

Saw-tooth tent (STT) cash-flow profile, 387
Scatter plot, 268
Schedule compression, 134–140
Schedule conflict, 90
Schedule project control system, 232
Scheduling heuristics, performance of, 163–164
Science, technology, engineering, and
     mathematics (STEM), 15–17
Scope management, 18
S-curve model, 400
Seasonal pattern modeling, 442–443
     beta distribution, 449–450
     cyclic distribution, 446–450
     cyclic pdf, 443–446
     expected value and variance, 445
     trigonometric sine function, 443
Second-order half-life, 47
Self-actualization needs, 64, 481
Sensitivity analysis, 435–436
Sequence-based scheduling, 471
Sequencing rules, 176
Sequential organization, 107–108
Serial priority scheduling process, 179
Serial scheduling heuristic method, 118
Services views (SVs), 15
Shortest-path problem, 341–343
SHRDLU program, 499
Silver-meal heuristic, 440–442
Simple average forecast method, 416
Simple interest rate, 355–357
Simplex communication, 84
Simplex method, 190
Simulated annealing, resource allocation using,
     347–349
Single-stage sampling, 263
SMART principles, 13–14
Social cooperation, 88
Social feasibility, 67
Social needs, 64, 481
Solar energy, 16
Span of control, wide and narrow, 99
SSE, *see* Error sum of squares (SSE)
SST, *see* Total sum of squares (SST)
Standard normal distribution, 270
Start critical (SC) activity, 157
State-space model, 42
State transformation, in project systems, 43–44
Statistical analysis, for project control, 261–262
     diagnostic tools, 266–269
     sampling techniques, 262–266
Step-by-step implementation, project
     management, 20–22
Stepping-stone method, 329

Stratified sampling, for project control, 263
STUDENT problem solver, 498
Subcontract management, 19–20
Subcritical paths analysis, critical path method,
     123–124, 128–130
Successor activity, network, 120
Supervised learning networks, 501
Supralevel planning, 60
SVM, *see* Systems value model (SVM)
Sympathy, 482–483
Systems constraints, 5–7
Systems control, 1–2
Systems decision analysis, 28–31
Systems engineering, definition of, 4
Systems influence philosophy, 7–8
Systems management, 475
Systems success, critical factors for, 13–14
Systems value model (SVM), 8–9
Systems-wide project planning
     guidelines for, 55
     implementation, 58–59
     knowledge areas and process cluster, 56
     project life cycle, 56–57

T

Takt time, for activity planning, 191–193
Task-combination Gantt chart, 133
Technical analysis
     conflict, 90–91
     feasibility, 67
     precedence, 121
Technical standards views (TVs), 15
Tent cash flows analysis, 382–383
     design and analysis of, 383–388
Theory X, 63
Theory Y, 64
Third-order half-life, 47
Time–cost–performance, for project planning,
     60
Time–cost trade-off model, 309–310
     assignment problem, in project
          optimization, 335–339
     knapsack activity scheduling, 322–324
     knapsack problem, 319–322
     maximum flow procedure, 310–311
     2-opt technique, 341
     procedural steps, 311–318
     sensitivity analysis for, 318–319
     shortest-path problem, 341–343
     transportation problem (*see* Transportation
          problem)
     transshipment formulation, 335

traveling resource formulation, 339–341
Time management, 18–19
Time resources (TIMRES), 180
    vs. ACTIM and ACTRES, 185–188
Time series analysis, 422–424
TIMRES, *see* Time resources (TIMRES)
Top-down budgeting, 73
Total relevant costs (TRC), 431
    calculation of, 433
        discount option, evaluation of, 433–435
        notations and variables, 437
        sensitivity analysis, 435–436
        silver-meal heuristic, 440–442
        Wagner–Whitin algorithm, 436–437
Total slack (TS), 125–127, 129
Total sum of squares (SST), 421
Tour construction techniques, 340
Tour improvement heuristics, 340
Traditional organization structure, 97–98
Transient data, 242, 283
Transportation method, 190, 324
Transportation problem
    algorithm, 329–330
    balanced *vs.* unbalanced, 326–327
    cost matrix for, 330, 331
    initial solution to, 327–329
    for project scheduling, 324–326
    tableau for, 330–335
Transportation tableau, 325
Transshipment formulation, 335
Traveling salesman problem (TSP), 339–341
TRC, *see* Total relevant costs (TRC)
Tree arc, 165
Triad approach, 12
Triangular distribution, 144–145
Triple C model, 79, 282
    communication, 79–87
    cooperation, 87–89
    coordination, 89–90
    resolving project conflicts with, 90–91
TSP, *see* Traveling salesman problem (TSP)
2-opt technique, 341
Two-stage sampling, 263

U

Uniform distribution, 145–146
Uniform series
    present worth factor, 363–364, 369

sinking fund factor, 365–366
Unit cost model, 211–212
United States Football League (USFL), 383
Unit utilization charts, 468
Unsupervised learning networks, 501
Urban environments, 16
U.S. Congress, 15
USFL, *see* United States Football League (USFL)
U.S. Justice Department, 75
Utility function, 395
    idle time reduction, 398
    for investment ROI, 396
    productivity and quality improvement, 397
    safety improvement, 398–399
    S-curve model for, 400
Utility theory, 390–393

V

Validation decision model, 30
Value vector modeling, 10–11
Variable interest rate, 377–378
Vertical cooperation, 88
Virtual reality, 17
Vogel's approximation, 328

W

Wagner–Whitin (W–W) algorithm, 436–437
    propositions for, 438–440
WBS, *see* Work breakdown structure (WBS)
Weighted average forecast method, 416–417
Weighted T-period moving average forecast, 417
Work breakdown structure (WBS), 74–75, 97, 240
worker assignment, quantitative modeling of, 188–191
Work rate analysis, 191, 193–196
    RS chart, 201–202
World economy , project systems view of, 473–474
W-W algorithm, *see* Wagner-Whitin (W-W) algorithm

Z

Zero-base budgeting, 73–74
Zero project slack convention, 122

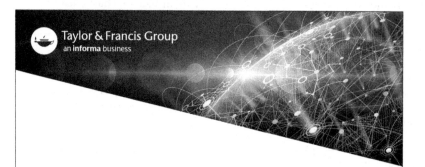